D1349357

CHALLENGES IN

Growth Hormone Therapy

CHALLENGES IN

Growth Hormone Therapy

EDITED BY

John P. Monson

MD, FRCP
Head of Division of General and Developmental Medicine
Reader in Endocrinology
Department of Endocrinology
Medical Unit
St Bartholomew's Hospital
and The Royal London School of Medicine and Dentistry
London

Blackwell
Science

Editorial Offices:
Osney Mead, Oxford OX2 0EL
25 John Street, London WC1N 2BL
23 Ainslie Place, Edinburgh EH3 6AJ
350 Main Street, Malden
MA 02148 5018, USA
54 University Street, Carlton
Victoria 3053, Australia
10, rue Casimir Delavigne
75006 Paris, France

Other Editorial Offices:
Blackwell Wissenschafts-Verlag GmbH
Kurfürstendamm 57
10707 Berlin, Germany

Blackwell Science KK
MG Kodenmacho Building
7–10 Kodenmacho Nihombashi
Chuo-ku, Tokyo 104, Japan

First published 1999

Set by Sparks Computer Solutions, Oxford
Printed and bound in Great Britain by MPG Books Ltd, Bodmin, Cornwall

A catalogue record for this title is available from the British Library

ISBN 0-632-05164-7

Library of Congress
Cataloging-in-publication Data

Challenges in growth hormone therapy / edited by John P. Monson
p. cm.
Includes bibliographical references and index
ISBN 0-632-05164-7
1. Somatotropin—Therapeutic use.
2. Dwarfism, Pituitary—Hormone therapy. I. Monson, John P.
RC658.7.C47 1999
616.4'7061—dc21

99-17635
CIP

DISTRIBUTORS

Marston Book Services Ltd
PO Box 269
Abingdon, Oxon OX14 4YN
(*Orders*: Tel: 01235 465500
Fax: 01235 465555)

USA
Blackwell Science, Inc.
Commerce Place
350 Main Street
Malden, MA 02148 5018
(*Orders*: Tel: 800 759 6102
781 388 8250
Fax: 781 388 8255)

Canada
Login Brothers Book Company
324 Saulteaux Crescent
Winnipeg, Manitoba R3J 3T2
(*Orders*: Tel: 204 837 2987)

Australia
Blackwell Science Pty Ltd
54 University Street
Carlton, Victoria 3053
(*Orders*: Tel: 3 9347 0300
Fax: 3 9347 5001)

For further information on Blackwell Science, visit our website www.blackwell-science.com

Contents

List of contributors

EDITOR

John P. Monson MD, FRCP, *Head of Division of General and Developmental Medicine, Reader in Endocrinology, Department of Endocrinology, Medical Unit, St Bartholomew's Hospital, London EC1A 7BE, UK*

CONTRIBUTORS

Roger Abs MD, PhD, *University Hospital Antwerp, Department of Endocrinology, Wilrijkstraat 10, B-2650 Edegem, Belgium*

Philippe Bareille MD, *The Institute of Child Health and Great Ormond Street Hospital, NHS Trust, London WC1N, UK*

Bengt-Åke Bengtsson MD, PhD, *Research Centre for Endocrinology and Metabolism, Sahlgrenska University Hospital, Sahlgrenska, S-413 45 Göteborg, Sweden*

Salem A. Beshyah PhD, MRCP, *Consultant Endocrinologist, Diabetes and Endocrine Unit, Princess Alexandra Hospital, Hamstel Road, Harlow CM20 1QX, UK*

Peter R. Betts MD, FRCP, FRCPCH, *Consultant Paediatrician, Department of Child Health, Southampton General Hospital, Tremona Road, Southampton SO16 6YD, UK*

Per Björntorp MD, PhD, *Department of Heart and Lung Diseases, Sahlgrenska University Hospital, Sahlgrenska, S-413 45 Göteborg, Sweden*

Nicola A. Bridges DM, MRCP, MRCPCH, *Consultant Paediatric Endocrinologist, Chelsea and Westminster Hospital, Fulham Road, London SW10 9NH, UK*

Pierre Chatelain MD, *Service de Pediatrie, INSERM U 307, Hôpital Debrousse, 29 Rue Soeur Bouvier, 69322 Lyon Cedex 05, France*

Lynda C. Doward BSc, *Senior Research Associate, Galen Research, Enterprise House, Manchester Science Park, Lloyd Street North, Manchester M15 6SE, UK*

William M. Drake MA, MRCP, *Department of Endocrinology, St Bartholomew's Hospital, London EC1A 7BE, UK*

David B. Dunger MD, FRCP, *Department of Paediatrics, Level 4, John Radcliffe Hospital, Oxford OX3 9QN, UK*

Fiona Frazer MB, BS, FRACP, *The Institute of Child Health and Great Ormond Street Hospital, NHS Trust, London WC1N, UK*

Peter C. Hindmarsh BSc, MD, FRCP, FRCPCH, *Reader in Paediatric Endocrinology, University College London, Cobbold Laboratories, Middlesex Hospital, Mortimer Street, London W1N 8AA, UK*

Anita C.S. Hokken-Koelega MD, PhD, *Consultant Pediatric-Endocrinologist, Sophia Children's Hospital, Erasmus University Rotterdam, PO Box 2060, 3000 CB Rotterdam, The Netherlands*

Jörgen Isgaard MD, PhD, *Research Center for Endocrinology and Metabolism, Gröna Stråket 8, Sahlgrenska University Hospital, S-413 45 Göteborg, Sweden*

Richard C. Jenkins BM, BS, MRCP, *Section of Medicine, Division of Clinical Sciences, Clinical Sciences Centre, Northern General Hospital, Sheffield S5 7AU, UK*

Gudmundur Johannsson MD, PhD, *Research Centre for Endocrinology and Metabolism, Sahlgrenska University Hospital, Sahlgrenska, S-413 45 Göteborg, Sweden*

Jean-Marc Kaufman MD, PhD, *Department of Endocrinology and Metabolic Bone Disease, University Hospital, Gent, Belgium*

Stephen P. McKenna BA, PhD, CPsychol, AFBPsS, *Director of Galen Research, Enterprise House, Manchester Science Park, Lloyd Street North, Manchester M15 6SE, UK*

Angelika Mohn MD, FRCP, *Department of Paediatrics, Level 4, John Radcliffe Hospital, Oxford OX3 9QN, UK*

Pierre Moulin MD, *Hôpital de Enfants, Unité d'Endocrinologie, CHU Purpan, Place du Docteur-Baylac, 31059 Toulouse Cedex, France*

Asad Rahim MBChB, MRCP, *Department of Endocrinology, Christie Hospital, Wilmslow Road, Manchester M20 4BX, UK*

Richard J.M. Ross MD, FRCP, *Reader in Endocrinology, Section of Medicine, Division of Clinical Sciences, Clinical Sciences Centre, Northern General Hospital, Sheffield S5 7AU, UK*

Stephen M. Shalet BSc (hons), MD, FRCP, *Professor of Medicine, Department of Endocrinology, Christie Hospital, Wilmslow Road, Withington, Manchester M20 4BX, UK*

Richard Stanhope BS, MD, DCH, FRCP, FRCPCH, *Consultant Paediatric Endocrinologist, Great Ormond Street Hospital for Children NHS Trust, London WC1N 3JH, UK*

Maïthé Tauber MD, *Hôpital des Enfants, Unité d'Endocrinologie, CHU Purpan, Place du Docteur-Baylac, 31059 Toulouse Cedex, France*

Tracy S. Tinklin BM, MRCP, MRCPCH, *Senior Registrar in Paediatric Endocrinology, Department of Child Health, Southampton General Hospital, Tremona Road, Southampton SO16 6YD, UK*

Andrew A. Toogood MB, CHB, MRCP, *Post-Doctoral Fellow, Department of Internal Medicine, Box 511, University of Virginia, Health Sciences Center, Charlottesville VA 22908, USA*

Mark Vandeweghe MD, PhD, *Department of Endocrinology, University Hospital, Gent, Belgium*

Johan Verhelst MD, *Middelheim Hospital Antwerp, Department of Endocrinology, Lindendreef 1, B-2020 Antwerpen, Belgium*

Christian Wüster MD, *University of Heidelberg, Department of Internal Medicine I, Endocrinology and Metabolism, Im Neuenheimer Feld 400, D-69120 Heidelberg, Germany*

Foreword

The availability of unlimited supplies of recombinant human GH has led, not surprisingly, to an extensive re-evaluation of clinical indications, both in the traditional endocrine setting and in the non-classical clinical areas. Dr John Monson has assembled an impressive array of practising clinicians and clinical scientists who have expertly reviewed some of the changes to practice, as well as extending our horizons.

One might have thought that the deliberations on GH treatment in children were settled but this still appears to be far from the case. GH is clearly indicated for treating children with GH deficiency (GHD), and there is agreement on its benefits in Turner's syndrome, but dosing schedules and effects in other syndromes still need resolving. In particular there is great interest in treating children with short stature who do not have classic GHD. The definition of short stature does, however, pose some problems and the results of GH responses to pharmacological stimuli may be ambiguous, or indeed, may be normal in children with abnormalities in the molecular structure of their GH or GH receptor. In addition the definition of GH *sufficiency* is arbitrary. Should auxological guidelines replace biochemical tests for the diagnosis of GHD and decisions regarding GH treatment? A simple definition from the American Academy of Pedatrics is that GH is recommended for very short children whose ability to participate in basic activities of daily living is limited as a result of their short stature and who have a condition for which GH is effective. How efficacious is GH therapy in increasing the final height of normal, short children? Studies have produced conflicting results and in few is there clear evidence that the final height of treated children was greater than the target height, nor that body image and self-esteem are improved. Part 1 provides some answers but raises even more questions.

There is an agreed and substantial body of evidence that mortality is increased in adults with hypopituitarism who are receiving conventional hormonal replacement with glucocorticoids, thyroid hormone and sex steroids. GHD is likely to be an important etiological factor but an additional contribution from unphysiological replacement by other pituitary hormones is possible. the availability of synthetically-derived GH has allowed us to explore these questions as never before. The supporting evidence for the detrimental effects of adult GHD is based primarily on the known adverse cardiovascular risk factors, the reduction of bone mass, and the poor quality of life in patients with adult GHD, much of which appears to be reversible with GH replacement. It thus could be argued that hormone deficiency automatically warrants hormone replacement and therefore GH replacement is justified for all adults with GHD.

More realistically, a selective approach to treatment of severe GHD is required, which after all is likely to be life-long, and thus priorities and guidelines need to be developed. Part 2 deals with these issues in a very professional and authoritative manner.

Part 3 identifies some of the difficulties associated with the transition from adolescence to adulthood, common to a number of endocrine disorders, but in this unusual circumstance the major indication for hormone replacement changes entirely. Finally some extremely challenging and exciting areas of non-traditional endocrinology where GH may have a role are discussed by experts in each field, completing an outstanding resume of the current status of treatment with GH.

Professor Michael Sheppard

Preface

The aim of this book is to pose questions, and attempt to provide answers, to fundamental questions relating to growth hormone (GH) deficiency and therapy across the whole age range and in the diverse situations in which treatment may be given in the future.

The utility of GH replacement in the paediatric setting of GH deficiency is well established spanning the use of cadaveric GH and recombinant human GH. The advent of the latter, with the subsequent increased availability of therapy, has expanded the potential paediatric clinical applications to include conditions associated with short stature, other than overt GH deficiency. More recently, the consequences of adult GH deficiency have been described and are increasingly recognized in clinical practice. As a result, GH replacement in the adult is now a licensed therapeutic indication in many countries and a large volume of literature testifies to its value, particularly in relation to body composition, bone mineral density and quality of life.

However, significant and fundamental questions remain regarding the use of GH therapy to treat short stature which is not specifically secondary to GH deficiency and it is this specific debate surrounding the use of GH in these situations which forms the basis for the first section of this book. In the context of adult GH deficiency, we have examined areas of current diagnostic and therapeutic controversy extending from the evidence for increased cardiovascular risk in hypopituitarism to safety data on long-term GH replacement. The transition from paediatric to adult care of pituitary disease and GH deficiency is assuming increasing importance and their fundamental issues are addressed in Part 3.

Non-classical uses of GH therapy currently under investigation in the adult include the management of osteoporosis, the 'metabolic syndrome' and cardiac failure. GH has also been used extensively in catabolic illness, burns and trauma but the adverse results of two recent trials of GH supplementation in the ICU setting prompt a re-evaluation of the rationale for this treatment. These issues have been defined and discussed in detail.

The production of this book has been dependent on the outstanding contributions of the chapter authors, all of whom are recognized authorities in their subject areas. I am grateful to them for the high quality of their work and scholarship. The reader will note some inevitable overlap in the subject matter of specific chapters. This is both inevitable and desirable, in that each chapter is designed to stand alone while obviously being linked with others.*

The expansion of clinical indications for GH therapy has been based on, and has given rise to, important advances in our understanding of the relationships between GH and GH-dependent factors and metabolism. I hope that the issues raised and discussed in this book will assist clinicians in the assessment and treatment of their patients.

John P. Monson

**Please note*: to assist you, specific references in each chapter have been denoted with one or two asterisks to indicate their importance as source material.

Part 1: GH Replacement in Childhood

1: GH replacement in children: what is the optimum dosing schedule?

Philippe Bareille, Fiona Frazer and Richard Stanhope

Introduction

Growth hormone (GH) deficiency is the usual indication for human GH (hGH) treatment in children. hGH is a single polypeptide of 191 amino acids synthesized in the anterior pituitary gland. It is secreted intermittently, mainly in a pulsatile fashion. Deficiency of this hormone causes severe growth failure with a final stature below 4 standard deviation (SD) scores. GH is not only important for growth, but also has additional major actions affecting glucose, lipids and protein metabolism.

GH-deficient children represent a heterogeneous group. GH deficiency can be either isolated or associated with other pituitary hormone deficiencies, and in either case it can be 'idiopathic' or 'organic'. Organic conditions include acquired GH deficiency (notably secondary to tumours and cranial irradiation) and congenital forms of GH deficiency (GH gene defect, congenital developmental abnormalities). Patients with GH deficiency may present clinically at any age from birth to adulthood.

Until the achievement of final stature, growth is the main parameter that is taken into consideration during GH replacement. The aim of growth hormone replacement is to help the child attain his or her target height by restoring a normal growth pattern. The achievement of that goal may depend on the underlying aetiology, the timing, dosage and duration of hGH treatment, and on whether the child is genuinely lacking GH. GH therapy was used initially in the late 1950s [1], and the hormone was extracted from the pituitary glands of cadavers. Treatment, with small doses being given three times a week, was suboptimal due to the limited supply. More than half of the children treated failed to exceed the third centile [2–5]. Moreover, adult heights were not always significantly higher than those reported in GH-deficient individuals who had never received GH substitution [6,7]. In 1984, the first cases of young patients affected by Creutzfeldt–Jakob disease were observed. A connection was rapidly established between this degenerative neurological disease and pituitary GH, which was then subsequently withdrawn. There was a gap of approximately nine months before biosynthetic GH became available. The advent of GH prepared by recombinant DNA techniques has allowed the use of higher and more appropriate doses, bringing about dramatic improvements in final height in GH-deficient individuals. In very young patients, GH deficiency may be responsible for fasting hypoglycaemia, which can justify early commencement of GH replacement even before the onset of growth failure.

Pathophysiology

GH is secreted in a pulsatile manner, with a dominant periodicity of 200 minutes [8]. Its secretion is under the control of at least two hypothalamic hormones: growth hormone–releasing hormone (stimulating effect) and somatostatin (inhibiting action). It is essential for growth as well as for normal carbohydrate, lipid, mineral and nitrogen metabolism. Growth effects are (mainly) mediated by insulin-like growth factor-I (IGF-I), a member of the insulin-like gene family. It is only recently that clear evidence has been obtained to show that GH affects somatic growth during intrauterine life [9–12]. GH receptors have been identified in fetal tissues, new-born infants with congenital GH deficiency are shorter than normal infants at birth, and postnatal growth is significantly impaired. However, the effect of GH on antenatal growth may be relatively minor [11], as the birth size of GH-deficient infants is reduced to a limited extent.

GH plays an important role both in gonadal development and in the maintenance of sexual maturation. GH acts on the gonads either by inducing local production of insulin-like growth factors [13] or by augmenting gonadotrophin-modulated gonadal function [14,15]. GH may facilitate the induction of ovulation by gonadotrophins in human females [16].

GH secretion increases significantly during puberty, with a GH secretion rate from mid-puberty to late puberty that is two to three times the secretion rate in the prepubertal period [17,18]. After that, physiological GH secretion decreases with age [19].

Current usage

GH is administered by daily single subcutaneous injections, usually in the evening to mimic the nocturnal secretion of GH, and also in young children to prevent possible nocturnal hypoglycaemia. Subcutaneous injections are better tolerated by patients and have a longer duration of action than intramuscular injections [20]. The average minimum weekly dose is 15 IU/m^2 (equivalent to 0.5 IU/kg or 0.19 mg/kg). GH therapy is commenced as early as possible when the diagnosis of GH deficiency has been ascertained, and is continued at least until the completion of growth. Doses are adjusted every six to 12 months according to the body surface area (square metre) or body weight (kilogram). A slight increase in the dose per kilogram body weight or dose per body surface area may be necessary when growth velocity decelerates, which may occur after several years of therapy. Weekly doses of up to 25 IU/m^2 (equivalent to 0.78 IU/kg or 0.3 mg/kg) have been proposed.

Auxological data are monitored every six months, and bone age is recorded every year. Follow-up is usually recommended in a specialized centre by a paediatric endocrinologist.

Initiation of GH therapy

There has been clear evidence that GH therapy should be initiated as soon as possible—even before the onset of growth failure—in order to optimize long-term growth (see below). Another important reason for early treatment is the prevention of fasting hypoglycaemia. In young children, GH plays an important role in the maintenance of normal plasma glucose levels—in contrast to adults, in whom GH and cortisol do not appear to play important roles in the prevention of short-term hypoglycaemia but are involved in the defence against prolonged hypoglycaemia [21]. Hopwood *et al.* [22] showed that symptomatic and asymptomatic hypoglycaemia occurred with equal frequency in children with isolated GH and multiple anterior pituitary deficiencies. Of 52 children studied, nine (17%) had symptomatic hypoglycaemia and 14 (27%) had asymptomatic hypoglycaemia. Spontaneous hypoglycaemia rarely occurred after the age of four years, and decreased with increasing body adiposity. Regular administration of GH to these patients is associated with improvement in clinically symptomatic hypoglycaemia, and usually results in complete amelioration of symptoms. However, the tendency towards hypoglycaemia, as detected by provocative testing, may persist for a period of many months after the initiation of hGH therapy.

The aetiology of hypoglycaemia in GH-deficient children remains obscure, mainly because of differences in study designs and because GH has only recently been changed to daily administration from a three times per week regimen. Possible aetiological factors for hypoglycaemia in young GH-deficient children include increased insulin sensitivity [23], hypoketonaemia [24] and reduced hepatic glucose production through diminished gluconeogenesis and/or abnormal glycogen utilization [25]. The exaggerated brain : body disproportion also plays a role by amplifying the normal tendency of all fasting children to become hypoglycaemic [23].

Optimum time and dosing schedule of GH therapy

Several factors have been identified to influence the growth outcome after initiation of GH therapy [26–29]. Early diagnosis and commencement of treatment, severity of GH deficiency, and GH dose have been found consistently to have a strong impact on the growth response. GH therapy started before the age of five years, i.e. after early diagnosis, appears to bring about complete catch-up and possibly ultimate achievement of the target height [12,29]. The duration of therapy seems to be an important factor—the longer the better. However, not all children may reach their target height centile [29,30]. Indeed, there seems to be a waning effect of GH therapy with time, with the occurrence of a plateau after four or five years before complete catch-up [29,30]. Even if the greatest response in the short term is observed in the most retarded children [29], only children treated early before the onset of significant growth failure may ultimately be able to attain their target height [29]. Recent final height data for individuals who had received higher GH doses (daily injection 0.3 mg/kg/week, or 25 IU/m^2/wk) clearly indicate that the most important factor influencing the final height outcome is the age at which treatment is initiated [27]. Thus, both a smaller height deficit at the start of GH therapy and longer duration of treatment appear to be prerequisites for the optimization of growth. Complete catch-up before the start of puberty seems important.

Controversial issues

Diagnosis of GH deficiency

It is established that the growth response depends on the severity of the GH deficiency, children with profound GH deficiency growing faster after initiation of GH treatment than those with less severe GH deficiency [31]. The difficulty in diagnosing GH deficiency may in part account for the variability in the growth response among individuals. The diagnosis of GH deficiency is commonly based on the GH response to a provocation test. A large number of stimuli are used in clinical practice, and are usually pharmacological. Insulin-induced hypoglycaemia, glucagon, arginine, clonidine and L-dopa are some of the commonest tests used in practice. The cut-off limit of adequate response depends on the assay used to measure plasma GH. These tests are considered to be the gold standard for diagnosing GH deficiency. Nevertheless, they are not without flaws: they may not reflect the physiological endogenous secretion; the development of different immunoassays of GH coupled with many different tests of various degrees of reliability render their interpretation difficult; their reproducibility is uncertain; and the threshold value between GH deficiency and normal response is purely empirical [32–38]. Serum IGF-I and insulin-like growth factor binding protein-3 (IGFBP-3) may be helpful in diagnosis, but normal values do not exclude the diagnosis [39,40]. A 24-hour GH profile (blood sampling every 15–20 minutes) is not easy to perform on a routine basis. However, the pattern of 24-hour GH secretion can be helpful in a short child with subnormal growth velocity but a normal response to provocative tests. Nevertheless, it should be pointed out that there is a significant overlap between the lower range in normal individuals and values reported in growth hormone deficiency. Hence, study results have been contradictory and have failed to establish a clear correlation between 24-hour GH secretion and auxological parameters [41,42].

Dose of GH (Table 1.1)

The optimum GH dose remains a controversial issue. Originally, the replacement dose was determined empirically, according to the growth response. Subsequently, the daily dose requirement was assessed by measuring plasma IGF-I, urine IGF-I, urine GH and plasma insulin [43,44], by measuring the plasma profile of GH after GH injections [20], or by carrying out a multiple-parameter deconvolution analysis of 24-hour serum GH concentrations [18]. A weekly dose of 15 IU/m^2 (0.5 IU/kg or 0.19 mg/kg) seems to be an adequate replacement dose before puberty. However, studies using higher doses (25 IU/m^2, equivalent to 0.3 mg/kg) have yielded promising results [27]. Moreover, some short-term studies have shown that the effects of GH therapy start to wane after the fourth year of treatment, even before the achievement of full catch-up (Fig. 1.1) [29,30]. It is unknown whether a slight increase in the GH dose or a longer period of therapy can be sufficient to normalize height in these children. On the other hand, an excessive increment in GH dose may advance bone age unduly, thereby compromising the final height outcome [45]. Individualized dosage of GH, titrated against GH-dependent serum markers, has not been undertaken in paediatric practice, in contrast to the evolving experience in adult GH replacement.

GH and puberty

It has been consistently demonstrated that delaying puberty improves final stature [28,46,47]. GH-deficient adults with induced puberty were found to be taller than those with spontaneous puberty. It is speculated that GH deficiency *per se* delays the onset of puberty [48], but that GH therapy accelerates the pubertal process [49]. Delaying the onset of puberty may

Table 1.1 Height outcome in children with growth hormone (GH) deficiency: results of four recent studies. The differences in the study design and numbers of patients do not allow strict comparisons between these data. Some patients were initially treated with pituitary GH therapy. In the studies by Blethen *et al.* [27] and by Price and Ranke [28] only data for boys are shown.

Author	Number of patients	GH dose IU/kg/wk	Duration GH (yr)	Age at enrolment (yr)	Baseline height SDS	Height SDS at onset of puberty	Final height SDS (apart from Rappaport study)	Target height SDS	Adult height SDS minus target height SDS	Comments
Blethen [28]	72	0.8	8 ± 3.2	11.3 ± 2.1	−3.0 ± 1.2	−1.9 ± 1.2	−0.7 ± 1.3	−0.2 ± 0.8	−0.6 ± 1.2	Multicentre prospective study
Coste [88]*	1152	0.4 ± 0.1	5.1 ± 2.2	9.7 ± 4	−3 ± 1.2		−1.9 ± 1	−0.8 ± 1.2		Register-based cohort study
Price [29]†	55	0.6	8.5 (7.1–11.4)	9.3 (4.6–12.9)	−2.8 (−4 to −1.7)	−1.59 (−2.6 to −0.4)	−1.03 (−2.4 to 0.3)	−0.4 (−1.5 to 1.1)	−0.7 (−2.6 to 0.5)	Multicentre prospective study
Rappaport [30]‡	25 (8)	0.6	3–5	1.4 ± 0.7 (0.7 ± 0.9)	−3.6 ± 1 (−1.1 ± 0.6)		−0.8 ± 1.2§ (0.35 ± 1)¶			Multicentre prospective study

*Data of children who started GH before puberty. Fifty-six were not GH-deficient but their removal does not change the results. †10th–90th centile given in brackets. ‡Two groups, one with growth retardation, the other (in brackets) without growth delay. §Height after five years of treatment. ¶Height after four years of treatment. SDS: standard deviation score.

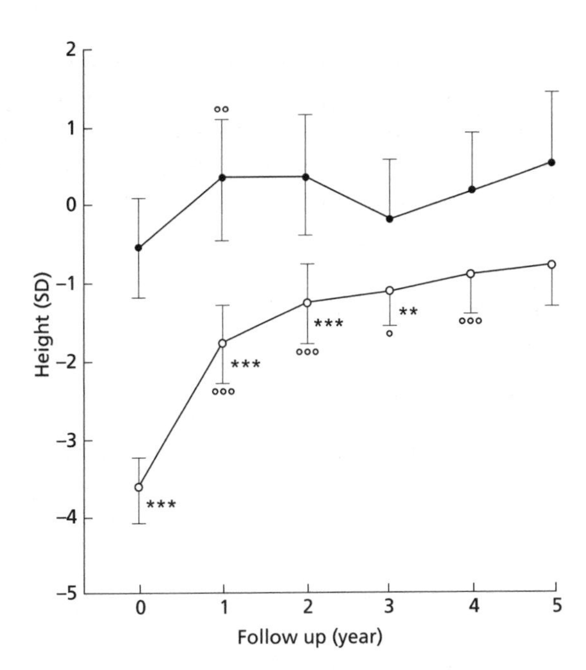

Fig. 1.1 Height standard deviation (SD) score during five years of growth hormone (GH) therapy. Comparison between patients without (group A, top) and with (group B, bottom) initial growth delay. Patients in group B exhibit a more dramatic response initially, but have incomplete catch-up due to waning GH effects. Reproduced with permission from Rappaport *et al.* [29].

therefore be an advantage in terms of final height outcome. However, the psychological sequelae of sexual immaturity may by far outweigh the benefits of the potential gains in final height. Moreover, a recent multicentre and prospective study did not corroborate the previous findings and showed no differences in final stature between patients with spontaneous puberty and those with gonadotrophin deficiency [27] as long as an adequate dosage regimen was used. This study therefore suggests that higher doses coupled with early treatment are sufficient to optimize growth without having recourse to delaying the onset of puberty.

Another important question is whether GH doses should be increased at puberty in order to mimic natural physiology. Interestingly, although we have a reasonable concept of GH dosage in prepuberty, there is very little agreement about the actual physiological importance of raised GH secretion at puberty. Undoubtedly, spontaneous GH secretion increases during puberty, presumably under the action of oestrogen derived from aromatization of androgen. However, whether the growth spurt is induced by sex steroids alone, or by a synergistic effect of raised GH and gonadal steroids, is not clear. In fact, it does not seem necessary to increase GH doses at puberty. Stanhope *et al.* [50] suggested that the short-term gain in height may eventually be offset by an acceleration of the pubertal process, but these findings are controversial. A temporary increase to twice-daily injections has also been proposed in order to maximize final height [19], but, to our knowledge, this hypothesis has not as yet been tested. Higher doses, bringing about complete normalization of growth before puberty, may be sufficient to optimize growth. Preliminary analysis of a recent study [51] suggests that higher GH doses during puberty (0.7 mg/kg/week) may increase final stature without unduly advancing bone maturation. These results need to be confirmed. The influence of gender on sensitivity to GH has not been addressed in paediatric practice, but the recent demonstration of increased GH responsiveness in adult males compared with females may have important implications for the peripubertal situation.

Frequency of injections

It is well established that the frequency of injections affects the growth response. GH-deficient children receiving daily injections exhibit more dramatic growth catch-up than those treated with three injections a week, even with a similar weekly dose [12,20,31,52]. Since GH levels return to baseline approximately 12 hours after a subcutaneous injection [44], it appears logical to administer GH in two doses at 12-hour intervals. However, in terms of anthropometric response there does not seem to be any advantage in doing this, at least in the short term [52,53]. In the study by Hindmarsh *et al.* [8], hGH was administered either in a single daily dose or three times daily, for a total weekly dose of 20 IU/m^2 in both groups. The growth response was not significantly greater in the twice-daily group after six months compared to the single daily group (Fig. 1.2), but the serum GH concentration profiles were closer to normal physiology.

Further evidence supporting the use of daily GH injections is the possible risk of hypoglycaemia with intermittent GH therapy. GH given three times per week has been associated with hypoglycaemia 36–60 hours after injections [54]. The signs and symptoms of hypoglycaemia resolved immediately when GH was given daily. GH raises plasma glucose by antagonizing the effects of insulin on both the production and uptake of glucose. In GH-deficient children, serum concentrations of GH peak at three to six hours after an injection and return to baseline within 24 hours [55], whereas circulating IGF-I levels do not peak until 19 hours after the injection, and then fall slowly with an apparent half-life of 20 hours [56]. Thus, on the second or third day after an injection, raised levels of insulin-like growth factors are not accompanied by elevated levels of GH. Under these circumstances, the insulin-like effect of endogenous somatomedins may become apparent, resulting in possible hypoglycaemia.

Houdijk *et al.* [57] documented asymptomatic nocturnal hypoglycaemia 8–12 hours after the GH injection in a group of 16 GH-deficient adolescents. One possible explanation the authors provided was the relatively insufficient amount of circulating GH to counteract the natural increase in

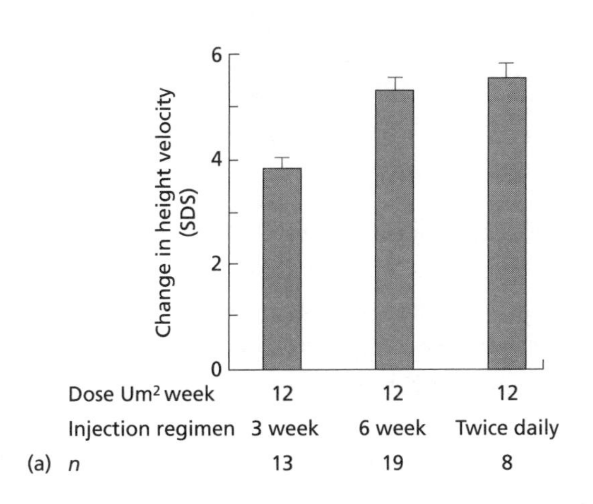

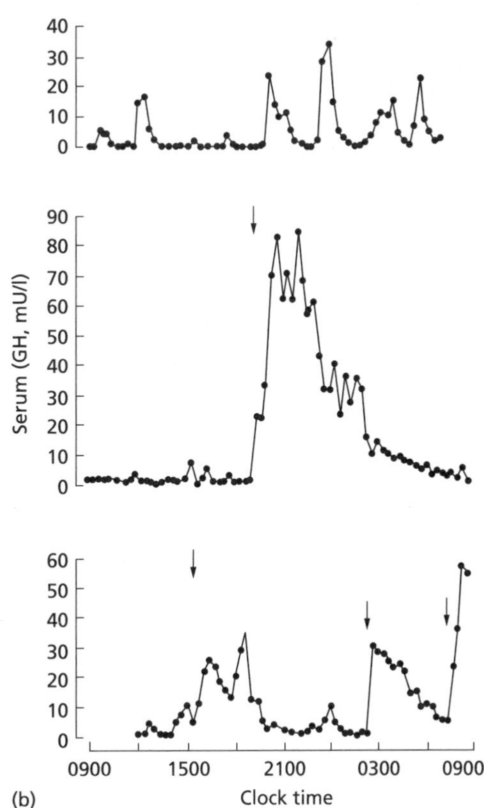

Fig. 1.2 (a) Growth effect of different frequencies of growth hormone (GH) administration; 3 times a week, 6 times a week and twice daily. (b) Twenty-four-hour GH profiles. *Upper panel*: normal profile from a seven-year-old boy. *Middle panel*: GH-deficient child receiving a subcutaneous recombinant human growth hormone (rhGH) injection of 2 IU (arrow). *Lower panel*: GH-deficient child receiving three rhGH injections of 0.75 IU. Reproduced with permission from Hindmarsh *et al.* [53].

insulin secretion from 3 AM to dawn. The authors were, however, puzzled by this unexpected observation. Acute and transitory insulin-like effects (independently of those induced by circulating IGF-I) of GH have been demonstrated *in vitro* [58] and experimentally *in vivo* [59]. In the latter case, infusion of GH induced greater hypoglycaemia than a saline infusion when the infusions were given along with somatostatin (to suppress GH, glucagon and insulin endogenous secretion) and insulin. GH administration was associated with greater suppression of hepatic glucose production and an inappropriate rate of glucose utilization. The insulin-like effects of GH disappeared after two hours. This experiment suggests that GH either has intrinsic insulin-like effects, or may increase insulin action. Changes in IGF-I were not observed within three hours during GH infusion, hence almost excluding the involvement of these factors in this study.

Several studies have assessed the growth effects of administering GH-releasing hormone [60–62] twice a day or every three hours using a pump. Preliminary results with this treatment up to 30 months were promising in some children. However, these trials were not pursued further, as there was no real advantage over GH treatment and there was the need for at least two injections a day or a pump.

Termination of GH therapy

While there is now little doubt about when GH therapy should be initiated, it is still unclear when it should be terminated. There has been increasing evidence over the past decade that GH continues to play an important role even after the cessation of growth [63]. However, most studies have been conducted in groups of patients with a wide range of ages in whom GH deficiency was mainly of adult onset. The benefits of GH therapy may not be the same in young adults with childhood-onset GH deficiency as they are in ageing individuals with adult-onset GH deficiency. A recent study [64] suggests that adults with childhood-onset GH deficiency and those with adult-onset GH deficiency represent two different entities. Those with adult-onset GH deficiency had lower plasma high-density lipoprotein–cholesterol, and were more obese and more affected by psychological distress than those with childhood-onset GH deficiency. By contrast to children, GH-deficient adults can develop insulin resistance through defects in glucose utilization [65]. Increased abdominal obesity is an important factor associated with insulin resistance, but other factors are likely to be involved as well. The marked reduction in insulin sensitivity, coupled with hyperlipidaemia [66–68], is responsible for an increased prevalence of cardiovascular disease in these individuals [69]. One study [70] was not able to demonstrate any significant alterations in the lipid profile (apart from elevated triglycerides) or gross insulin resistance in a group of young adults with childhood onset of GH deficiency. However, there was a significant increase in the intima–media thickness of the carotid arteries compared to controls. Whether the continuation of GH treatment after cessation of growth in GH-deficient adolescents can prevent cardiovascular diseases in the long term is still unknown. Optimization of treatment in terms of dose and frequency therefore remains an open issue. For example, it may be advantageous to pursue GH treatment as long as bone density is increasing, which may still occur after the cessation of growth, and thereby maximize the peak bone density. These questions are further discussed in Chapters 14 and 15 of the present volume.

GH secretion should be rechecked in late adolescence to confirm or reject the diagnosis of GH deficiency, especially in idiopathic, partial and isolated GH deficiency. Indeed, the majority of patients with childhood-onset GH deficiency have a normal GH status on retesting at completion of height [71]. Tauber *et al.* [71] reassessed 131 young adults with childhood-onset GH deficiency. Seventy-one per cent of patients with partial GH deficiency had a normalized GH test, compared to 36% with complete GH deficiency. Normalization occurred in 67% of patients with idiopathic GH deficiency, but only in 10% of those with organic GH deficiency. These findings have been corroborated by other authors [72].

Side-effects

The use of biosynthetic growth hormone has definitely averted the risk of the fatal Creutzfeldt–Jacob disease [73], but the absolute safety of GH therapy remains to be firmly established. The possibility of inducing leukaemia and tumours has not been confirmed [74,75]. The enormous benefits of GH therapy have been confirmed in classical GH deficiency (at least in the paediatric age group), certainly outweighing untoward reactions. We can postulate that side-effects are more likely to occur during high-dosage regimens. It is beyond the scope of this chapter to discuss in detail the 'potential' side-effects of GH therapy. However, the potential risk of glucose intolerance or diabetes should be mentioned. Excessive GH secretion, as seen in acromegaly or gigantism, is commonly associated with hyperinsulinaemia, insulin resistance and glucose intolerance [76,77]. The diabetogenic nature of GH has been demonstrated experimentally. Most experimental studies were performed in healthy adults given continuous GH infusions for several hours. Insulin sensitivity and glucose production and utilization were assessed using clamp studies (hyperglycaemic and euglycaemic hyperinsulinaemic clamps). GH induced a shift in the dose–response curve towards higher values indicating insulin resistance. Insulin resistance can be induced by 'physiological' levels of GH (equivalent to nocturnal peaks of GH). The antagonistic effect of GH on insulin action is mainly exerted on glucose utilization (peripheral glucose uptake), but a moderate effect on hepatic glucose production is also observed [78,79]. Hence careful monitoring of glucose homeostasis in all children treated with GH therapy is recommended, especially in susceptible individuals such as those with intrauterine growth retardation [80,81] and Turner syndrome [82].

The only proven 'serious' adverse event is the occurrence of 'benign cranial hypertension' [83]. However, this is extremely rare and is reversible on cessation of GH therapy. It is dose-dependent, and usually does not recur after restarting GH therapy at a lower dose [84].

Other indications for GH therapy

The development of hGH produced by recombinant DNA technology has dramatically broadened the spectrum of indications for hGH therapy. GH is notably used in Turner syndrome, intrauterine growth retardation, renal failure and 'idiopathic' growth failure [85]. These indications are discussed in detail elsewhere in this book.

Summary

GH therapy, beginning as early as possible before the onset of significant height deficit, can normalize growth, with ultimate achievement of the genetic target height in GH-deficient children. The minimum weekly dose required is 15 IU/m^2/day (approximately equivalent to 0.2 mg/kg). However, the use of higher doses (up to 25 IU/m^2/day, equivalent to 0.3 mg/kg) may be necessary to optimize the final height outcome. Normalization of height should have occurred before the onset of puberty. Any increment in the dosage should be introduced cautiously, to detect any undue advancement of bone age. It is now well recognized that GH therapy ought to be administered daily. It is still unclear whether augmenting GH doses during puberty is beneficial, or deleterious in terms of final height as a result of accelerating the pubertal process. It seems that continuation of GH therapy after the completion of growth is beneficial in some patients, but this remains an area of intensive research. In summary, very early and prolonged treatment are the two main key factors for optimizing GH therapy in GH-deficient children.

Acknowledgement

We are grateful to the Child Growth Foundation in the United Kingdom for funding P.B. and F.F., and to Serono UK for secretarial support.

References

1 Raben MS. Treatment of a pituitary dwarf with human growth hormone. *J Clin Endocrinol Metab* 1958; **18**: 901–3.
2 Burns EC, Tanner JM, Preece MA, Cameron N. Height and pubertal development in 55 children with idiopathic growth hormone deficiency treated for between 2 and 15 years with human growth hormone. *Eur J Pediatr* 1981; **137**: 155–63.
3 Joss E, Zuppinger K, Schwarz HP, Roten H. Final height of patients with pituitary growth failure and changes in growth variables after long term hormonal therapy. *Pediatr Res* 1983; **17**: 676–9.
4 Bourguignon JP, Vandeweghe M, Vanderschueren-Lodeweyckx M *et al.* Pubertal growth and final height in hypopituitary boys: a minor role of bone age at onset of puberty. *J Clin Endocrinol Metab* 1986; **63**: 376–82.
5 Hibi I, Tanaka T. Final height of patients with idiopathic growth hormone deficiency after long term growth hormone treatment. *Acta Endocrinol* 1989; **120**: 409–15.
6 Van der Werff ten Bosch JJ, Bot A. Treatment with hGH has no effect on adult height in hypopituitary children. *Pediatr Res* 1988; **23**: 114.
7 Van der Werff ten Bosch JJ, Bot A. Growth of males with idiopathic hypopituitarism without growth hormone treatment. *Clin Endocrinol* 1990; **32**: 707–17.

* 8 Hindmarsh PC, Matthews DR, Brook CGD. Growth hormone secretion in children determined by time series analysis. *Clin Endocrinol* 1988; **29**: 35–44.
9 Hill DJ, Riley SC, Bassett NS, Waters MJ. Localization of thr growth hormone receptor, identified by immunocytochemistry in second trimester human fetal tissues and in placenta throughout gestation *J Clin Endocrinol Metab* 1992; **75**: 646–650.
10 Wit JM, van Unen H. Growth of infants with neonatal growth hormone deficiency. *Archives Dis Chil* 1992; **67**: 920–924.
11 Gluckman PD. Growth hormone deficiency diagnosed and treated in the first two years of life: evidence of the role of growth hormone in human perinatal growth In: Ranke MB, Gunnarson R, eds *Progress in Growth Therapy: 5* Years of KIGS. Mannheim: J & J-Verlag, 1994: 88–96.
12 Boersma B, Rikken B, Wit JM. Catch-up in early treated patients with growth hormone deficiency (Dutch Growth Hormone Working Group). *Arch Dis Child* 1995; **72**: 427–31.
13 Davoren JB, Hsueh AJ. Growth hormone increases ovarian levels of immunoreactive somatometin C/insulin-like growth factor I *in vivo*. *Endocrinology* 1986; **118**: 888–890.
14 Adashi EY, Resnick CE, D'Ercole AJ, Svoboda ME, Van-Wyk JJ. Insulin-like growth factors as intraovarian regulators of granulosa cell growth and function. *Endocrine Review* 1985; **6**: 400–420.
15 Mason HD, Martikainen H, Beard RW, Anyaoku V, Franks S. Direct gonadotrophic effect of growth hormone on oestradiol production by human granulosa cells *in vitro*. *Journal of Endocrinology* 1990; **126**: R1–R4.
16 Homburg R, Eshel A, Abdalla HI, Jacobs HS. Growth hormone facilitates ovulation induction by gonadotrophins. *Clinical Endocrinology* 1988; **29**: 113–117.
17 Mauras N, Blizzard RM, Link K *et al.* Augmentation of growth hormone secretion during puberty: evidence for a pulse amplitude–modulated phenomenon. *J Clin Endocrinol Metab* 1987; **64**: 596–601.
** 18 Martha PM, Gorman KM, Blizzard RM, Rogol AD, Veldhuis JD. Endogenous growth hormone secretion and clearance rates in normal boys, as determined by deconvolution analysis: relation to age, pubertal status, and body mass. *J Clin Endocrinol Metab* 1992; **74**: 336–44.
19 Ho KKY, Hoffman DM. Aging and growth hormone. *Horm Res* 1993; **40**: 80–6.
** 20 Albertsson-Wikland K, Westphal A, Westgren U. Daily subcutaneous administration of human growth hormone in growth hormone-deficient children. *Acta Paediatr Scand* 1986; **75**: 89–97.
21 Boyle PJ, Cryer PE. Growth hormone, cortisol, or both are involved in defense against, but are not critical to recovery from, hypoglycemia. *Am J Physiol* 1991; **260**: E395–E402.
* 22 Hopwood NJ, Forsman PJ, Kenny FM, Drash AL. Hypoglycemia in hypopituitary children. *Am J Dis Child* 1975; **129**: 918–29.
23 Soman V, Tamborlane W, DeFronzo R, Genel M, Felig P. Insulin binding and insulin sensitivity in isolated growth hormone deficiency. *N Engl J Med* 1978; **299**: 1025–30.
* 24 Wolfsdorf JI, Sadeghi-Nejad A, Senior B. Hypoketonemia and age-related fasting hypoglycemia in growth hormone deficiency. *Metabolism* 1983; **32**: 457–62.
25 Haymond MW, Karl I, Weldon VV, Pagliara AS. The role of GH and cortisone on glucose and gluconeogenic substrate regulation in fasted hypopituitary children. *J Clin Endocrinol Metab* 1976; **42**: 846–56.
26 Vanderschueren-Lodewyckx M, Van Den Broeck J, Wolter R, Malvaux P. Early initiation of growth hormone treatment: influence on final height. *Acta Paediatr Scand Suppl* 1987; **337**: 4–11.
** 27 Blethen SL, Baptista J, Kuntze J *et al.* Adult height in growth hormone (GH)-deficient children treated with biosynthetic GH (Genentech Growth Study Group). *J Clin Endocrinol Metab* 1997; **82**: 418–20.
* 28 Price DA, Ranke MB. Final height following growth hormone treatment. In: Ranke MB, Gunnarson R, eds. *Progress in Growth Therapy: 5 Years of*

KIGS. Mannheim: J & J-Verlag, 1994: 129.

29 Rappaport R, Mugnier E, Limoni C *et al.* A 5-year prospective study of growth hormone (GH)-deficient children treated with GH before the age of 3 years. *J Clin Endocrinol Metab* 1997; **82: 452–6.

30 Arrigo T, Bozzola M, Cavallo L *et al.* Growth hormone deficient children treated from before two years old fail to catch-up completely within five years of therapy. *J Pediatr Endocrinol Metab* 1998; **11**: 45–50.

31 Blethen SL, Compton P, Lippe BM, Rosenfeld RG, August GP, Johanson A. Factors predicting the response to growth hormone (GH) therapy in prepubertal children with GH deficiency. *J Clin Endocrinol Metab* 1993; **76**: 574–579.

32 Bercu BB, Shulman D, Root AW, Spiliotis BE. Growth hormone (GH) provocative testing frequently does not reflect endogenous GH secretion. *J Clin Endocrinol Metab* 1986; **63**: 709–16.

33 Reiter EO, Morris AH, McGillivray MH, Weber D. Variable estimates of serum growth hormone concentrations by different radioassay systems. *J Clin Endocrinol Metab* 1988; **66**: 68–71.

34 Ceniker AC, Chen AB, Wert RM, Sherman BM. Variability in the quantification of circulating growth hormone using commercial immunoassays. *J Clin Endocrinol Metab* 1989; **68**: 469.

35 Rosenfeld RG, Albertsson-Wikland K, Cassorla F *et al.* Diagnostic controversy: the diagnosis of childhood growth hormone deficiency revisited. *J Clin Endocrinol Metab* 1995; **80**: 1532–40.

*36 Ghigo E, Bellone J, Aimaretti G *et al.* Reliability of provocative tests to assess growth hormone secretory status: study in 472 normal children. *J Clin Endocrinol Metab* 1996; **81**: 3323–7.

37 Carel JC, Tresca JP, Letrait M *et al.* Growth hormone testing for the diagnosis of growth hormone deficiency in childhood: a population register-based study. *J Clin Endocrinol Metab* 1997; **82**: 2117–21.

38 Rosenfeld RG. Is growth hormone deficiency a viable diagnosis? [editorial]. *J Clin Endocrinol Metab* 1997; **82**: 349–51.

39 Juul A, Skakkeboek NE. Prediction of the outcome of growth hormone provocative testing in short children by measurement of serum levels of insulin-like growth factor I and insulin-like growth factor binding protein 3. *J Pediatr* 1997; **130**: 197–204.

40 Tillmann V, Buckler JMH, Kibirige MS *et al.* Biochemical tests in the diagnosis of childhood growth hormone deficiency. *J Clin Endocrinol Metab* 1997; **82**: 531–5.

41 Albertsson-Wikland K, Rosberg S, Karlberg J, Groth T. Analysis of 24-hour growth hormone profiles in healthy boys and girls of normal stature: relation to puberty. *J Clin Endocrinol Metab* 1994; **78**: 1195–201.

42 Pozo J, Argente J, Barrios V *et al.* Growth hormone secretion with normal variants of short stature. *Horm Res* 1994; **41**: 185–92.

43 Hibi I, Tanaka T, Yano H *et al.* An attempt to assess the replacement dose of human growth hormone in the treatment of growth hormone deficient children. *Acta Paediatr Scand Suppl* 1987; **337**: 87–92.

44 Jørgensen JO, Flyvbjerg A, Lauritzen T *et al.* Dose–reponse studies with biosynthetic human growth hormone (GH) in GH-deficient patients. *J Clin Endocrinol Metab* 1988; **67**: 36–40.

45 Zadik Z, Zung A, Chen M, Fink A. Dose-related differences of growth hormone (GH) effect on bone age [abstract]. *Horm Res* 1997; **48** (Suppl 2): 81.

*46 Hibi I, Tanaka T, Tanae A *et al.* The influence of gonadal function and the effect of gonadal suppression treatment on final height in growth hormone (GH)-treated GH-deficient children. *J Clin Endocrinol Metab* 1989; **69**: 221–6.

47 Tanaka T, Hibi I. Bone maturation in growth hormone treated growth hormone deficient boys with and without gonadal suppression treatment. *Clin Pediatr Endocrinol* 1992; **1**: 101.

48 Tanaka T. Pubertal aspects of idiopathic growth hormone deficiency. In: Ranke MB, Gunnarsson R, eds. *Progress in Growth Hormone Therapy: 5 Years of KIGS*. Mannheim: J & J-Verlag, 1994: 112–28.

*49 Darendelier F, Hindmarsh PC, Preece MA, Cox L, Brook CGD. Growth hormone increases rate of pubertal

maturation. *Acta Endocrinol (Copenh)* 1990; **122**: 414.
50 Stanhope R, Uruena M, Hindmarsh P, Leiper AD, Brook CGD. Management of growth hormone deficiency through puberty. *Acta Paediatr Scand Suppl* 1991; **372**: 47–52.
51 Saenger P, Mauras N, Reiter EO *et al.* Effects of high-dose rhGH therapy in adolescent children with GH deficiency: a randomized, multicenter study [abstract]. *Horm Res* 1998; **49**: O15.
*52 Smith PJ, Hindmarsh PC, Brook CGD. Contribution of dose and frequency of administration to the therapeutic effect of growth hormone. *Arch Dis Child* 1988; **63**: 491–4.
*53 Hindmarsh PC, Stanhope R, Preece MA, Brook CGD. Frequency of administration of growth hormone: an important factor in determining growth response to exogenous growth hormone. *Horm Res* 1990; **33** (Suppl 4): 83–9.
*54 Press M, Notarfrancesco A, Genel M. Risk of hypoglycaemia with alternate-day growth hormone injections. *Lancet* 1987; **i**: 1002–4.
55 Parker ML, Utiger RD, Daughaday WH. Studies on human growth hormone: the physiological disposition and metabolic fate of human growth hormone in man. *J Clin Invest* 1962; **41**: 262–8.
56 Blethen SL, Daughaday WH, Weldon VV. Kinetics of the somatomedin C/insulin-like growth factor I response to exogenous growth hormone in growth hormone deficient children. *J Clin Endocrinol Metab* 1982; **54**: 986–90.
57 Houdijk EC, Herdes E, Delemarre-Van de Waal HA. Pharmacokinetics and pharmacodynamics of recombinant human growth hormone by subcutaneous jet- or needle-injection in patients with growth hormone deficiency. *Acta Paediatr Scand* 1997; **86**: 1301–7.
58 Birnbaum RS, Goodman HM. Comparison of several insulin-like effects of growth hormone. *Horm Metab Res* 1979; **11**: 136–42.
59 MacGorman LR, Rizza RA, Gerich JE. Physiological concentrations of growth hormone exert insulin-like and insulin antagonistic effects on both hepatic and extrahepatic tissues in man. *J Clin Endocrinol Metab* 1981; **53**: 556–9.
60 Smith PJ, Brook CGD, River J, Vale W, Thorner MO. Nocturnal pulsatile growth hormone releasing hormone treatment in growth hormone deficiency. *Clin Endocrinol* 1986; **25**: 35–44.
61 Ross RJM, Tsagarakis S, Grossman A *et al.* Treatment of growth hormone deficiency with growth hormone releasing hormone. *Lancet* 1987; **i**: 5–8.
62 Thorner MO, Rogol AD, Blizzard RM. Acceleration of growth rate in growth hormone-deficient children treated with growth hormone–releasing hormone. *Pediatr Res* 1988; **24**: 145–51.
63 Carroll PV, Christ ER, Growth Hormone Research Society Scientific Committee. Growth hormone deficiency in adulthood and the effects of growth hormone replacement: a review. *J Clin Endocrinol Metab* 1998; **83**: 382–95.
64 Attanasio AF, Lamberts SWJ, Matranga AMC *et al.* Adult growth hormone (GH)-deficient patients demonstrate heterogeneity between childhood onset and adult onset before and during human GH treatment (Adult Growth Hormone Deficiency Study Group). *J Clin Endocrinol Metab* 1997; **82**: 82–8.
65 Hew FL, Oschmann M, Christopher M *et al.* Insulin resistance in growth hormone-deficient adults: defects in glucose utilization and glycogen synthase activity. *J Clin Endocrinol Metab* 1996; **81**: 555–64.
66 Cuneo RC, Salomon F, Watts GF, Hesp R, Sonksen PH. Growth hormone treatment improves serum lipids and lipoproteins in adults with growth hormone deficiency. *Metabolism* 1993; **42**(12):1519–23.
67 De Boer H, Blok GJ, Voerman HJ, Phillips M, Schouten JA. Serum lipid levels in growth hormone-deficient men. *Metabolism* 1994; **43**(2):199–203.
68 Al-Shoumer KAS, Gray R, Anyaoku V *et al.* Effects of four years' treatment with biosynthetic human growth hormone (GH) on glucose homeostasis, insulin secretion and lipid metabolism in GH-deficient adults. *Clin Endocrinol* 1998; **48**: 795–802.
*69 Rosén T, Bengtsson BÅ. Premature mortality due to cardiovascular disease in hypopituitarism. *Lancet* 1990; **336**: 285–8.
70 Capaldo B, Patti L, Oliviero U *et al.* Increased arterial intima–media thickness in childhood-onset growth

hormone deficiency. *J Clin Endocrinol Metab* 1997; **82**: 1378–81.
*71 Tauber M, Moulin P, Pienkowski C, Jouret B, Rochiccioli P. Growth hormone (GH) retesting and auxological data in 131 GH-deficient patients after completion of treatment. *J Clin Endocrinol Metab* 1997; **82**: 352–6.
72 Wacharasindhu S, Cotterill AM, Camacho-Hubner C, Besser GM, Savage MO. Normal growth hormone secretion in growth hormone insufficient children retested after completion of linear growth. *Clin Endocrinol* 1996; **45**: 553–6.
73 Preece M. Human pituitary growth hormone and Creutzfeldt–Jakob disease. *Horm Res* 1993; **39**: 95–8.
74 Allen DB. Risk of leukemia in children treated with human growth hormone: review and reanalysis. *J Pediatr* 1997; **131**: S32–S36.
*75 Allen DB. Safety of human growth hormone therapy: current topics. *J Pediatr* 1996; **128**: S8–S13.
76 Muggeo M, Bar RS, Roth J, Kahn R, Gorden P. The insulin resistance of acromegaly: evidence for two alterations in the insulin receptor on circulating monocytes. *J Clin Endocrinol Metab* 1979; **48**: 17–25.
77 Trimble ER, Atkinson AB, Buchanan KD, Hadden DR. Plasma glucagon and insulin concentrations in acromegaly. *J Clin Endocrinol Metab* 1980; **51**: 626–31.
78 Bratusch-Marrain R, Smith D, DeFronzo RA. The effect of growth hormone on glucose metabolism and insulin secretion in man. *J Clin Endocrinol Metabol* 1982; **55**: 973–82.
79 Fowelin J, Attvall S, Von Schenck H, Smith U, Lager I. Characterization of the insulin-antagonist effect of growth hormone in man. *Diabetologia* 1991; **34**: 500–6.
80 Phillips DIW, Barker DJP, Hales CN, Hirst S, Osmond C. Thinness at birth and insulin resistance in adult life. *Diabetologia* 1994; **37**: 150–4.
81 Leger J, Levy-Marchal C, Bloch J *et al.* Reduced final height and indications for insulin resistance in 20-year-olds born small for gestational age: regional cohort study. *Br Med J* 1997; **315**: 341–7.
82 Wilson DM, Frane JW, Sherman B *et al.* Carbohydrate and lipid metabolism in Turner syndrome: effect of therapy with growth hormone, oxandrolone and a combination of both. *J Pediatr* 1988; **112**: 210–17.
83 Malozowski S, Tanner LA, Wysowski D, Fleming GA, Stadel BV. Benign intracranial hypertension in children with growth hormone deficiency treated with growth hormone. *J Pediatr* 1995; **126**: 996–9.
84 Crock PA, McKenzie JD, Nicoll AM *et al.* Benign intracranial hypertension and recombinant growth hormone therapy in Australia and New Zealand. *Acta Paediatr* 1998; **87**: 381–6.
85 Coste J, Letrait M, Carel JC *et al.* Long-term results of growth hormone treatment in France in children of short stature: population register-based study. *Br Med J* 1997; **315**: 708–13.

2: What is the current status of GH therapy in children with Turner syndrome?

Tracy S. Tinklin and Peter R. Betts

Introduction

Adult women with Turner syndrome are noticeably shorter than their peers, and many families of children with this condition are willing to try treatment that might improve the final adult height. Several combinations of hormonal preparations have been tested for maximizing growth, but the variability of individual growth and conflicting evidence on the benefits of such preparations have made it difficult to counsel families about treatment during childhood.

Growth hormone has been licensed for use in the treatment of Turner syndrome for the past 10 years in the United Kingdom. Initial information on short-term growth looked promising, and this led to the widespread use of growth hormone internationally. However, as final height data are becoming available, controversy remains about whether this type of treatment can significantly alter adult height, and there has been much discussion about the optimal treatment regimen for growth promotion in Turner syndrome [1–3]. This chapter will discuss the benefits and risks of growth hormone treatment peculiar to Turner syndrome.

Are there psychosocial consequences of short stature in Turner syndrome?

It is generally assumed that short stature affects the quality of life. Busschbach investigated the effect of stature on aspects of normal life, looking at adults with a variety of conditions, including Turner syndrome, that cause short stature [4]. Those with Turner syndrome admitted to experiencing difficulty in finding employment, although the majority had eventually been successful. They were less likely to have found a partner than other short people, but while height was a concern, problems regarding their insecurity about infertility were considered more important. The majority of this selected group of women wanted to be taller, and an interview technique designed to justify the cost of treatment was used to assess the importance of stature. The women with Turner syndrome were asked to 'trade off' the number of years that they would be willing to lose at the end of their life in order to be taller. Forty-four per cent of this population were prepared to lose some years, but the amount of time was small, suggesting that the impact of short stature on the quality of life might be less significant than supposed. Adults with Turner syndrome express low self-esteem, and many admit to being dissatisfied with their lives [5,6], but features of the syndrome other than short stature may account for this lack of contentment.

If an improvement in final adult height might resolve some of the perceived problems affecting the quality of life, would the cost of growth hormone treatment for 10 years be justified?

What is the untreated growth pattern in Turner syndrome?

The mean reduction in adult height potential for a woman with Turner syndrome is in the order of 20 cm [7]. Parental height is the most important factor determining final adult height for the individual girl [8–11]. As with all children, there is a strongly positive correlation between mid-parental height and the height of the child, so a girl with Turner syndrome with tall parents may be expected to reach an adult height that is at the lower end of the normal range.

The ethnic origin of a child will also affect height [9], with the mean adult height of an untreated girl with Turner syndrome in Japan being only 138 cm [12], compared with a mean height in Northern Europe of 147 cm [7].

The majority of affected girls have a fairly normal growth rate during early childhood. Unless the diagnosis is made antenatally, or there are phenotypic features suggestive of Turner syndrome associated with low birth weight in the neonatal period, the diagnosis may not be made until growth failure becomes apparent in mid-childhood [13,14].

All children in the United Kingdom are screened for height at school entry to identify growth problems. Fifty per cent of girls with Turner syndrome will be below the 0.4th centile on the UK 1990 growth charts [15] at this age, and they should then be identified if not already diagnosed. An additional proportion of girls with Turner syndrome will be identified if the target height is taken into consideration, as they will be small for family size. However, the remainder will not be picked up from a single height measurement, and although their growth rate is below that of normal girls, it may take up to four years to fall by as much as one centile band [16].

Absence of the adolescent growth spurt in girls with Turner syndrome accentuates their difference from the normal population. In normal puberty, endogenous oestrogen production leads to the growth spurt and also accelerates maturation of the cartilage, causing fusion of the epiphysis. In Turner syndrome, the failure of oestrogen production causes growth failure and marked delay in the closure of the epiphyseal plate, leading to a prolonged growth phase, with the final height often not being reached until nearly 20 years of age [13].

It has been suggested that the chromosomal pattern of an individual girl with Turner syndrome may have an effect on spontaneous growth [14]. Although most studies have shown no difference in the growth of girls with Turner syndrome when comparing the 45XO karyotype to all others [8–10,17], one showed that those with X mosaicism have a growth spurt between the age of eight and 12 years [18]. The same girls had spontaneous breast development and less delay in bone age than the girls without mosaicism, suggesting the presence of residual functioning ovarian tissue. Even those girls with no visible ovaries on pelvic ultrasound seemed to have a growth spurt. Despite this 'pubertal' growth spurt, the girls with mosaicism reached a shorter adult height than the 45XO girls, suggesting that a surge of endogenous oestrogen may be detrimental to final height.

Growth standards

Ranke's and Lyon's groups have designed growth charts specific to Turner syndrome, using cross-sectional height data for European girls. The Ranke chart [13] uses data from 150 German girls and gives the mean height plus or minus two standard deviations (Fig. 2.1). The height data used by Lyon [17] were collated from four studies of 366 untreated girls, including the girls from the Ranke study. There were between 24 and 138 height measurements used at different ages. The chart gives a range of heights between the third and 97th centile superimposed on Tanner–Whitehouse standards for normal girls (Fig. 2.2). The height standards show that, as in the normal population, there is a wide variation in height for the Turner population. Although the Ranke and Lyon data are derived from relatively small numbers of girls with Turner syndrome, larger populations have since been compared to the standards, and have been shown to have similar heights [9].

The mean height at different ages differs between the two charts. This is illustrated in Tables 2.1 and 2.2, from which it can be seen that while the mean height of girls at the age of five years is comparable, the older Lyon girls are shorter than the Ranke girls, finishing at a mean adult height of 142.9 cm at the age of 20 years, compared to 146.8 cm for the Ranke girls. This difference may potentially be due to the relatively small number of cases in each group, or may be the result of the varying heights of the population groups studied.

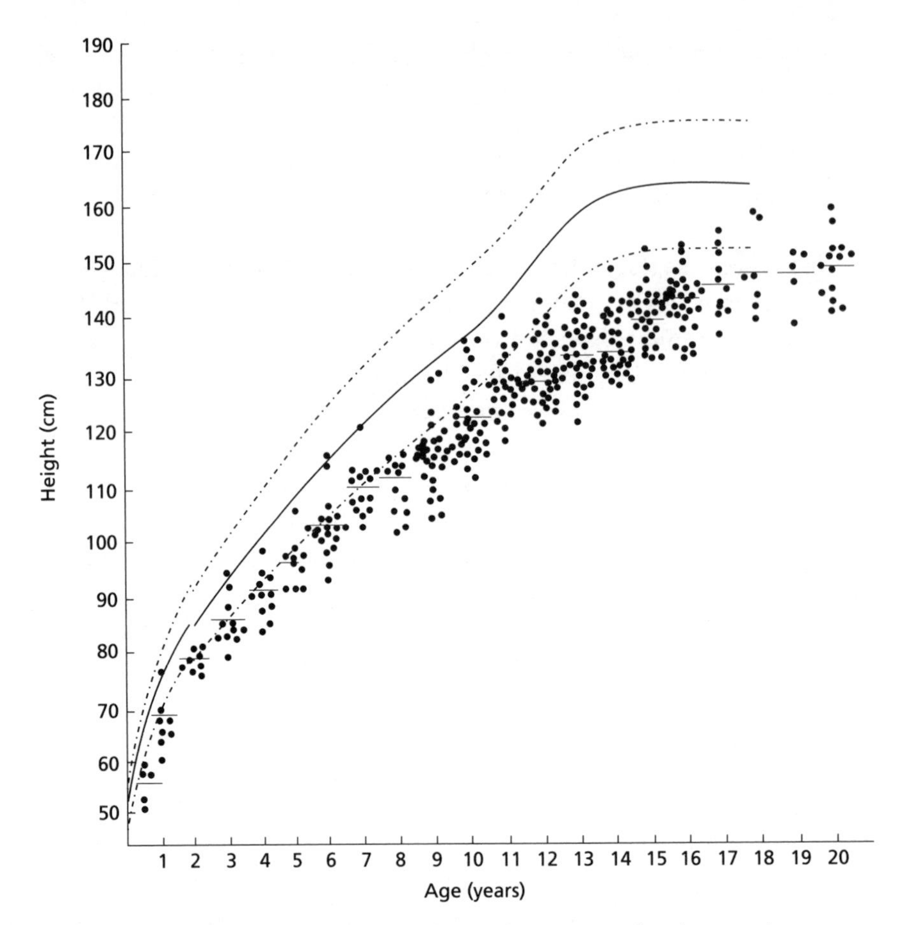

Fig. 2.1 Height in patients with Turner syndrome (dots) compared to the normal range (lines). Reproduced with permission from Ranke *et al.* [13].

The difference between the two growth charts is very important when interpreting final height data in growth studies. Using the Lyon data, the treated girls will appear to have gained more height than if the Ranke data are used [3,19,20].

Predicting adult height

The Lyon standards were used to derive a regression equation from the relationship between the height standard deviation score (SDS) during childhood and at final adult height for 29 British girls who had completed growth [17]. This method has since been used to predict adult height by projecting the current height centile, or more specifically the SDS, into maturity. This Lyon method of 'projected adult height' has since been validated by other

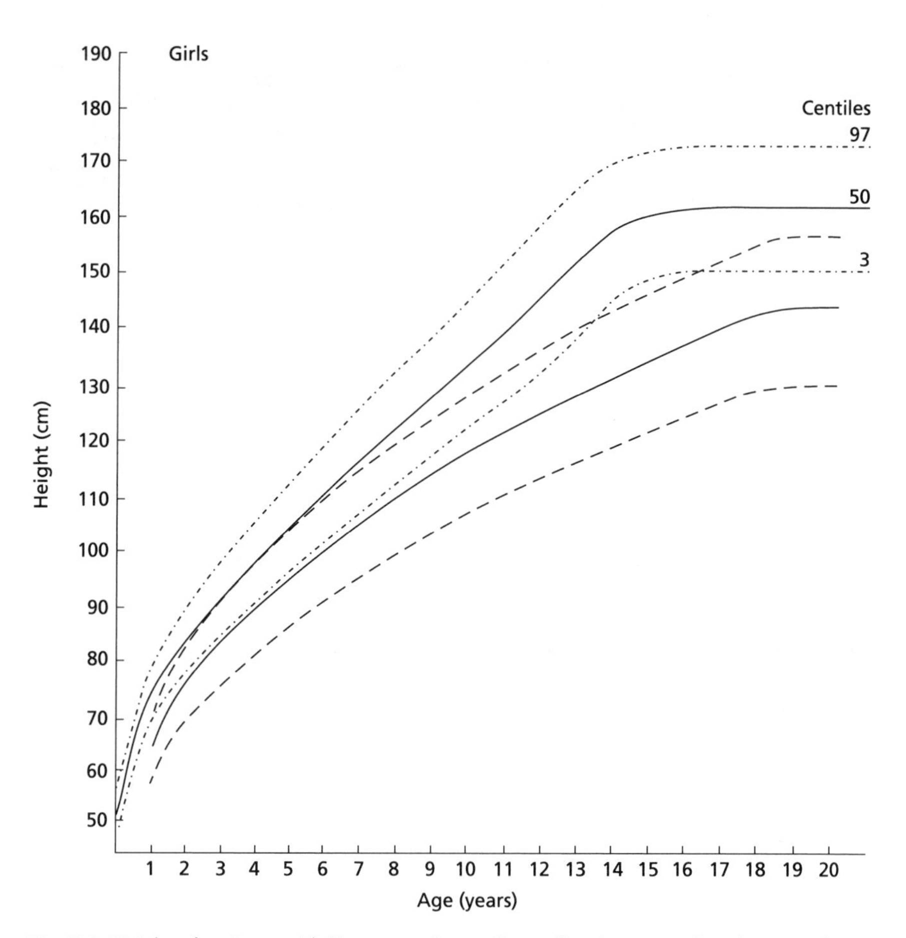

Fig. 2.2 Height of patients with Turner syndrome (heavy lines) compared to the normal range (dashed lines). Reproduced with permission from Lyon *et al.* [17].

Table 2.1 Height data used for Turner-specific standards (from Lyon *et al.* [17]).

Age (years)	Number of patients	Mean (cm)	SD (cm)
5	66	96.5	4.5
10	93	119.5	5.6
15	84	135.8	6.4
20	138	142.9	6.7

Table 2.2 Height data used for Turner-specific standards (from Ranke *et al.* [13]).

Age (years)	Number of patients	Mean (cm)	SD (cm)
5	10	95.7	4.6
10	33	121.1	6.4
15	31	138.7	4.9
20	14	146.8	5.8

researchers [21], but the predictive value of this method has a large confidence interval that makes precision for the individual girl difficult.

A prediction of final adult height using bone age determination ('predicted adult height') has also been used to assess final height [22]. However, this model of prediction was based on normal children, and since bone mineralization in Turner syndrome is abnormal, the methodology remains questionable.

Comparison of the actual final adult height to the projected or predicted adult height is used to assess the effect of growth hormone treatment in Turner syndrome. This may be applicable for study groups, but it must be remembered that, due to interindividual variability, this difference is purely theoretical, since the actual untreated adult height of the treated girl is unknown. The majority of papers published on final adult height have included the difference between final adult height and projected adult height as an outcome measure, but this method relies on historical data.

The use of historical control data in growth studies

The majority of studies use the Lyon or Ranke standards, or height data specific to the country of origin [12,21,23] as controls, because it is difficult to find concurrent children matched for age, current height, mid-parental height, bone age and karyotype. However, historical controls have disadvantages, because the girls measured for standards were those who had been younger and were therefore the shortest patients with Turner syndrome. This means that girls in treatment groups may look as though they have had a beneficial effect from the treatment when compared to the standard charts. Another concern about the use of historical controls is the trend towards increasing height in successive generations in the general population [14]. This may have a small but significant effect on current trial data, and it prevents historical data from being directly comparable.

Practicalities of growth hormone treatment

Growth hormone secretion

The majority of girls with Turner syndrome are able to make adequate amounts of growth hormone on formal provocative testing, but smaller than normal peaks of growth hormone secretion have been measured on 24-hour sampling during mid-childhood and puberty [24,25]. This may reflect the lack of natural circulating oestrogens, which would be expected to increase growth hormone secretion in normal girls. However, it is unlikely to be simply the lower concentration of growth hormone production that explains the poor growth pattern in Turner syndrome. Early reports showed that girls with Turner syndrome do not respond to exogenous growth hormone as well as those children who are growth hormone-deficient, implying that there may be relative end organ resistance [24,26]. As a result, higher doses than the 15 IU/m^2/week used in growth hormone-deficient children have been proposed to maximize growth velocity in Turner syndrome.

Calculation of growth hormone dose

Growth hormone dose may be calculated according to weight or body surface area. The majority of centres in Britain now prescribe a dosage of ≈ 30 IU/m^2/week, while the standard dosage outside the UK is 1 unit/kg/week. The two methods of dosage calculation result in different weekly doses depending on the girl's age. Using Ranke's height and weight data for girls with Turner syndrome in Germany and Austria [7], the dosage calculated by weight and surface area is shown in Table 2.3. A girl aged five years will receive 24% more growth hormone when the dosage is calculated by body surface area than when it is calculated by weight, and at age 15, she would receive 9.5% less.

The differing dosages may have a significant effect on final height. A higher dosage at the start of treatment in the younger girl might improve the height gained during the initial years of treatment, and supports the use of surface area-calculated dosage. However, girls with Turner syndrome gain weight easily during adolescence, and they would then receive a lower dosage of growth hormone when calculated on body surface area during puberty.

Number of injections

Growth hormone was traditionally given by injection three times a week until the 1980s, but it was then shown that growth velocity improved if the same dose was given by six or seven daily injections per week [27,28]. Growth

Table 2.3 Height and weight data and GH dose calculated on the basis of weight and body surface area (from Ranke *et al.* [7]).

Age (years)	Height (cm)	Weight (kg)	Body surface area (m^2)	Dose/week (IU/kg/week)	Dose/week (30 IU/m^2/week)
5	97.0	14.4	0.62	14.4	18.6
10	120.8	26.5	0.93	26.5	27.9
15	139.0	41.8	1.27	41.8	38.1
> 18	146.9	46.8	1.38	46.8	41.4

hormone is therefore usually prescribed for use once daily before bedtime, to mimic the nocturnal peak of normal physiological secretion. Growth response does not improve with a twice-daily regimen in girls with Turner syndrome [23].

When should treatment be started?

Evidence suggests that the growth response is better if growth hormone is started early in childhood in Turner syndrome. Data from final height studies show significantly improved height gain in those girls started on treatment before the age of 10 years [21,29,30]. Once a decision has been made to treat a girl with Turner syndrome using growth hormone, there would appear to be no advantage in waiting to start. The age of commencing treatment is falling, and many patients now start around the age of five years, especially if they are particularly short.

How do oestrogens affect growth in Turner syndrome?

Ovarian dysgenesis is one of the most constant features of Turner syndrome, although a proportion of girls will still have some oestrogen production. Oestrogen therapy is therefore necessary to promote breast development in puberty, and in the long term to reduce the prevalence of osteoporosis, cardiovascular disease and cerebrovascular accidents. These conditions currently contribute to the reduced life expectancy in women with Turner syndrome [31].

There is controversy about the appropriate age to introduce oestrogen supplements. They promote skeletal growth, and have previously been given from the age of nine years in an attempt to increase final height. However, even in low doses, oestrogens advance skeletal maturation and shorten the time available for growth [32]. It is now accepted that final height is improved if their introduction is delayed, and starting oestrogen supplements at the age of 13–14 years on a low dosage of 50–100 ng/kg daily seems appropriate. The failure of pubertal development is a major problem during adolescence for girls with Turner syndrome, and some may wish to commence treatment earlier than might be advised on the grounds of optimal growth in order to match their peers.

Is oxandrolone useful as a growth-promoting agent in Turner syndrome?

It has been shown that girls with Turner syndrome produce lower amounts of androgens than normal girls [33]. Oxandrolone, which is a synthetic anabolic agent with few androgenic effects, has been used in low doses in Turner syndrome as a growth-promoting agent. It has been given at a dosage of 0.05–0.2 μg/kg/day alone and in combination with growth hormone. The higher dosage will accelerate bone age [34] and therefore reduce ultimate height, while the lower dosage appears to improve final height [35] without significant virilization. The effect of oxandrolone is greatest on prepubertal children and those with significant bone age delay. Since children have a surge of endogenous androgens at the age of 6–8 years during adrenarche, it is reasonable to introduce oxandrolone just after this time at a dosage of 0.0625 mg/kg/day.

What is the effect of growth hormone on short-term growth?

Various outcome measures have been used to assess the initial effect of growth hormone on height. Change in height velocity was used in earlier studies, but the assessment of short-term velocity may be inaccurate due to the effect of measurement error [36]. Change in height SDS is a more useful outcome measure, but cannot be used to compare studies unless the same height standards are used.

Growth hormone has been shown to accelerate growth during the first year of treatment in a number of studies [12,28,34]. Some used growth hormone alone, while others showed the results of growth hormone in combination with oestrogens or oxandrolone. Some of the most influential work on the early effects of growth hormone was reported by Rosenfeld *et al.* [27,34]. This group started a randomized controlled study that compared four groups of patients: those on no treatment; those on oxandrolone alone; those on growth hormone alone; and those on a combination of growth hormone and oxandrolone. The girls receiving growth hormone alone initially grew a mean of 6.6 cm per year; those on oxandrolone alone grew a mean of 7.9 cm; and those on combination treatment grew a mean of 9.8 cm per year. This compared to a mean height velocity of 4.3 cm per year prior to treatment.

Although growth velocity improves during the first year of treatment with growth hormone, the rate of growth is not sustained, and it has been suggested that step-wise dosage increments might avoid this apparent 'tolerance', thereby maximizing the effect from growth hormone. Van Teunenbroek *et al.* compared the effect of accelerating the growth hormone dosage to maintaining a constant dosage over a four-year period, and found a significant increase in height velocity during the first year of treatment as expected, but no continuing improvement of growth velocity [29].

What is the effect of growth hormone on final height?

Many studies have now reported data on final height. Interpretation of these studies is complicated by many factors, particularly the change in the treatment dose that occurred once further information became available on dose response. Many of the studies involved small numbers of patients, and meaningful meta-analysis has therefore not been possible due to the differences in height standards and outcome measures used.

Definition of final adult height

The untreated girls measured for the Ranke and Lyon standards reached final height by the age of 20 years. For this reason, final height at the end of any study should be compared to height standards at this age.

Final height has been defined as a height velocity of < 0.5 cm per year, while near-final height is a height velocity of < 1 cm per year [1]. In studies of final height, the growth hormone treatment was stopped when growth had slowed to between 0.5 and 2.5 cm per year, or in some cases when bone age had reached 13 or 14 years [21,37]. Consequently, the height gain published in many studies may be a conservative estimate, as growth may not have completely stopped.

Treatment with growth hormone alone

The published studies on final height show a range of height gain over projected adult height from –0.2 cm to 8.4 cm. The results of the most directly comparable studies are shown in Table 2.4. Most reports are of girls who started growth hormone treatment later than would now be recommended, so this may limit the final height achieved. Certainly the best results were from the Rosenfeld group, in which 17 girls started growth hormone at a mean age of 9.3 years [21] on a dosage of 1.14 IU/kg/week and achieved a mean gain in height of 8.4 cm according to the Lyon data. Oestrogens were started at a mean age of 15.2 years, and would therefore not be expected to have a detrimental effect on final height. The mean final adult height was 150.4 cm, with 16 of the 17 patients reaching a height above their projected adult height.

Takano published data on girls who were started on growth hormone at 10 years of age [12]. They were treated for a mean of six years, and those on 1 IU/kg/week reached a height 6.3 cm taller than historical adult controls.

Taback compared a group of 17 girls on growth hormone to 14 contemporary patients who had by choice not received growth hormone [38]. The mean final height of the treated girls was 148 cm, compared to 140.7 cm,

Table 2.4 Summary of studies of final height in girls with Turner syndrome treated with growth hormone alone.

First author	Year	Number of patients	Age at start of treatment (years)	Projected adult height (cm)	Final height (cm)	Height 'gain' (cm)
Heinrichs [19] (Belgium)	1995	46	13.1	144.2 (L) 146.2 (R)	152.1	7.9 (L) 5.9 (R)
Ranke [3] (KIGS, world-wide)	1995	82	12	142.4 (L) 145.9 (R)	149.6	6.1 (L) 2.5 (R)
Rochiccioli [37] (France)	1995	117	12.9	144.1 (L)	147.7	3.6 (L)
Van den Broeck [40] (Europe)	1995	56	12.9	147.8 (R)	150.7	2.9 (R)
Taback [38] (Canada)	1996	17	12.4	148.2 (L)	148	–0.2 (L)
Takano [12] (Japan)	1997	15* 15†	10	138‡	142.2 144.3	
Chu [39] (Scotland)	1997	11	12.5	140.9 (R)	141.3	0.4 (R)
Rosenfeld [21] (USA)	1998	17	9.3	142 (L)	150.4	8.4 (L)

Please note: (L) according to Lyon data, (R) according to Ranke data.
*patients on 0.5 IU/kg/week.
†patients on 1 IU/kg/week.
‡compared to mean height of untreated Japanese girls with Turner syndrome.

but they were not matched controls and had a 4-cm taller projected adult height at referral. Disappointingly, only eight of the 17 patients achieved a final adult height above their projected adult height.

Chu also showed little effect from growth hormone, with a mean height gain above the projected adult height of 0.4 cm in 11 girls who received growth hormone with or without oestrogens [39]. Only five reached a height above their projected adult height—the same proportion as would be expected by chance. The low dosage of growth hormone used (22.5 IU/m^2/week) may partly explain the lack of effect on final adult height.

Rochiccioli and Chaussain [37] and Van den Broeck *et al.* [40] reported multicentre data, and were therefore able to include larger patient numbers. Both groups were started on ≈ 0.8 IU/kg/week of growth hormone, with a similar moderate height gain. Van den Broeck's group found that height gain was strongly related to growth response in the first two years of treatment, and the authors suggest that if there is no catch-up growth in the first years of treatment, it may never occur. This raises the interesting possibility that growth hormone treatment should be reassessed and stopped in a girl with Turner syndrome if there has been no significant change in growth rate after two years of treatment.

When analysing the results of final height gained on treatment, it is very important to consider which of the published charts or data have been used to predict the untreated final height. As mentioned previously, there is a marked difference in the mean heights of girls at different ages using either the Ranke or Lyon data. This will significantly alter the estimated benefit of growth hormone therapy, with the Lyon data giving better results, as has been demonstrated by the studies by Heinrichs *et al.* [19] and Ranke *et al.* [3].

Treatment with growth hormone and oxandrolone

The addition of oxandrolone appears to improve the benefit of growth hormone on final height, as shown in Table 2.5. Although 15 girls in the Chu study [39] who received oxandrolone in addition to growth hormone only 'gained' 0.6 cm, the Nilsson group showed a mean height gain over the Lyon projected adult height of 8.5 cm, giving a mean adult height of 154.2 cm [32]. Forty-three patients in the Rosenfeld study [21] were on a combination of growth hormone and oxandrolone, and their mean final adult height was 152.1 cm, with all patients achieving a final adult height above their projected adult height and a mean gain of 10.3 cm.

Table 2.5 Summary of studies of final height in girls with Turner syndrome treated with growth hormone and oxandrolone.

First author	Year	Number of patients	Age at start of treatment (years)	Projected adult height (cm)	Final height (cm)	Height 'gain' (cm)
Heinrichs [19] (Belgium)	1995	46	13.1	144.2 (L) 146.2 (R)	152.1	7.9 (L) 5.9 (R)
Rochiccioli (France)	1995	24	12.9	144.1 (L)	148.1	4 (L)
Nilsson (Sweden)	1996	17	12.2	145.7 (K)	154.2	8.5 (K)
Chu (Scotland)	1997	15	12.5	142.9 (R)	143.5	0.6 (R)
Rosenfeld (USA)	1998	43	9.3	141.8 (L)	152.1	10.3 (L)

Please note: (L) according to Lyon data, (R) according to Ranke data, (K) according to Karlberg data.

Variability of response to growth hormone

Although there is improved growth when groups of girls with Turner syndrome are given growth hormone, it has become apparent that there is a heterogeneous response to treatment. The reasons behind this variability are not yet clear, but growth hormone seems to have a greater effect on growth in those girls who are the shortest prior to treatment [19,30,38,39]. There is a significant negative correlation between the pretreatment height standard deviation scores and height gain. Hofman *et al.* also found that an improved growth response was associated with an increased weight-for-height index, and suggested that it is the heavier girls who have a better response to growth hormone. Other factors associated with a good response to treatment were tall parents and a delayed bone age [30].

Poor adherence to the treatment regimen will have a negative effect on the growth response. Previous work has shown that nearly 50% of patients treated with growth hormone were not complying fully with treatment, either because of a poor understanding of the treatment or difficulties with mixing growth hormone and accurate dosage [41]. Simplification of injecting devices should improve the injection technique, and education by nurse specialists has been shown to improve compliance [42].

Are there any ill effects due to treatment with growth hormone?

The Kabi International Growth Study (KIGS) database has accumulated adverse event reports on over 20 000 patients treated with growth hormone over the last decade, and has demonstrated a good safety record [43]. Complications of treatment that are unique to Turner syndrome are uncommon. An increased incidence of hypertension and impaired glucose tolerance in untreated women with Turner syndrome is well recognized, but there is no apparent further increase in either on growth hormone therapy. Studies have shown that there is no difference between fasting or stimulated glucose levels in treated and untreated girls during an oral glucose tolerance test, and glycosylated haemoglobin does not increase during treatment [12,44]. However, there is a tendency for insulin levels to increase in the fasting state and with stimulation, suggesting that insulin resistance may be increased by growth hormone treatment [45].

There is also concern about the increased incidence of idiopathic intracranial hypertension on treatment. This has been reported in five patients with Turner syndrome included in the KIGS database, but it resolved in all cases when treatment was stopped.

What is the cost of growth hormone treatment?

The financial implications of prescribing growth hormone to every girl with Turner syndrome must be taken into consideration. Treatment from the age of 5–16 years of age will cost approximately £122 000, when calculated at a dosage of 30 IU/m^2/week according to published growth data [7]. On a limited health budget, it is important to be able to justify the height gained at this cost.

What does the future hold?

Growth hormone given at a dosage of 30 IU/m^2/week in daily injections will increase adult height in groups of patients with Turner syndrome. Although individual responses are hard to predict, many patients should now reach the normal range of adult height. The gain in height is correlated with the dosage given, with a maximum gain occurring during the initial years of treatment. As more results on final height become available, it is expected that the benefit of starting treatment at an earlier age will be supported. The addition of oxandrolone may further increase final height, whereas the early introduction of oestrogens will not. Further trials are necessary to determine the optimal dosage and age of commencement of oxandrolone and oestrogens.

The next important challenge will be to identify which factors predict a maximal growth response in girls with Turner syndrome receiving growth hormone therapy, so that treatment can be targeted appropriately.

References

** 1 Donaldson MDC. Growth hormone therapy in Turner syndrome: current uncertainties and future strategies. *Horm Res* 1997; **48** (Suppl): 35–44.

* 2 Donaldson MDC. Jury still out on growth hormone for normal short stature and Turner's syndrome. *Lancet* 1996; **348**: 3–4.

* 3 Ranke MB, Price DA, Maes M *et al.* Factors influencing final height in Turner syndrome following GH treatment: results of the Kabi International Growth Study. In: Albertsson-Wikland K, Ranke MB, eds. *Turner Syndrome in a Life-Span Perspective: Research and Clinical Aspects. Proceedings of the 4th International Symposium on Turner Syndrome, Gothenburg, Sweden, 18–21 May, 1995.* Amsterdam: Elsevier, 1995: 161–5.

** 4 Busschbach JJ, Rikken B, Grobbee DE, De Charro FT, Witt JM. Quality of life in short adults. *Horm Res* 1998; **49**: 32–8.

* 5 Delooz J, Van den Berghe H, Swillen A, Kleczkowska A, Fryns JP. Turner syndrome patients as adults: a study of their cognitive profile, psychosocial functioning and psychopathological findings. *Genet Couns* 1993; **4**: 169–79.

* 6 McCauley E, Sybert VP, Ehrhardt AA. Psychosocial adjustment of adult women with Turner syndrome. *Clin Genet* 1986; **29**: 284–90.

* 7 Ranke MB, Chavez-Meyer H, Blank B, Frisch H, Hausler G. Spontaneous growth and bone age development in Turner syndrome: results of a multicen-

tric study. In: Ranke MB, Rosenfeld RG, eds. *Turner Syndrome: Growth-Promoting Therapies. Proceedings of the 2nd International Symposium on Turner Syndrome, Frankfurt/Main, 1990.* Amsterdam: Excerpta Medica, 1991: 101–6.

* 8 Ranke MB, Grauer ML. Adult height in Turner syndrome: results of a multinational survey 1993. *Horm Res* 1994; **42**: 90–4.

* 9 Price DA, Albertsson-Wikland K. Demography, auxology and response to recombinant human growth hormone treatment in girls with Turner syndrome in the Kabi Pharmacia International Growth Study. *Acta Paediatr* 1993; **391** (Suppl): 69–74.

*10 Rochiccioli P, David M, Malpuech G, *et al.* Study of final height in Turner's syndrome: ethnic and genetic influences. *Acta Paediatr* 1994; **83**: 305–8.

*11 Naeraa RW, Eiken M, Legarth EG, Nielsen J. Spontaneous growth, final height and prediction of final height in Turner syndrome. In: Ranke MB, Rosenfeld RG, eds. *Turner Syndrome: Growth-Promoting Therapies. Proceedings of the 2nd International Symposium on Turner Syndrome, Frankfurt/Main, 1990.* Amsterdam: Excerpta Medica, 1991: 113–16.

12 Takano K, Ogawa M, Tanaka T *et al.* Clinical trials of growth hormone treatment in Turner syndrome in Japan: a consideration of final height. *Eur J Endocrinol* 1997; **137: 138–45.

13 Ranke MB, Pfluger H, Rosendahl W *et al.* Turner syndrome: spontaneous growth in 150 cases and review of the literature. *Eur J Pediatr* 1983; **141: 81–8.

**14 Karlberg J, Albertsson-Wikland K. Natural growth and aspects of growth standards in Turner Syndrome. In: Albertsson-Wikland K, Ranke MB, eds. *Turner Syndrome in a Life-Span Perspective: Research and Clinical Aspects. Proceedings of the 4th International Symposium on Turner Syndrome, Gothenburg, Sweden, 18–21 May, 1995.* Amsterdam: Elsevier, 1995: 75–85.

*15 Cole TJ. Do growth chart centiles need a face lift? *Br Med J* 1994; **308**: 641–2.

*16 Cole TJ, Hall DMB. Screening for growth: towards 2000 [letter]. *Arch Dis Child* 1996; **74**: 183.

17 Lyon AJ, Preece MA, Grant DB. Growth curve for girls with Turner syndrome. *Arch Dis Child* 1985; **60: 932–5.

*18 Mazzanti L, Nizzoli G, Tassinari D *et al.* Spontaneous growth and pubertal development in Turner's syndrome with different karyotypes. *Acta Paediatr* 1994; **83**: 299–304.

**19 Heinrichs C, De Schepper J, Thomas M *et al.* Final height in 46 girls with Turner syndrome treated with growth hormone in Belgium: evaluation of height recovery and predictive factors. In: Albertsson-Wikland K, Ranke MB, eds. *Turner Syndrome in a Life-Span Perspective: Research and Clinical Aspects. Proceedings of the 4th International Symposium on Turner Syndrome, Gothenburg, Sweden, 18–21 May, 1995.* Amsterdam: Elsevier, 1995: 137–47.

*20 Tillman V, Price DA, Bucknall JL, Clayton PE. Experience within the Manchester Growth clinic of growth hormone treatment of girls with Turner syndrome: the influence of duration of treatment on final height. In: Albertsson-Wikland K, Ranke MB, eds. *Turner Syndrome in a Life-Span Perspective: Research and Clinical Aspects. Proceedings of the 4th International Symposium on Turner Syndrome, Gothenburg, Sweden, 18–21 May, 1995.* Amsterdam: Elsevier, 1995: 149–54.

21 Rosenfeld RG, Attie KM, Frane J *et al.* Growth hormone therapy of Turner syndrome: beneficial effect on adult height. *J Pediatr* 1998; **132: 319–24.

*22 Bayley NB, Pinneau SR. Tables for predicting adult height from skeletal age: revised for use with the Greulich–Pyle hand standards. *J Pediatr* 1952; **40**: 423–41.

*23 Van Teunenbroek A, de Muinck Keizer-Schrama S, Stijnen T *et al.* Growth response and levels of growth factors after two years' growth hormone treatment are similar for a once and twice daily injection regimen in girls with Turner syndrome (Dutch Working Group on Growth Hormone). *Clin Endocrinol* 1997; **46**: 451–9.

*24 Masarano AA, Brook CGD, Hindmarsh PC *et al.* Growth hormone secretion in Turner's syndrome and influence of oxandrolone and ethinyloestradiol. *Arch Dis Child* 1989; **64**: 587–92.

*25 Albertsson-Wikland K, Rosberg S. Pattern of spontaneous growth hormone secretion in Turner syndrome. In: Ranke MB, Rosenfeld RG, eds. *Turner syndrome: growth promoting therapies. Proceedings of the 2nd International Symposium on Turner Syndrome, Frankfurt/Main, 1990.* Amsterdam: Excerpta Medica, 1991: 23–8.

*26 Hochberg Z, Pollack S, Aviram M. Resistance to insulin-like growth factor 1 in Turner syndrome. In: Hibi I, Takano K, eds. *Basic and clinical approach to Turner syndrome. Proceedings of the 3rd International Symposium on Turner Syndrome, Chiba, Japan, 8–10 July 1992.* Amsterdam: Excerpta Medica, 1993: 233–7.

*27 Rosenfeld RG, Frane J, Attie KM *et al.* Six-year results of a randomised, prospective trial of human growth hormone and oxandrolone in Turner syndrome. *J Pediatr* 1992; **121**: 49–55.

*28 Rongen-Westerlaken C, Wit JM. Growth hormone therapy in Turner syndrome: the Dutch experience. In: Ranke MB, Rosenfeld RG, eds. *Turner Syndrome: Growth-Promoting Therapies. Proceedings of the 2nd International Symposium on Turner Syndrome, Frankfurt/Main, 1990.* Amsterdam: Excerpta Medica, 1991: 225–9.

*29 Van Teunenbroek A, de Muinck Keizer-Schrama SM, Stijnen T *et al.* Yearly step-wise increments of the growth hormone dose results in a better growth response after four years in girls with Turner syndrome (Dutch Working Group on Growth Hormone). *J Clin Endocrinol Metab* 1996; **81**: 4013–21.

*30 Hofman P, Cutfield WS, Robinson EM *et al.* Factors predictive of response to growth hormone in Turner's syndrome. *J Pediatr Endocrinol* 1997; **10**: 27–33.

*31 Gravholt CH, Juul S, Naeraa RW, Hansen J. Morbidity in Turner syndrome. *J Clin Epidemiol* 1998; **51**: 147–58.

32 Nilsson KO, Albertsson-Wikland K, Alm J *et al.* Improved final height in girls with Turner's syndrome treated with growth hormone and oxandrolone. *J Clin Endocrinol Metab* 1996; **81: 635–40.

*33 Apter D, Lenko HL, Perheentupa J, Soderholm A, Vihko R. Subnormal pubertal increases of serum androgens in Turner's syndrome. *Horm Res* 1982; **16**: 164–73.

34 Rosenfeld RG, Hintz RL, Johanson AJ *et al.* Methionyl human growth hormone and oxandrolone in Turner syndrome: preliminary results of a prospective randomized trial. *J Pediatr* 1986; **109: 936–43.

*35 Crock P, Werther GA, Wettenhall HNB. Oxandrolone increases adult stature in Turner syndrome: a study to final height. In: Ranke MB, Rosenfeld RG, eds. *Turner Syndrome: Growth-Promoting Therapies. Proceedings of the 2nd International Symposium on Turner Syndrome, Frankfurt/Main, 1990.* Amsterdam: Excerpta Medica, 1991: 189–94.

*36 Voss LD, Wilkin TJ, Bailey BJR, Betts PR. The reliability of height and weight velocity in the assessment of growth (Wessex Growth Study). *Arch Dis Child* 1991; **66**: 833–7.

**37 Rochiccioli P, Chaussain JL. Final height in patients with Turner syndrome treated with growth hormone. In: Albertsson-Wikland K, Ranke MB, eds. *Turner Syndrome in a Life-Span Perspective: Research and Clinical Aspects. Proceedings of the 4th International Symposium on Turner Syndrome, Gothenburg, Sweden, 18–21 May, 1995.* Amsterdam: Elsevier, 1995: 123–8.

38 Taback SP, Collu R, Deal CL *et al.* Does growth hormone supplementation affect adult height in Turner's syndrome? *Lancet* 1996; **348: 25–7.

39 Chu CE, Paterson WF, Kelnar CJ *et al.* Variable effect of growth hormone on growth and final adult height in Scottish patients with Turner's syndrome. *Acta Paediatr* 1997; **86: 160–4.

40 Van den Broeck J, Massa GG, Attansio A *et al.* Final height after long-term growth hormone treatment in Turner syndrome (European Study Group). *J Pediatr* 1995; **127: 729–35.

*41 Smith SL, Hindmarsh PC, Brook CGD. Compliance with growth hormone treatment: are they getting it? *Arch Dis Child* 1993; **68**: 91–3.

*42 Smith SL, Hindmarsh PC, Brook CGD.

Compliance with growth hormone treatment: are they getting it? *Arch Dis Child* 1995; 73: 277.

* 43 Wilton P. KIGS adverse events report no. 9. Pharmacia and Upjohn International Growth Database Report No. 15. Biannual report, 1997.

* 44 Stahnke N, Attanasio A, van den Broeck J, Partsch CJ, Zeisel HJ. GH treatment studies to final height in girls with Turner syndrome: the German experience. In: Albertsson-Wikland K, Ranke MB, eds. *Turner Syndrome in a Life-Span Perspective: Research and Clinical Aspects. Proceedings of the 4th International Symposium on Turner Syndrome, Gothenburg, Sweden, 18–21 May, 1995.* Amsterdam: Elsevier, 1995: 95–103.

* 45 Wilson DM, Rosenfeld RG. Effect of GH and oxandrolone on carbohydrate and lipid metabolism. In: Ranke MB, Rosenfeld RG, eds. *Turner Syndrome: Growth-Promoting Therapies. Proceedings of the 2nd International Symposium on Turner Syndrome, Frankfurt/Main, 1990.* Amsterdam: Excerpta Medica, 1991: 269–74.

3: Is there a role for GH therapy in Noonan syndrome?

David B. Dunger and Angelika Mohn

Noonan syndrome

Noonan syndrome is a common form of familial short stature, with cases reported world-wide and without racial predilection. Although its incidence has been estimated to be between 1 : 1000 and 1 : 2500 live births [1], others [2] have argued that mild expression of the disease could occur in up to 1 : 100 live births. In the past, cases of Noonan syndrome may have remained underdiagnosed, because the similarity of the phenotype with Turner syndrome led to confused terminology.

The first patient with Noonan syndrome was probably described by Koblinsky, a medical student at the University of Dorpat in Estonia (then in Russia) in 1883, and a number of case reports of patients of both sexes with webbed neck, short stature, micrognathia and other abnormalities followed [2]. In 1938, Turner reported a series of older female patients with webbed neck, short stature, cubitus valgus and sexual infantilism, leading to the term 'Turner syndrome' [3]. Ullrich reported independently on a number of patients of both sexes with a similar phenotype, which he termed 'Bonnevie–Ullrich syndrome', after Bonnevie's work on mice [4], and more confusion was created by Flavell [5], who introduced the term 'male Turner syndrome' when reporting on a male patient with similar abnormalities and small testes. It was only after the identification by Ford *et al.* [6] of abnormal chromosomes in a patient with the diagnosis of Turner syndrome that the convention became established that female patients with webbed neck, ovarian dysgenesis and a 45XO karyotype would be described as having Turner syndrome. Finally, in 1963, Noonan and Ehmke [7] defined a specific group of 9 patients (6 males and 3 females) with clinical features of Turner syndrome and normal chromosomes, but with valvular pulmonary stenosis, which they considered to represent a new syndrome. To avoid the previous misunderstanding, Opitz *et al.* [8] in 1965 proposed that the term 'Noonan syndrome' should be used for this condition. Although Turner syndrome and Noonan syndrome undoubtedly have remarkable similarities, there are important differences between the two clinical conditions.

Clinical features

The diagnosis of Noonan syndrome is still made purely on clinical grounds, the key to the diagnosis being the typical facies, right-sided cardiac abnormalities and normal chromosomes.

The most striking clinical features are the facial abnormalities, characterized by hypertelorism, epicanthic folds, ptosis, down-slanting palbebral fissures with high-arched eyebrows, depressed nasal root with a wide nasal base and low-set ears. These abnormalities are best seen during early childhood, but may alter later, making the diagnosis more difficult [9], although other features may still be evident.

About 50–88% of patients have cardiac problems. The right side is more frequently involved and a stenotic pulmonary valve is the most characteristic lesion, but virtually every type of cardiac defect has been described [10]. In contrast, in Turner syndrome the typical cardiological defects are left-sided, and coarctation of the aorta and bicuspid aortic valves are the most typical lesions. In Noonan syndrome, an unusual electrocardiogram with an indeterminate or left-axis deviation and a dominant S wave over the entire precordium is frequently found, but this is not clearly related to any specific cardiac malformation, and the cause is unknown [11]. Hypertrophic cardiomyopathy (HCM) is found in 20–30% of patients, either presenting at birth or developing later in childhood [12]. Unlike idiopathic HCM, patients with Noonan syndrome may have involvement of both the right and the left ventricle. The natural history of HCM in children with Noonan syndrome has not been determined. However, Skinner *et al.* [13] studied the outcome in 29 patients with HCM, eight of whom were affected by Noonan syndrome. The children with Noonan syndrome fared less well than those in the idiopathic group, with HCM resolving in only one child in contrast to 38% of the idiopathic group. The high incidence of cardiac abnormalities means that every patient who is suspected of having Noonan syndrome should undergo careful cardiac evaluation.

Other abnormalities seen in Noonan syndrome have been summarized in a number of excellent clinical reviews [1,2,11,14–17]. The main features of the patients included in seven large clinical studies are summarized in Table 3.1, and include skeletal, haematological, neurodevelopmental and ocular abnormalities. Variability of expression was noted in all of these studies, and this can make clinical diagnosis difficult. Different scoring systems have been proposed [18,19] in which the symptoms and features of Noonan syndrome are divided into major and minor features (Table 3.2).

Table 3.1 Phenotypic features of Noonan syndrome.

Feature	Frequency (%)
Short stature	50–71
Delayed puberty	100
Undescended testes	72–77
Facies	100
Hypertelorism	74–98
Down-sloping palpebral apertures	38
Epicanthic folds	39
Ptosis	48–58
Low-set ears	78–90
Cardiac defect	55–88
Pulmonary stenosis	22–66
Hypertrophic cardiomiopathy	20–30
Atrial septal defect, ventricular defect, atrioventricular dissociation	5–13
Skeletal abnormalities	
Pectus carinatum sup./excavatum inf.	70–95
Cubitus valgus	47–54
Joint hyperextensibility	50
Neck webbing	22–50
Clinobrachydactyly	25
Haematological abnormalities	
Lymphatic dysplasia	20
Bleeding abnormalities	33–65
Neurodevelopmental findings	
Poor feeding in infancy	95
Attending normal school	84
Behaviour problems	52
Convulsion	13
Ocular abnormalities	58–95
Strabismus	48
Refractive errors	61
Prominent corneal nerves	46
Other	
Skin changes	27–67
Hearing loss (otitis media)	38
Hepato/splenomegalia	26–53

Information summarized from: Nora *et al.* [1], Mendez *et al.* [2], Burch *et al.* [10], Charr *et al.* [14], Allanson [15], Sharland *et al.* [16], Lee *et al.* [17].

Table 3.2 Clinical scoring system for diagnosis.

Feature	A = major	B = minor
1 Facial	Typical facies	Suggestive facies
2 Cardiac	Valvular pulmonary stenosis Hypertrophic cardiomiopathy Typical electrocardiogram	Other defects
3 Height	< 3rd centile	< 10th centile
4 Chest wall	Pectus carinatum/excavatum	Broad thorax
5 Family history	First-degree relative definite for Noonan syndrome	First-degree relative suggestive of Noonan syndrome
6 Other	All three (male): Mental retardation Cryptorchidism Lymphatic dysplasia	One of: Mental retardation Cryptorchidism Lymphatic dysplasia

Diagnosis if 1A + one of 2A–6A or two of 2B–6B
or 1B + two of 2A–6A or three of 2B–6B.
Adapted with permission from De Sanctis *et al.* [19].

While these proposed scoring systems are useful research tools, they have never been widely used in clinical practice, and the diagnosis still rests on clinical pattern recognition.

The differential diagnosis of Noonan syndrome includes: Turner syndrome, trisomy 8p and trisomy 22 mosaicism. It also includes the phenotype resulting from fetal effects of certain teratogens, such as alcohol and primidone. The association of neurofibromatosis type 1 and Noonan syndrome has been recognized in several case reports, and has become known as neurofibromatosis–Noonan syndrome [20]. It has been suggested that the syndrome is a nosologically discrete biological entity caused by a mutation at a locus that is different from either the neurofibromatosis or the Noonan syndrome loci. The alternative possibility, that the genes of the two diseases are contiguous and that a large deletion involving both genes produces the combined phenotype, has been excluded [21]. It has also been argued that cardiofaciocutaneous syndrome and Noonan syndrome are variable manifestations of the same entity, or examples of contiguous gene syndromes [22]. The debate is still ongoing, and will only be resolved by cloning the gene of the Noonan and/or cardiofaciocutaneous syndrome.

Genetics

Because of its striking similarities with Turner syndrome and the direct familial transmission, an X-linked dominant inheritance was first proposed

[23], but this was subsequently excluded because of reports of male-to-male transmission [24], and an autosomal dominant mode of inheritance became widely accepted. As in other autosomal dominant conditions, Noonan syndrome may show great variability in expression. In addition, since the clinical features become less obvious with age [9], affected adults may not be recognized. If both parents have only one possible sign or no signs of Noonan syndrome and have had one affected child, the empirical recurrence risk is still around 5% [25]. It must be borne in mind, however, that half of the cases of Noonan syndrome seem to be sporadic.

In 1994, using genome-wide linkage analysis of a large three-generation Dutch family with autosomal dominant Noonan syndrome and 20 smaller two-generation families, the gene was located on the distal part of the long arm of chromosome 12, and this locus is thought to be responsible for ≈ 80% of familial cases [26,27]. Despite this advance, the gene has not yet been identified, and the diagnosis is still clinical. Patients with typical cardiological problems are more likely to be diagnosed as having Noonan syndrome, while patients with normal cardiological function may not. These considerations need to be recalled when data relating to growth and response to recombinant human growth hormone (rhGH) treatment are analysed.

Growth

Short stature is found in about 50–70% of cases, according to different clinical studies [1,2,14–16], and Noonan syndrome may be one of the commonest forms of familial short stature.

Prenatal growth

Accurate data on growth before birth are not available, although pregnancy is complicated in one-third of cases by polyhydramnios [16], and ultrasound may detect fetal oedema [28]. This may explain reports of excessive weight loss during the first week of life [16]. Fetal oedema could lead to an overestimation of birth weight, which has been reported to be comparable to that of the normal population with an average weight of 3450 ± 510 g (both sexes, mean ± SD) by Sharland *et al.* [16] and 3182 ± 1052 g for males and 3219 ± 745 g for females by Ranke *et al.* [29]. Studies reporting size at birth in children born at term indicate that the average length is no different from that in the normal population, with values of 51.0 ± 1.9 cm in boys and 51.1 ± 2.4 cm in girls [29].

Postnatal growth and puberty

Postnatal growth of children with Noonan syndrome has been described in a number of studies, and two groups [29,30] have constructed syndrome-specific growth charts. In both studies, the data were obtained in a retrospective manner, mixing longitudinal and cross-sectional measurements, and may have been influenced by the age at diagnosis of the patients reported. In addition, the numbers of patients were small—particularly in contrast to studies of normal populations [31], but also in relation to studies defining growth in Turner syndrome [32]. Although these charts give an overall impression of growth in Noonan syndrome, they still need to be treated with caution.

Witt *et al.* [30] studied 112 patients (64 of whom were male), ranging in age from birth to 64 years. Although normal at birth, children of both sexes dropped off the fifth centile by three months of age, and remained below this centile until adulthood (Figs. 3.1, 3.2). In contrast, Ranke *et al.* [29] studied 144 patients (89 males), 83 of whom were seen because of a congenital heart defect. He showed a growth curve that ran along the third centile during childhood up to 12 years in boys and 10 years in girls; puberty was then delayed by about two years and an insufficient growth spurt was observed, with the heights of children of both sexes falling below the normal range. The total growth phase was prolonged, with a delayed bone age of about two years and subsequent delayed epiphyseal closure. The final height was not reached until the end of the second decade of life (Figs. 3.3, 3.4). In both studies, the final height was close to the third centile for normal males and females: in Witt *et al.* [30], the male final height was 161.0 ± 8.5 cm and the female final height 150 ± 6.2 cm; in Ranke *et al.* [29], males were 162.5 ± 5.4 and females 152.7 ± 5.7 cm (mean ± SD).

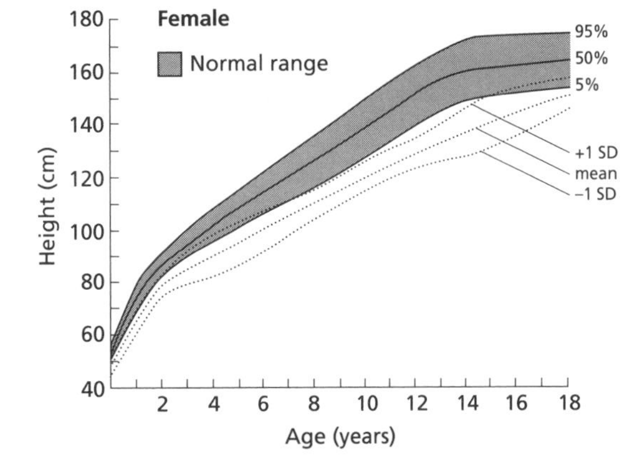

Fig. 3.1 Growth curve for height from birth to 18 years in females with Noonan syndrome. Reproduced with permission from Witt *et al.* [30].

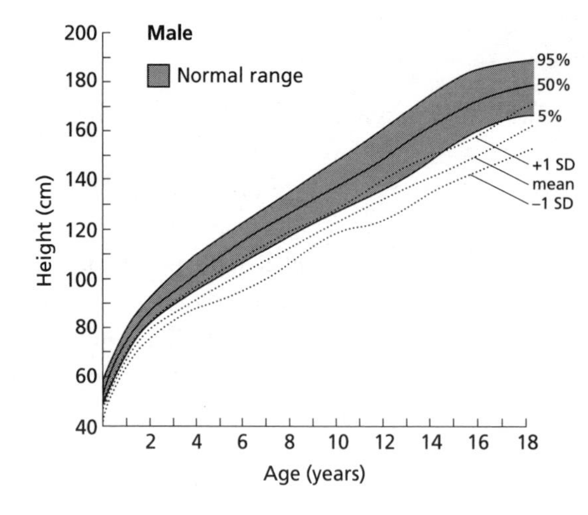

Fig. 3.2 Growth curve for height from birth to 18 years in males with Noonan syndrome. Reproduced with permission from Witt *et al.* [30].

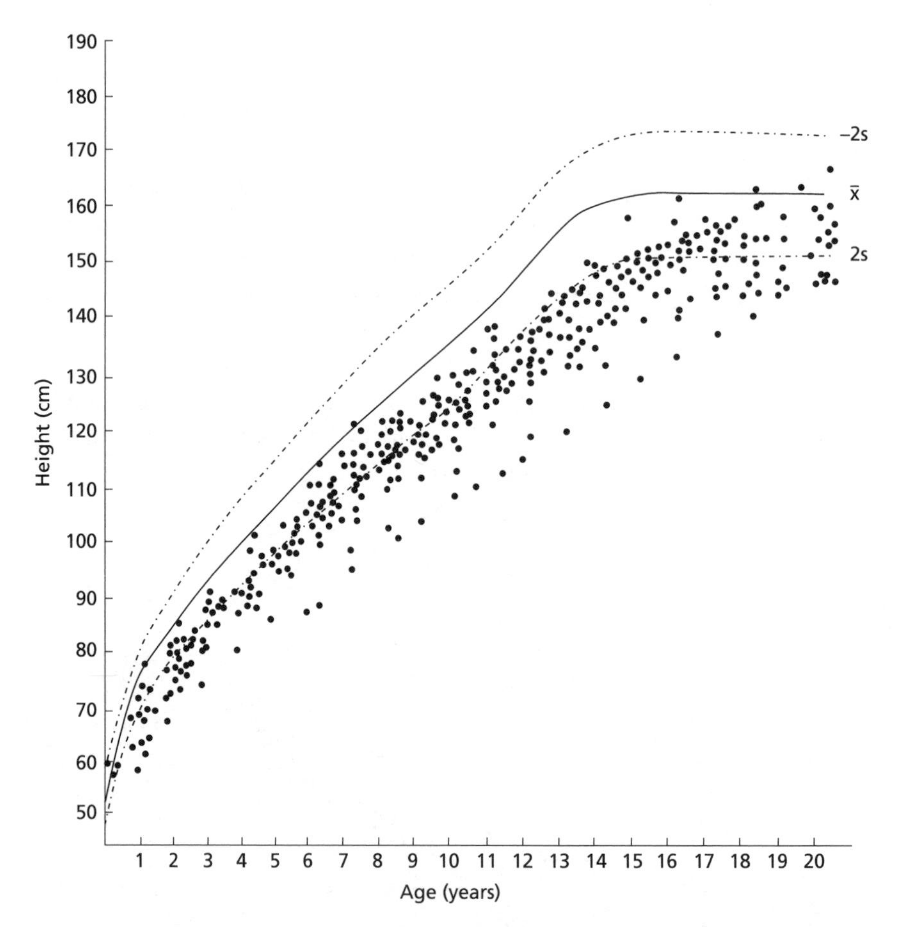

Fig. 3.3 Growth curve for height from birth to 18 years in females with Noonan syndrome. Reproduced with permission from Ranke *et al.* [29].

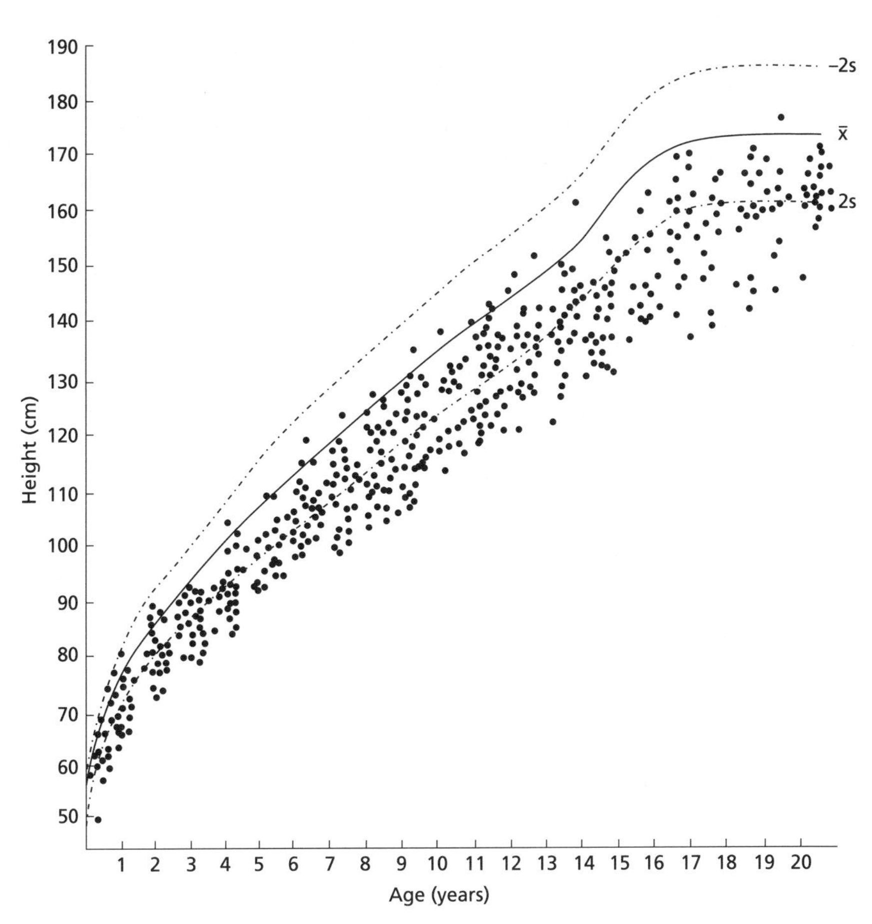

Fig. 3.4 Growth curve for height from birth to 18 years in males with Noonan syndrome. Reproduced with permission from Ranke *et al.* [29].

A delay in puberty has been consistently reported in a number of studies [1,2,15,16]. Sharland *et al.* [16] showed a mean age of menarche of 14.6 ± 1.7 years (1.6 years later than normal girls) and a significantly delayed onset of puberty in boys. Accurate data relating to onset of puberty in boys are not available, but most studies indicate an approximate delay of two years [16,29,33]. Despite the delayed puberty, gonadotrophin deficiency has never been documented [33,34]. Up to 77% of boys present with undescended testes. One longitudinal study [33] of pubertal development in boys indicated that boys with normal descent of the testis developed late, but progressed normally through puberty, while in boys with undescended testes, puberty needed to be induced, suggesting a degree of primary gonadal failure. Fertility in patients with Noonan syndrome and undescended testes has been assessed in a different study using semen sample analysis, and azoospermia and oligospermia were found in four out of six patients, suggesting impairment of fertility [34]. In contrast, female fertility seems to be normal. The difference in fertility in Noonan syndrome between the two sexes has been suggested by all the large family studies, in which maternal transmission has been reported more often than paternal [25].

Pathogenesis of short stature

The cause of the short stature in Noonan syndrome has still not been completely elucidated. The possibility of a disorder in the GH–IGF-I axis has been investigated in the various studies summarized in Table 3.3 [35–42]. The results of growth hormone (GH) provocation tests and spontaneous

Table 3.3 Published data on GH secretion.

Author/year (Ref.)	GH secretion status	Findings	Comment
Elders 1976 [42]	15 patients, standard provocation testing, IGF-I levels	Normal GH secretion, high IGF-I levels	No alteration in the GH–IGF-I axis, peripheral resistance to IGF-I
Spadoni 1990 [35]	8 patients (4 M), spontaneous secretion overnight	1 prepubertal patient with reduced secretion	Normal somatotrophic function
Ahmed 1991 [36]	6 patients (3 M) standard provocation testing, 4 patients spontaneous secretion overnight, IGF-I levels	1 patient with partial GH deficiency, low IGF-I levels	Degree of GH resistance or presence of neurosecretory dysfunction
Bernadini 1991 [37]	8 patients (4 M), clonidine or arginine provocation testing, spontaneous GH secretion overnight, IGF-I levels	Normal GH secretion with provocation testing, reduced spontaneous GH secretion in 3 patients and low IGF-I levels in 2	Possible neurosecretory dysfunction
Tanaka 1992 [38]	1 prepubertal boy, spontaneous GH secretion overnight	Insufficient GH secretion	Neurosecretory dysfunction
Thomas 1993 [39]	5 prepubertal patients (4 M), glucagon test	1 patient with GH < 20 mU/L	GH deficiency
Romano 1996 [40]	150 patients (97 M) standard GH provocation testing	45% peak levels < 25 mU/L	No difference in growth response to rhGH between GH sufficient and GH insufficient patients
Cotterill 1996 [41]	30 patients (19 M), glucagon test	10 patients with GH peak < 32 mU/L	Prompt rise of IGF-I with therapy and low GH on stimulation test may point to impairment of GH release

GH: growth hormone; IFG-I: insulin-like growth factor-I; rhGH: recombinant human growth hormone.

overnight secretion are contradictory. While some authors have reported occasionally poor GH secretion [40,41], others report normal or increased GH levels after standard provocation tests and overnight [35,37]. The same is true of insulin-like growth factor-I (IGF-I) levels, which were reported as normal or low in some studies [36,41] and normal or high in another study [42]. Two studies documented a discrepancy between GH and IGF-I concentrations [36,37], which may suggest a degree of GH receptor resistance, or alternatively the presence of GH neurosecretory dysfunction [36].

As has been suggested in Turner syndrome, the possibility of an abnormality of the skeleton and/or growth plate as a cause of the short stature is an attractive hypothesis. Skeletal defects are documented in both syndromes, and in Noonan syndrome they include cubitus valgus, clinobrachydactyly and pectus carinatus/excavatus. However, these defects would not limit growth. Modelling and tubulation defects of the metacarpals or metatarsals have been documented in over 50% of patients, but skeletal surveys do not show any convincing evidence of an underlying skeletal dysplasia [14]. The growth response to high-dose, but not low-dose, GH in both Turner and Noonan syndromes suggests that there may be a relative resistance to GH at the growth plate [43].

The congenital heart disease associated with Noonan syndrome could have a negative effect on growth. However, in the study conducted by Ranke *et al.* [29], the pattern of growth in patients with heart disease did not differ from that in patients with no heart disease—suggesting that the impact of heart problems on overall growth is small, although this does not exclude a role in individual patients.

Growth-promoting therapy

The similarities between Noonan syndrome and Turner syndrome have led to the expectation that rhGH treatment in Noonan syndrome might have the same beneficial effect on final height, irrespective of the biochemical demonstration of GH insufficiency or sufficiency. As in every rhGH trial, the goal of treatment with rhGH is to improve height, without causing significant adverse events. Although the prevalence of Noonan syndrome may be greater than that of Turner syndrome, only a few studies on the growth-promoting effects of rhGH have been published (Table 3.4) [36,37,39–41,44–48].

Clinical studies of rhGH

The very first study of rhGH treatment in Noonan syndrome was conducted by Cianfarani *et al.* [44] in 1987 on three patients. The rhGH dosage used

Table 3.4 Published studies on recombinant human growth hormone (rhGH) treatment.

Author/year (Ref)	Population studied	Treatment schedule	Variables controlled	Findings
Cianfarani 1987 [44]	3 patients	6 IU/m²/week for 6–9 months	Growth velocity	No improvement
Wales 1990 [47]	2 prepubertal patients	10 IU/m²/week for 12 months	Growth velocity	Increment of growth velocity from 4.5 to 7.5 and 5.8 to 9.5 cm/year
Ahmed 1991 [36]	6 patients (3 M); 3 prepubertal	12 IU/m²/week for 12 months	Growth velocity	Increment of growth velocity from 4.8 ± 1.09 to 7.4 ± 0.63 cm/year
			SDS growth velocity	Increment of SDS growth velocity from 0.93 ± 0.67 to 1.48 ± 1.40
Bernardini 1991 [37]	2 prepubertal male patients	12 IU/m²/week for 6 months	Growth velocity	Increment of growth velocity of 3.9 and 4.0 cm/year
Lu 1993 [48]	7 patients (5 M); age mean 11.3; range 6.5–17.5	10 IU/m²/week for 12 months	SDS height	Increment of SDS height of 0.3 ± 0.3
			Growth velocity	Increment of growth velocity of 2.8 ± 0.8 cm/year
			SDS growth velocity	Increment of SDS growth velocity of 3.8 ± 1.6
Thomas 1993 [39]	5 prepubertal patients (4 M); age mean 3.9; range 2.5–5.9	30 IU/m²/week for a mean of 2.9 years (range 1.8–4.6)	SDS height	Increment of SDS height from –3.3 to –2.4 (1st year), to –2.1 (3rd year)
			SDS growth velocity	Increment of SDS growth velocity from –2.1 to +3.1 (1st year) to –0.4 (2nd year) to –2.0 (3rd year)

Otten 1994 [45]	50 patients (34)	12 IU/m²/week for 3 years	SDS height	Increment of SDS height from –3.1 to –2.6 (1st year), to –2.5 (2nd year), to –2.4 (3rd year)
			Growth velocity	Increment of growth velocity from 4.3 to 7.0 (1st year), to 6.0 (2nd year), to 4.8 (3rd) cm/year
Municchi 1995 [49]	4 prepubertal girls; age range 12.3–15.1	10 IU/m²/week for 1st year; 12 IU/m²/week for 2nd 3rd 4th years	SDSR final height (Ranke standards)	Increment of SDSR final height in 3 patients
Cotterill 1996 [41]	30 patients (19 M); mean age 8.9; range 4.8–13.7	28 IU/m²/week for 12 months	SDS height	Increment of SDS height from –3.01 ± 0.10 to –2.36 ± 0.10
			Growth velocity	Increment of growth velocity from 4.9 ± 0.2 to 8.1 ± 0.4 cm/year
			SDS growth velocity	Increment of SDS growth velocity from –0.7 ± 0.15 to 2.42 ± 0.32
Romano 1996 [40]	150 patients (97 M); mean age 10.6 ± 3.8	30 IU/m²/week for 4 years	SDS height	Increment of SDS height from –3.3 ± 0.9 to –2.8 ± 1.1 (1st year), to –2.6 ± 1.1 (2nd year), to –2.4 ± 1.1 (3rd year), to –2.1 ± 1.2 (4th year)
			Growth velocity	Increment of growth velocity (cm/year) from 4.3 ± 2.3 to 8.0 ± 2.0 (1st year), to 6.9 ± 1.7 (2nd year), to 6.3 ± 1.5 (3rd year), to 5.7 ± 1.9 (4th year)
De Schepper 1997 [46]	23 prepubertal patients (18 M); mean age 9.4; range 5.4–14.3	30 IU/m²/week for 12 months	SDS height	Increment of SDS height from –2.28 ± 0.68 to –1.78 ± 0.76
			SDS growth velocity	Increment of growth velocity from 4.5 ± 1.0 to 8.5 ± 1.6 cm/year

was 6 IU/m²/week, divided into three doses weekly over a short period of six to nine months. The results were disappointing, as no improvement in growth velocity was observed, but this was most likely due to the low dosage of rhGH administered and the short study period. Using a similar daily dose, but administering it every day (12 IU/m²/week), Ahmed *et al.* [36] observed an increase in mean growth velocity in five children from 4.8 ± 1.09 before treatment to 7.4 ± 0.63 cm/year (mean ± SD) over a study period of one year. Height prediction based on the Tanner–Whitehouse radius, ulnar, and short bones (TW2 RUS) method indicated a mean increase of 3.0 cm after 12 months of rhGH therapy. The same rhGH treatment schedule of 12 IU/m²/week was used a few years later in a larger cohort of 50 patients in the Kabi International Growth Study (KIGS) [45]. The study design used cross-sectional and longitudinal data, combined for analysis. They also demonstrated an improvement in growth velocity (4.3 cm/year before treatment to 7.0 cm/year) over one year of treatment. However, with prolonged treatment for up to three years, growth velocity decreased during the second and third year of treatment (6.0 and 4.8 cm/year, respectively). The height standard deviation score (SDS) increased over the whole study period (from –3.1 SDS before treatment to –2.4 SDS after three years of treatment).

It was later argued that if, as in Turner syndrome, the underlying growth disorder is a partial resistance to GH, then higher doses of rhGH might be more effective. Cotterill *et al.* [41] studied 30 patients using a dose of 28 IU/m²/week over one year of treatment. They observed an increase in height SDS from –3.01 to –2.36 and an increase in growth velocity from 4.9 to 8.9 cm/year. De Schepper *et al.* [46], in a one-year study of 23 patients using 30 IU/m²/week, showed a similar increase in growth velocity of 4.0 ± 1.6 cm/year combined with an increase in height SDS of 0.5 ± 0.46 (mean ± SD). The increment in growth with these higher doses of GH was therefore more consistent, and this induced Thomas *et al.* [39] to use a high rhGH dosage over a longer period of three years. As with the low-dose rhGH treatment schedule, a significant increase in growth velocity SDS over the first year of treatment was observed (–2.1 to –3.1), followed by a marked and continuous decrease in growth velocity SDS over the next two years of treatment (–0.4 SDS at the end of the second year and –2.0 SDS at the end of the third year). Nevertheless, height SDS showed a persistent improvement compared to the pretreatment value (from –3.3 to –2.1 SDS at the end of the third year). The same growth pattern was confirmed by Romano *et al.* [40] in 150 patients over a treatment period of four years, indicating that despite the slowing down of the growth rate after the first year of treatment, encouraging and sustained improvements in height SDS can be obtained with high-dose rhGH therapy in Noonan syndrome.

Final height after rhGH therapy

Although the results of short-term studies of rhGH therapy in Noonan syndrome are encouraging, as yet there are few data concerning final height.

In many of the studies, height prediction or change in height SDS for bone age are used as short-term surrogates. Improvements in height SDS for bone age have been observed in several studies [39–41,46], and these have been interpreted as showing that rhGH is not associated with an undesirable advance in bone maturation. Ahmed *et al.* [36] observed an improvement in height prediction based on bone age (TW2 RUS) after treatment with rhGH. Bone age is usually delayed in children with Noonan syndrome, but it cannot be assumed that height prediction is reliable. The alternative strategy of looking at the projected final height using Noonan-specific growth charts may also be misleading, as these charts were constructed using relatively small populations.

There are few studies in which final height data are available for children with Noonan syndrome treated with rhGH. Municchi *et al.* [49] studied four girls who were treated with rhGH (10 IU/m^2/week) for up to four years or until reaching final height. Final height was compared with the projected height based on the Ranke–Noonan growth charts [29]. The authors observed an improved final height in three girls (patient 1 from –0.7 to 1.1, patient 2 from –1.9 to –0.9 and patient 3 from 0.1 to 0.4). Two of the patients exceeded their corrected mid-parental height: patient 1, 149.5 cm vs. 156.0 cm and patient 3, 145 vs. 151.8 cm (mid-parental height vs. final height). In the study conducted by Romano *et al.* [40], in which a higher rhGH dose (30 IU/m^2/week) was used, six of the 150 patients reached final height during the study period. The mean duration of treatment (SEM) was 4.6 ± 0.7 years, and three of these patients reached heights greater than their Bayley–Pinneau-predicted heights.

Complications of rhGH

None of the above-mentioned studies have reported any important adverse events. However, a major concern regarding the use of rhGH in Noonan syndrome is a possible risk of alteration in cardiac function, particularly the development or deterioration of HCM. Even without the addition of rhGH, HCM does progress in some patients, leading to heart failure. Recombinant GH has been used in idiopathic dilated cardiomyopathy, in which it has been shown to increase left ventricular wall thickness and to reduce chamber size significantly [50], resulting in improved cardiac function. Although this represents an important achievement in patients with cardiac failure due to idiopathic dilated cardiomyopathy, the same effects might be harmful in patients with underlying HCM.

A recent study investigated the long-term cardiac effects of rhGH treatment in short normal children [51]. During this randomized controlled study, no important changes in cardiac size or function were observed in any of the 15 treated and 13 untreated patients over a study period of four years. However, a tendency towards increased left ventricular mass was documented, and was interpreted as being a reflection of the increase in lean body mass during treatment. Reassuringly, recent data from Saenger *et al.* [52] indicate that children with Turner syndrome treated with rhGH do not develop any short-term cardiac complications. However, GH-sufficient children with underlying cardiac abnormalities such as HCM may constitute a special group of patients in whom extra caution is required.

In their study of rhGH treatment, Romano *et al.* [40] included patients with a congenital heart defect (42%); no adverse events due to rhGH were recorded, but a detailed cardiological assessment during therapy was not performed. The effect of rhGH on cardiac function in Noonan syndrome was investigated more closely by Cotterill *et al.* [41] in a study of 30 children. Echocardiograms were taken at the start and the end of rhGH therapy, and were recorded and interpreted by the same operator. There was no increase in the mean maximal left ventricular wall thickness during the study. However, none of the patients had shown any indication of incipient ventricular hypertrophy or any features of HCM at the beginning of the study. The effect of GH in patients with early evidence of HCM has not been specifically studied, but there have been anecdotal reports of deterioration in cardiac function with rhGH treatment.

Limitations of rhGH treatment studies

The greatest limitation in all of the studies of rhGH in Noonan syndrome is that none of them have been randomized or controlled. Cotterill *et al.* [41] compared their study growth data with data from 10 control individuals in whom consent for rhGH had not been obtained or the inclusion criteria were not met. In the untreated group, height SDS and growth velocity did not change over the study period of one year, indicating a beneficial effect of rhGH on growth in children with Noonan syndrome. De Schepper *et al.* [46] compared their data for rhGH treatment in children with Noonan syndrome with a comparable group of girls with Turner syndrome, and found that the increase in growth velocity and height SDS was similar in the two groups.

A further limiting factor in the interpretation of many studies is the fact that Noonan syndrome is diagnosed clinically, introducing the possibility of misdiagnosis or underdiagnosis of the condition. The possible inclusion of patients with other similar syndromes could alter the outcome and interpretation of any study. Although all the patients were examined by a single geneticist in one of the studies [41], misdiagnosis will only be completely excluded when a specific genetic test is developed.

Conclusion

Noonan syndrome is a common cause of familial short stature, and its true prevalence may only become evident when the dominant gene is identified. Short stature occurs in 70% of children with Noonan syndrome, and the growth pattern is similar to that observed in Turner syndrome. Bone age is often delayed and the growth period may be prolonged, with final heights of around 161 cm in boys and 151 cm in girls being reached towards the end of the second decade. As in Turner syndrome, the cause of the short stature is unknown, but it may result from a mild bony dysplasia or resistance to the effects of GH at the growth plate. High-dose, but not low-dose, rhGH improves the short-term growth velocity, but there are as yet no definitive data on final height. HCM is relatively common in patients with Noonan syndrome, and as the effects of rhGH on this complication are unknown, cardiac assessment is indicated before and during rhGH treatment.

There is an urgent need for a large prospective randomized study of the effects of rhGH in Noonan syndrome, in order to determine the gains in final height and the incidence of adverse events. In the absence of a prospective study, attempts should be made to match study patients with comparable historical controls. Previous studies indicate that gains in growth velocity may decline with consecutive years of treatment, and there may be a place for high-dose intermittent rhGH therapy. Overall, further study is necessary before the role of rhGH in Noonan syndrome can be firmly established.

References

** 1 Nora JJ, Nora AH, Sinha AK, Spangler RD, Lubs HA. The Ullrich–Noonan syndrome (Turner phenotype). *Am J Dis Child* 1974; **127**: 48–55.

** 2 Mendez HM, Opitz JM. Noonan syndrome: a review. *Am J Med Genet* 1985; **21**: 493–506.

3 Turner HH. A syndrome of infantilism, congenital webbed neck, and cubitus valgus. *Endocrinology* 1938; **23**: 566–74.

4 Ullrich O. Turner's syndrome and status Bonnevie–Ullrich. *Am J Hum Genet* 1949; **1**: 179–200.

5 Flavell G. Webbing of the neck with Turner's syndrome in the male. *Br J Surg* 1943; **31**: 150–3.

6 Ford CE, Jones KW, Polani PE, de Almeida JC, Briggs JH. A sex chromosome anomaly in a case of gonadal dysgenesis (Turner syndrome). *Lancet* 1959; **i**: 711–13.

** 7 Noonan JA, Ehmke DA. Associated noncardiac malformations in children with congenital heart disease. *J Pediatr* 1963; **63**: 468–70.
8 Opitz JM, Summitt RL, Sarto GE. Noonan syndrome in girls: a genocopy of the Ullrich–Turner syndrome. *J Pediatr* 1965; **67**: 968.
9 Sharland M, Morgan M, Patton MA. Photoanthropometric study of facial growth in Noonan syndrome. *Am J Med Genet* 1993; **45**: 430–6.
** 10 Burch M, Sharland M, Shinebourne E *et al.* Cardiologic abnormalities in Noonan syndrome: phenotypic diagnosis and echocardiographic assessment of 118 patients. *J Am Coll Cardiol* 1993; **22**: 1189–92.
** 11 Noonan JA. Noonan syndrome: an update and review for the primary pediatrician. *Clin Pediatr (Phila)* 1994; **33**: 548–55.
12 Burch M, Mann JM, Sharland M *et al.* Myocardial disarray in Noonan syndrome. *Br Heart J* 1992; **68**: 586–8.
* 13 Skinner JR, Manzoor A, Hayes AM, Joffe HS, Martin RP. A regional study of presentation and outcome of hypertrophic cardiomyopathy in infants. *Heart* 1997; **77**: 229–33.
** 14 Char F, Rodriquez-Fernandez HL, Scott CI *et al.* The Noonan syndrome: a clinical study of forty-five cases. *Birth Defects Orig Artic Ser* 1972; **8**: 110–18.
** 15 Allanson JE. Noonan syndrome. *J Med Genet* 1987; **24**: 9–13.
** 16 Sharland M, Burch M, McKenna WM, Patton MA. A clinical study of Noonan syndrome. *Arch Dis Child* 1992; **67**: 178–83.
* 17 Lee NB, Kelly L, Sharland M. Ocular manifestations of Noonan syndrome. *Eye* 1992; **6**: 328–34.
18 Duncan WJ, Fowler RS, Farkas LG *et al.* A comprehensive scoring system for evaluating Noonan syndrome. *Am J Med Genet* 1981; **10**: 37–50.
19 De Sanctis V, Pinamonti A. *Manual of Disease-Specific Growth Charts and Body Standard Measurements.* Pisa: Pacini Editori, 1997: 86.
20 Borochowitz Z, Berant N, Dar H, Berant M. The neurofibromatosis–Noonan syndrome: genetic heterogeneity versus clinical variability—case report and review of the literature. *Neurofibromatosis* 1989; **2**: 309–14.
21 Sharland M, Taylor R, Patton MA, Jeffery S. Absence of linkage of Noonan syndrome to the neurofibromatosis type 1 locus. *J Med Genet* 1992; **29**: 188–90.
22 Fryer AE, Holt PJ, Hughes HE. The cardio-facio-cutaneous (CFC) syndrome and Noonan syndrome: are they the same? *Am J Med Genet* 1991; **38**: 548–51.
23 Nora JJ, Sinha AK. Direct familial transmission of the Turner phenotype. *Am J Dis Child* 1968; **116**: 343–50.
24 Nora JJ, Sinha AK. Direct male to male transmission of the XY Turner phenotype. *Lancet* 1970; **i**: 250.
25 Sharland M, Morgan M, Smith G, Burch M, Patton MA. Genetic counselling in Noonan syndrome. *Am J Med Genet* 1993; **45**: 437–40.
** 26 Jamieson CR, van der Burgt I, Brady AF *et al.* Mapping a gene for Noonan syndrome to the long arm of chromosome 12. *Nat Genet* 1994; **8**: 357–60.
** 27 Brady AF, Jamieson CR, van der Burgt I *et al.* Further delineation of the critical region for Noonan syndrome on the long arm of chromosome 12. *Eur J Hum Genet* 1997; **5**: 336–7.
28 Witt DR, Hoyeme E, Zonana J *et al.* Lymphedema in Noonan syndrome: clues to pathogenesis and prenatal diagnosis and review of the literature. *Am J Med Gen* 1987; **27**: 841–56.
** 29 Ranke MB, Heidemann P, Knupfer C *et al.* Noonan syndrome: growth and clinical manifestations in 144 cases. *Eur J Pediatr* 1988; **148**: 220–7.
** 30 Witt DR, Keena BA, Hall JG, Allanson JE. Growth curves for height in Noonan syndrome. *Clin Genet* 1986; **30**: 150–3.
31 Tanner JM, Whitehouse RH, Takaishi M. Standards from birth to maturity for height, weight, height velocity, and weight velocity. *Arch Dis Child* 1966; **41**: 454–71.
** 32 Lyon AJ, Preece MA, Grant DB. Growth curve for girls with Turner syndrome. *Arch Dis Child* 1985; **60**: 932–5.
33 Theintz G, Savage MO. Growth and pubertal development in five boys with Noonan's syndrome. *Arch Dis Child* 1982; **57**: 13–17.
* 34 Elsawi MM, Pryor JP, Klufio G, Barnes C, Patton MA. Genital tract function in men with Noonan syndrome. *J Med Genet* 1994; **31**: 468–70.
* 35 Spadoni GL, Bernadini S, Cianfarani S

et al. Spontaneous growth hormone secretion in Noonan's syndrome. *Acta Paediatr Scand* 1990; **367** (Suppl): 157.

** 36 Ahmed ML, Foot AB, Edge JA *et al.* Noonan's syndrome: abnormalities of the growth hormone/IGF-I axis and the response to treatment with human biosynthetic growth hormone. *Acta Paediatr Scand* 1991; **80**: 446–50.

** 37 Bernadini S, Spadoni GL, Cianfarani S *et al.* Growth hormone secretion in Noonan's syndrome. *J Pediatr Endocrinol Metab* 1991; **4**: 217–21.

* 38 Tanaka K, Sato A, Naito T *et al.* Noonan syndrome presenting growth hormone neurosecretory dysfunction. *Intern Med* 1992; **31**: 908–11.

** 39 Thomas BC, Stanhope R. Long-term treatment with growth hormone in Noonan's syndrome. *Acta Paediatr* 1993; **82**: 853–5.

** 40 Romano AA, Blethen SL, Dana K, Noto RA. Growth hormone treatment in Noonan syndrome: the National Cooperative Growth Study experience. *J Pediatr* 1996; **128**: S18–21.

** 41 Cotterill AM, McKenna WJ, Brady AF *et al.* The short-term effects of growth hormone therapy on height velocity and cardiac ventricular wall thickness in children with Noonan's syndrome. *J Clin Endocrinol Metab* 1996; **81**: 2291–7.

* 42 Elders MJ, Char F. Possible etiologic mechanisms of the short stature in the Noonan syndrome. *Birth Defects Orig Artic Ser* 1976; **12**: 127–33.

43 Saenger P. Turner's syndrome. *N Engl J Med* 1996; **335**: 1749–54.

* 44 Cianfarani S, Spadoni GL, Finocchi G *et al.* Trattamento con ormone della crescita (GH) in tre casi di sindrome di Noonan. *Minerva Pediatr* 1987; **39**: 281–4.

** 45 Otten BJ. Short stature in Noonan syndrome: demography and response to growth hormone treatment in the Kabi International growth study. In: Ranke MB, Gunnarsson R, eds. *Progress in Growth Hormone Therapy: 5 Years of KIGS.* Mannheim: J & J-Verlag, 1994: 206–15.

** 46 De Schepper J, Otten BJ, Francois I *et al.* Growth hormone therapy in prepubertal children with Noonan syndrome: first year growth response and comparison with Turner syndrome. *Acta Paediatr Scand* 1997; **86**: 943–6.

47 Wales JKH, Milner RDG. Successful therapy with recombinant somatotropin for short stature in Noonan's syndrome. *Acta Paediatr Scand* 1990; **366** (Suppl): 144.

48 Lu PW, Cowell CT, Moore B, Craighead A, Horward N. Growth hormone therapy in Noonan syndrome. *Acta Paediatr Scand* 1991; **377** (Suppl): 167.

** 49 Municchi G, Pasquino AM, Pucarelli I, Cianfarani S, Passeri F. Growth hormone treatment in Noonan syndrome: report of four cases who reached final height. *Horm Res* 1995; **44**: 164–7.

50 Fazio S, Sabatini D, Capaldo B *et al.* A preliminary study of growth hormone in the treatment of dilated cardiomyopathy. *N Engl J Med* 1996; **334**: 809–14.

* 51 Daubeney PEF, McCaughey ES, Chase C *et al.* Cardiac effects of growth hormone in short normal children: results after four years of treatment. *Arch Dis Child* 1995; **72**: 337–9.

52 Saenger P, Wesoly S, Glickstein J, Appel P, Issenberg H. No evidence for ventricular hypertrophy in Turner syndrome after growth hormone therapy. In: Albertsson-Wikland K, Ranke M, eds. *Turner Syndrome in a Life-Span Perspective: Research and Clinical Aspects. Proceedings of the 4th International Symposium on Turner Syndrome, Gothenburg, Sweden, 18–21 May, 1995.* Amsterdam: Elsevier, 1995: 259–62.

4: Does GH therapy influence final height in skeletal dysplasia syndromes?

Nicola A. Bridges

Introduction

Individuals with skeletal dysplasia are considerably disabled by their short stature, and it is hard to imagine that a potential therapy such as growth hormone (GH) administration should be available and not be tried, even if the use of GH is not logical in these conditions. We now have a wide range of publications concerning the use of GH in skeletal dysplasia. However, the small number of individuals included in these studies and the fact that there are no controlled trials mean that we are not yet able to give realistic information to the parents of children with this group of disorders.

Skeletal dysplasias are growth disorders characterized by an abnormal appearance at X-ray. The central control and secretion of growth hormone is normal, but the response at the level of the growth plate is abnormal. The growth patterns described in a number of other disorders, such as Turner syndrome, Down syndrome and Noonan syndrome, are similar to those seen in skeletal dysplasia, and the situation is comparable in that these are genetic disorders in which GH secretion is normal but bone growth is abnormal. GH treatment in these disorders is discussed elsewhere in the present volume (Chapters 2 & 3). The genetic defects causing many of the commoner skeletal dysplasias have now been identified.

The growth problem and clinical picture in skeletal dysplasia

Despite the wide genetic variation, the pattern of growth is similar in most skeletal dysplasias, with cumulative loss in all phases of growth (Fig. 4.1). There is a loss in fetal growth, with reduced birth length and then diminished growth velocity during childhood, with an attenuated pubertal growth spurt (the pattern is similar in Turner, Noonan and Down syndromes). This pattern is important in assessing the effect of therapy where there is an inevitable loss of growth before treatment can be started.

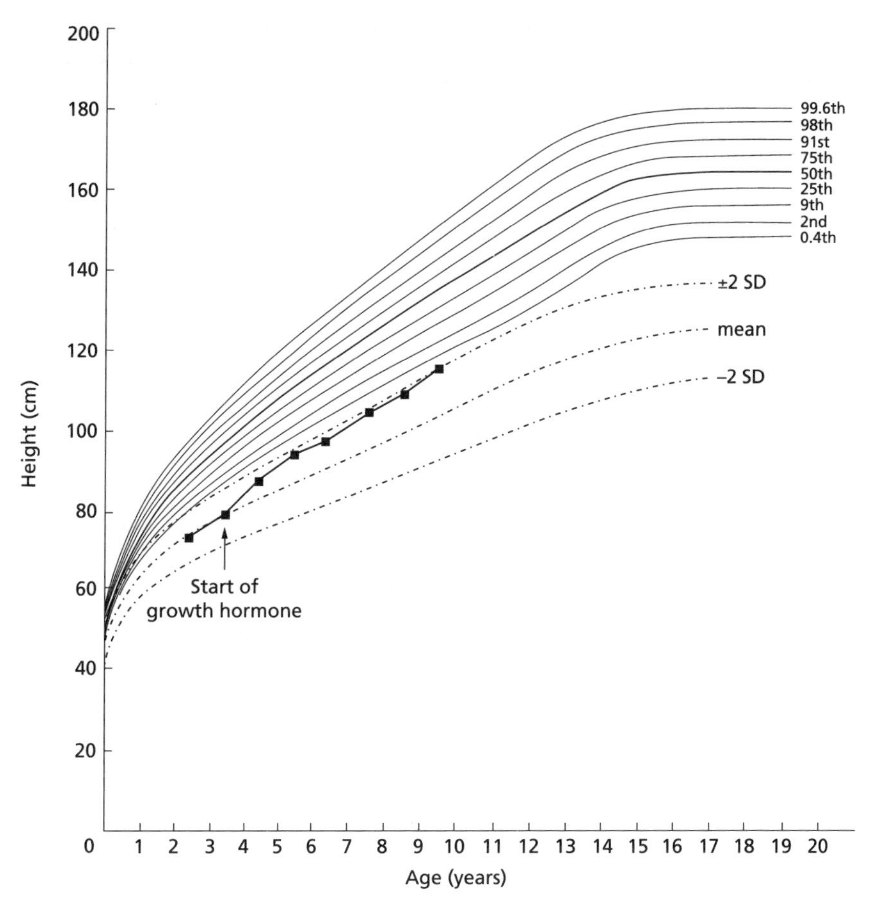

Fig. 4.1 Centiles for normal girls (UK standards, 1990 [47]) (thin lines, 0.4th to 99.6th centiles) compared to achondroplasia (Horton *et al.* [1]) (dotted lines, mean ± 2 SD). Superimposed on this is the growth pattern of an individual girl with achondroplasia (thick line, solid squares), starting treatment with recombinant growth hormone at 3.2 years and continuing for six years.

There are specific growth charts for a range of skeletal dysplasias [1,2]. The numbers of patients are very small compared to those used in the 'normal' population charts. The data may be neither up to date nor specific to the ethnic group. The value of charts may be limited where there is genetic heterogeneity (for example hypochondroplasia). For these reasons, the value of these growth data in interpreting the effect of treatment may be limited.

The genetics of skeletal dysplasia

The genetic basis of many skeletal dysplasias have been identified in the last few years. These defects affect a range of factors important for the control of bone growth (Table 4.1).

Table 4.1 Mutations identified in the commoner skeletal dysplasias.

Mutation identified	Skeletal dysplasia	Comments
Fibroblast growth factor receptors		
FGFR2	Craniosynostosis (Apert, Crouzon, Pfeiffer, Jackson Weiss) [7]	Activating mutations of the receptor [12]
FGFR3	Thantophoric dysplasia [5] achondroplasia [4] hypochondroplasia [6]	Activating mutations of the receptor [8]. Not all children with hypochondroplasia have FGFR3 mutations
Collagen gene mutations		
Type 1 collagen COL1A1, COL1A2	Osteogenesis imperfecta type 1 and neonatal lethal forms [3, 18]	
Type 2 collagen COL2A1	Achondrogenesis type 2 Hypochondrogenesis Spondyloepiphyseal dysplasia Kneist dysplasia Stickler syndrome [3, 35, 36]	Different mutations result in skeletal dysplasias of varying severity—mildest is 'late onset' SED with osteoarthritis, most severe is achondrogenesis
Type 9 collagen COL9A2	Multiple epiphyseal dysplasia [37]	
Type 10 collagen	Schmidt type metaphyseal dysplasia [19]	
Other mutations		
SOX9	Camptomelic dysplasia [21]	SOX9 is also important in sex determination
DTDST (transmembrane sulphate transporter)	Achondrogenesis type 1 b Diastrophic dysplasia [38]	
G protein related to parathyroid hormone/ parathyroid hormone-related peptide receptor	Jansen type metaphyseal dysplasia [20]	G protein mutation resulting in constitutive activation
COMP (cartilage oligomeric matrix protein)	Pseudoachondroplasia Multiple epiphyseal dysplasia (Fairbanks type) [39]	

This is not an exhaustive list but covers the most common skeletal dysplasias.

Mutations of fibroblast growth factor receptors

Mutations of the fibroblast growth factor receptors have been identified as the cause of two different groups of skeletal abnormalities—mutations of the fibroblast growth factor receptor type 3 gene (FGFR3) are responsible for thanatophoric dysplasia (TD) [3], achondroplasia [4] and most (but not all) cases of hypochondroplasia [5,6], and mutations of the fibroblast growth factor receptor type 2 gene (FGFR2) are responsible for inherited forms of craniosynostosis (Apert, Pfeiffer, Jackson, Weiss and Crouzon types [7]). The mutations in FGFR3 that cause TD, achondroplasia and hypochondroplasia are heterozygous mutations that constitutively activate the FGFR3 receptor. In TD, there is evidence that the receptor abnormality results in tyrosine kinase activity (with activation of transcription factors such as *Stat 1* [8]) and increased receptor dimerization [9,10]. Activation of the FGFR3 receptor appears ultimately to inhibit bone growth [8,11]. The FGFR2 receptor mutations responsible for craniosynostosis are also constitutively activating mutations [7,12].

Studies in most populations have demonstrated that nearly all individuals with achondroplasia have the same mutation (Gly 380 → Arg) of the transmembrane domain of FGFR3 [4,9]. All TD type 2 individuals have been reported to have the same mutation of the second tyrosine kinase domain of FGFR3 [13,14], while TD type 1 and hypochondroplasia are more genetically heterogeneous. In TD type 1, a range of FGFR3 mutations have been reported [5]. In hypochondroplasia, a range of FGFR3 mutations have been described, but some families with typical features (who cannot be differentiated using clinical and radiological features alone) have not been demonstrated to have a defect in FGFR3 [5,15].

Mutations of collagen genes

Numerous mutations of the gene for type 2 collagen (COL2 A1) have been identified, and result in skeletal problems of varying severity. Mild ('late-onset') spondyloepiphyseal dysplasia with osteoarthritis is the mildest manifestation. The most severe forms are achondrogenesis type 2 and hypochondrogenesis, and lying between these in severity are congenital spondyloepiphyseal dysplasia, Kneist dysplasia and Stickler syndrome [16,17].

Most cases of osteogenesis imperfecta are caused by defects in the production of type 1 collagen [3,18]. Abnormalities in other collagen genes have been identified as the cause of multiple epiphyseal dysplasia and Schmidt metaphyseal dysplasia [3,19].

Other causes

An activating mutation of the G protein–associated receptor for parathyroid hormone and parathyroid hormone–related peptide (PTHrp) has been identified in Jansen-type metaphyseal chondrodysplasia [20]. Abnormalities in the SOX9 gene have been identified in camptomelic dysplasia. SOX9 is a transcription factor that regulates the type 2 collagen gene and is also involved in gender determination [3,21].

Therapies for skeletal dysplasia

The mutations underlying the more common skeletal dysplasias have now been identified, and this knowledge has advanced the study both of the way in which these mutations result in abnormal bone growth and of the patterns of normal bone growth. Our increasing knowledge emphasizes the difficulties of producing potential therapies directed at the defect. The discovery of the genetic basis for achondroplasia might eventually lead to a disorder-specific therapy, but a source of recombinant fibroblast growth factor would be unlikely to improve the growth of children with achondroplasia. There are no data concerning the way in which GH acts to accelerate growth in children with skeletal dysplasia. Given the diversity of genetic defects that result in skeletal dysplasia, it is not surprising that the effects vary between different disorders [22] and between the different genetic variants of the same disorder [23].

While the genetic basis of the most common disorders has been identified, there remain many individuals with rarer skeletal abnormalities that may not fall into any diagnostic category and which may be limited to one individual or family. These individuals are often the subjects for 'one-off' trials of GH, the results of which are unlikely to add to our understanding of the role of GH in skeletal dysplasia.

Leg lengthening

The disproportion associated with skeletal dysplasia makes leg lengthening an attractive treatment strategy, and there is now experience of this in a number of centres, particularly with achondroplasia. Leg lengthening requires prolonged immobilization and intensive hospital supervision, and there is a significant risk of complications (pin site infections, fracture, angulation of the elongated segment [24]). Not all patients are suitable for this procedure. However, many patients regard the risk as acceptable given the potential increment in height (Saleh and Burton reported increases of 60–120 mm in length in the femur and 60–145 mm in the tibia in a series of 94 patients [25]), and this remains the only proven treatment to increase height in skeletal dysplasia. The long-term results of GH treatment must be compared with those achieved by leg lengthening.

Growth hormone

In Turner syndrome, GH treatment has been shown to increase final height [26] even though there is normal GH secretion, and this provides a model for trials of GH in skeletal dysplasia. Similar doses of GH, which are greater than those used for GH insufficiency, have been used. Although in Turner syndrome several studies have used oxandrolone as well [26], there is only one report of the use of oxandrolone in skeletal dysplasia, with a short-term increase in growth velocity [27].

The potential benefits of GH treatment in children with skeletal dysplasia would be increased height in childhood, increased final height and changes in proportion. In osteogenesis imperfecta, there is the additional potential benefit of decreased fracture frequency.

Achondroplasia and hypochondroplasia

There are now many published trials of GH in achondroplasia and hypochondroplasia. However, the total patient numbers are very small. The results of these trials are summarized in Table 4.2, in which the most recent paper is cited when an individual trial has been reported on several occasions.

In hypochondroplasia and achondroplasia, GH treatment for one or two years increases growth velocity over pretreatment values. Data after this are too limited for conclusions to be drawn—one study suggests a longer-

Table 4.2 Summary of results of GH studies in skeletal dyplasia.

Source	N	Pretreatment growth velocity (cm/yr)	1st year growth velocity (cm/yr)	2nd year growth velocity (cm/yr)	Comments
Achondroplasia					
Okabe *et al.* 1991 [40]	3	3.8	7.6		
Horton *et al.* 1992 [41]	6	3.4	6.0		
Nishi *et al.* 1993 [42]	6	3.7	6.0		
Bridges *et al.* 1994 [22]	17	Change in height SDS from –5.0 to –4.3			
Key *et al.* 1996 [43]	5	4.6	5.5		
Shohat *et al.* 1996 [29]	11	4.0	5.3		No change in disproportion
Weber *et al.* 1996 [44]	6	5.0	5.8		
Stamouyannou *et al.* 1997 [45]	15	3.2	8.3	6.3 ($n = 9$)	
Tanaka *et al.* 1998 [46]	35	3.9	6.5	4.6	
Hypochondroplasia					
Allen *et al.* 1994 [28]					Apparent difference between those starting prepubertally and in established puberty
Starting prepubertally	39	Change in ht SDS from –2.8 to –2.5		ht SDS = –2.5	
Starting in puberty	33	Change in ht SDS from –2.5 to –2.2		ht SDS = –1.9	
Key *et al.* 1996 [43]	10	5.1	7.5		
Shohat *et al.* 1996 [29]	4	3.7	7.0		

SDS: standard deviation score.

term benefit in children with hypochondroplasia who are started on treatment with signs of puberty (rather than prepubertally), and there are individual reports of longer treatment in achondroplasia [28]. Investigators who have reported on body segment disproportion in achondroplasia have found no change with GH treatment [22,29]. There are no final height data.

Other skeletal dysplasias

There are very limited data concerning other skeletal dysplasias. Burren and Werther report a poor short-term response in children with metaphyseal dysplasia and dyschondrosteosis (Léri–Weill syndrome) [30]. Thuestad *et al.* report an increase in growth velocity in five children with dyschondrosteosis after one year of treatment [31]. There are some individual case reports of GH treatment in skeletal dysplasias in which an abnormality of growth hormone secretion has been identified, for example Robinow syndrome and pycnodysostosis [32,33].

Osteogenesis imperfecta

In osteogenesis imperfecta, growth hormone treatment has been studied as a way of preventing short stature and also as a potential treatment to increase bone density and reduce fracture frequency. A short-term increase in growth velocity has been reported [18,34], with increased bone density and decreased fractures demonstrated by one group [18].

Conclusions

It is now clear that the short-term treatment of achondroplasia and hypochondroplasia (and possibly some other disorders) with GH increases growth velocity over pretreatment values, and it could be argued that this is of psychological benefit to the children, without consideration of final height. However, there are few data on treatment after the first year, and there is no information concerning final height. None of the studies has included a control group, and predicting the expected final height is difficult; bone ages cannot be used in disorders causing an abnormal skeleton; there may be genetic heterogeneity within a disorder; and—as discussed above—diagnosis-specific growth charts may be unhelpful. Many trials involve small numbers (or only one subject). It may be difficult to assess final height results even when they are available. Judgements as to the benefits of treatment will be difficult in a situation in which the therapy cannot be expected to return the patient to complete normality.

All these uncertainties mean that final height data from the trials reported to date (which will take many years) may not give us the information needed to assess the advisability of treatment for an individual patient. Clinicians at present have little clinical evidence to give families who are understandably keen to try any therapy that might help their child. In the long term, research into the genetics of skeletal dysplasia and the mechanisms of the abnormal bone growth is perhaps more likely to clarify the situation than further trials of GH.

References

1 Horton WA, Rotter JI, Rimoin DL, Scott CI, Hall JG. Standard growth curves of achondroplasia. *J Pediatr* 1978; **93**: 435–8.

2 Horton WA, Hall JG, Scott CI, Pyeritz RE, Rimoin DL. Growth curves for height for diastrophic dysplasia, spondyloepiphyseal dysplasia congenita and pseudoachondroplasia. *Am J Dis Child* 1982; **136**: 316–19.

* 3 Horton WA. Molecular genetic basis of the human chondrodysplasias. *Endocrinol Metab Clin North Am* 1996; **25**: 683–97.

4 Bellus GA, Hefferon TW, de Ortiz Luna RI *et al.* Achondroplasia is caused by recurrent G380R mutations of FGFR3. *Am J Hum Genet* 1995; **56**: 368–73.

5 Bonaventure J, Rousseau F, Legeai-Mallet L *et al.* Common mutations in the fibroblast growth factor receptor 3 (FGFR3) gene account for achondroplasia, hypochondroplasia and thanatophoric dwarfism. *Am J Med Genet* 1996; **63**: 148–54.

6 Rousseau F, Bonaventure J, Legeai-Mallet L *et al.* Clinical and genetic heterogeneity of hypochondroplasia. *J Med Genet* 1996; **33**: 749–52.

7 Holloway GE, Suthers GK, Haan EA *et al.* Mutation analysis in FGFR2 craniosynostosis syndromes. *Hum Genet* 1997; **99**: 251–5.

8 Su WC, Kitagawa M, Xue N *et al.* Activation of Stat 1 by mutant fibroblast growth factor receptor on thanatophoric dysplasia type 2 dwarfism. *Nature* 1977; **386**: 288–92.

9 Horton WA. Fibroblast growth factor receptor 3 and the human chondrodysplasias. *Curr Opin Pediatr* 1997; **9**: 437–42.

10 D'Avis PY, Robertson SC, Meyer AN *et al.* Constitutive activation of fibroblast growth factor receptor 3 by mutations responsible for the lethal skeletal dysplasia thanatophoric dysplasia type 1. *Cell Growth Differ* 1998; **9**: 71–8.

11 Nguyen HB, Estacion M, Gargus JJ. Mutations causing achondroplasia and thanatophoric dysplasia alter bFGF induced calcium signals in human diploid fibroblasts. *Hum Mol Genet* 1997; **6**: 681–8.

12 Galvin BD, Hart KC, Meyer AN, Webster MK, Donoghue DJ. Constitutive receptor activation by Crouzon syndrome mutations in fibroblast growth factor receptor (FGFR) 2 and FGFR2/Neu chimeras. *Proc Natl Acad Sci USA* 1996; **93**: 7894–9.

13 Bonaventure J, Rousseau F, Legeai-Mallet L, Le Merrer M, Munnich A. Common mutations in the gene encoding fibroblast growth factor 3 account for achondroplasia, hypochondroplasia and thanatophoric dysplasia. *Acta Pediatr (Suppl)* 1996; **417**: 33–8.

14 Pokharel RK, Alimsardjono H, Takeshima Y *et al.* Japanese cases of type 1 thanatophoric dysplasia exclusively carry a C to T transition at nucleotide 742 of the fibroblast growth factor 3 gene. *Biochem Biophys Res Commun* 1996; **227**: 236–9.

15 Prinster C, Carrera P, Del Maschio M *et al.* Comparison of clinical radiological and molecular biological findings in hypochondroplasia. *Am J Med Genet* 1998; **75**: 109–12.

16 Jacenko O, Olsen BR, Warman ML. Of mice and men: heritable skeletal disorders. *Am J Med Genet* 1994; **34**:

163–8.
17 Francomano CA, McIntosh I, Wilkin DJ. Bone dysplasias in man: molecular insights. *Current Opin Genet Dev* 1996; **6**: 301–8.
*18 Antoniazzi F, Bertholdo F, Mottes M *et al.* Growth hormone treatment in osteogenesis imperfecta with quantitative defect of type 1 collagen synthesis. *J Pediatr* 1996; **129**: 432–9.
19 Wallis GA, Rash B, Sykes B *et al.* Mutations within the gene encoding the alpha 1 (X) chain of type X collagen (COL10A1) cause metaphyseal chondrodysplasia type Schmid but not several other forms of metaphyseal chondrodysplasia. *J Med Genet* 1996; **33**: 450–7.
20 Schipani E, Kruse K, Juppner H. A constitutively active mutant PTH-PTHrP receptor in Jansen-type metaphyseal chondrodysplasia. *Science* 1995; **268**: 98–100.
21 Bell DM, Leung KK, Wheatley SC *et al.* SOX9 directly regulates the type-II collagen gene. *Nature Genet* 1997; **16**: 174–8.
*22 Bridges NA, Brook CGD. Progress report: growth hormone in skeletal dysplasia. *Horm Res* 1994; **42**: 231–4.
23 Mullis PE, Patel MS, Brickell PM, Hindmarsh PC, Brook CGD. Growth characteristics and response to growth hormone therapy in patients with hypochondroplasia: genetic linkage of the insulin like growth factor 1 gene at chromosome 12q23 to the disease in a subgroup of these patients. *Clin Endocrinol* 1991; **34**: 285–74.
24 Faber FW, Keessen W, van Roermund PM. Complications of leg lengthening: 46 procedures in 28 patients. *Acta Orthop Scand* 1991; **62**: 327–32.
25 Saleh M, Burton M. Leg lengthening: patient selection and management in achondroplasia. *Orthop Clin North Am* 1991; **22**: 589–99.
26 Rosenfeld RG, Attie KM, Frane J *et al.* Growth hormone therapy of Turner's syndrome: beneficial effect on adult height. *J Pediatr* 1998; **132**: 319–24.
27 Buyukgebiz A, Kovanlikaya I. Oxandrolone therapy in skeletal dysplasia. *Turk J Pediatr* 1993; **35**: 189–96.
28 Allen DB, Brook CGD, Bridges NA *et al.* Therapeutic controversies: growth hormone treatment of non–GH-deficient subjects. *J Clin Endocrinol Metab* 1994; **79: 1239–48.
*29 Shohat M, Tick D, Barakat S *et al.* Short-term human growth hormone treatment increases growth rate in achondroplasia. *J Clin Endocrinol Metab* 1996; **81**: 4033–7.
*30 Burren CP, Werther GA. Skeletal dysplasias: response to growth hormone therapy. *J Pediatr Endocrinol Metab* 1996; **9**: 31–40.
31 Thuestad IJ, Ivarsson SA, Nilsson KO, Wattsgard C. Growth hormone treatment in Léri–Weill syndrome. *J Pediatr Endocrinol Metab* 1996; **9**: 201–4.
32 Kawai M, Yorifuji T, Yamanaka C *et al.* A case of Robinow syndrome accompanied by partial growth hormone insufficiency treated with growth hormone. *Horm Res* 1997; **48**: 41–3.
33 Soliman AT, Rajab A, Al Sami I, Darwish A, Asfour M. Defective growth hormone secretion in children with pycnodysostosis and improved linear growth after growth hormone treatment. *Arch Dis Child* 1996; **75**: 242–4.
34 Marini JC, Bordenick S, Heavner G *et al.* The growth hormone and somatomedin axis in short children with osteogenesis imperfecta. *J Clin Endocrinol Metab* 1993; **76**: 251–6.
35 Winterpacht A, Superti-Furga A, Schwarze U *et al.* The deletion of six amino acids at the C-terminus of the alpha 1 (11) chain causes overmodification of type II and type XI collagen: further evidence for the association between small deletions in COL2A I and Kniest dysplasia. *J Med Genet* 1996; **33**: 649–54.
36 Williams CJ, Ganguly A, Considine E *et al.* A-2→G transition at the 3′ acceptor splice site of IVS17 characterizes the COL2A1 gene mutation in the original Stickler syndrome kindred. *Am J Med Genet* 1996; **14**: 461–7.
37 Muragak Y, Mariman EC, van-Beersum SE *et al.* A mutation in the gene encoding the alpha 2 chain of the fibril-associated collagen IX, COL9A2, causes multiple epiphyseal dysplasia (EDM2). *Nat Genet* 1996; **12**: 103–5.
38 Superti-Furga A, Rossi A, Steinmann B, Gitzelmann R. A chondrodysplasia family produced by mutations in the diastrophic dysplasia sulfate transporter

gene: genotype/phenotype correlations. *Am J Med Genet* 1996; **63**: 144–7.

39 Cohn DH, Briggs MD, King LM *et al.* Mutations in the cartilage oligomeric matrix protein (COMP) gene in pseudoachondroplasia and multiple epiphyseal dysplasia. *Ann N Y Acad Sci* 1996; **785**:188–94.

*40 Okabe T, Nishikawa K, Miyamori C, Sato T. Growth promoting effect of human growth hormone on patients with achondroplasia. *Acta Paediatr Jpn* 1991; **33**: 357–62.

*41 Horton WA, Hecht JT, Hood OJ, *et al.* Growth hormone therapy in achondroplasia. *Am J Med Genet* 1992; **42**: 667–70.

*42 Nishi Y, Kajiyama M, Miyagawa S, Fujiwara M, Hamamoto K. Growth hormone therapy in achondroplasia. *Acta Endocrinol* 1993; **128**: 394–6.

43 Key LL, Gross AJ. Response to growth hormone in children with chondrodysplasia. *J Pediatr* 1996; **125: S14–S17.

*44 Weber G, Prinster C, Meneghel M et al. Human growth hormone treatment in prepubertal children with achondroplasia. Am J Med Genet 1996; **61**: 396–400.

*45 Stamouannou L, Karachaliou F, Neou P *et al.* Growth and growth hormone therapy in children with achondroplasia: a two-year experience. *Am J Med Genet* 1997; **72**: 71–6.

*46 Tanaka H, Kubo T, Yamate T *et al.* Effect of growth hormone therapy in children with achondroplasia: growth pattern, hypothalamic pituitary function, and genotype. *J Clin Endocrinol Metab* 1998; **138**: 275–80.

47 Freeman JV, Cole TJ, Chinn S *et al.* Cross-sectional stature and weight reference curves for the UK, 1990. *Arch Dis Child* 1995; **73**: 17–24.

5: Is there a role for GH therapy in short normal children?

Peter C. Hindmarsh

Introduction

The increased availability of biosynthetic recombinant human growth hormone (rhGH) in the mid-1980s led to increased interest in the wider application of this agent for promoting growth in non-GH-deficient individuals. Several candidate conditions emerged, including Turner syndrome and intrauterine growth restriction, as well as the conditions in patients labelled as being either of idiopathic short stature, short non-GH-deficient or short normal. Even prior to the introduction of rhGH, several groups had explored the use of pituitary-derived growth hormone (GH) in this group of individuals [1–6]. The majority of studies were short-term (6–12 months), and only one study used a double-blind, placebo-controlled approach [7]. Overall, a growth-accelerating effect was observed, but the heterogeneity of the patients in the studies made extrapolation difficult.

In contrast to GH deficiency, in which a clear abnormality—a deficit—is replaced, these studies had as their ultimate goal the aim of improving final height. This did not involve hormone replacement, but rather the addition of more GH in an already replete or near-replete individual. It was thought that this type of improvement in final height would benefit the recipient, as it was generally considered at that time that short stature at these extremes was a disadvantage. Testing this particular hypothesis—i.e. that short individuals are psychologically disadvantaged—came much later in the whole programme. The primary aim in the majority of studies was to demonstrate short- to medium-term growth acceleration, in the hope that this would translate into an improvement in final height. Out of necessity, this type of study has to be conducted over a long period of time, and it is only after 10 years or more that final height data are beginning to emerge. This chapter reviews the background to these studies, the problems with the clinical trials themselves and the outcome of studies reporting final height information.

Types of short stature

The purpose of this chapter is to consider two types of short stature—patients in whom there is a familial component to the short stature (FSS) and those in whom the short stature (perhaps transient) is associated with delayed skeletal maturation and puberty. The latter group of individuals are often referred to as having constitutional delay of growth and puberty (CDGP), and they may include patients with FSS. The definition of short stature is to some extent arbitrary, but for the purpose of this chapter we are effectively talking of individuals with a height between the second and 0.4th centiles of the 1990 UK Growth Standards [8]. Heights below the 0.4 centile have a high probability of being associated with pathology. It is assumed for the purpose of this discussion that conditions such as Turner syndrome, intrauterine growth restriction and to a large extent the skeletal dysplasias have been excluded, although milder forms can masquerade as idiopathic short stature. It is unlikely that patients with these conditions will be labelled as having FSS, since those with the above conditions will have average-sized parents, whereas FSS patients will have short parents and the height of the child will be appropriate for the target height of the parent. This is certainly not the case in Turner syndrome, intrauterine growth restriction, or the mild skeletal dysplasias. Bone age is not usually delayed in FSS unless it is associated with CDGP, and growth velocity is near normal, averaging on a year-by-year basis on the 25th height velocity centile, or between 0 and –0.8 standard deviation scores (SDS).

In short children with CDGP, growth is similar to that in those with FSS, but there is an accumulating deficit of bone age compared to chronological age. Height for the parents is short, but it normalizes after correction for the bone age deficit. The growth rate is the same as in the FSS group up to the point at which puberty becomes apparent. The importance of making this distinction is that there is a considerable body of data to suggest that the final height outcome (natural history) of the two groups differs. In FSS, the majority of individuals achieve the mid-parental height, as might be expected. In the CGDP group, many individuals fall short of this target, and end up shorter than might be expected given the parental heights and bone age delay [9,10]. This is particularly the case if puberty is delayed up to 15 or 16 years of age. The more delayed the bone maturation, the less confident one can be about the final height outcome. Severe delay in bone age may lead to an over-prediction of the final height of the individual. The reason for this is not entirely clear, but it may in part reflect the fact that the height prediction systems used are not designed for individuals with large delays in skeletal maturation, and as these systems also assume a normal timing and magnitude of the pubertal growth spurt, it is likely that the over-performance reflects inherent problems in the prediction systems [11]. Whether these observations pertain to the general population is unclear. Greco *et al.* [12] suggested that the problems observed with the CDGP population were less obvious in larger, community-based studies. These important caveats need to be considered in the interpretation of final height data and in determining how well-balanced the clinical trials in this area are.

The demonstration of a normal growth rate and pattern implies that the endocrine axis is probably functioning intact. Many of the clinical trials have also attempted to define the GH secretory status of individuals. Several studies have used pharmacological methods of assessing GH secretory status, and very few have supported these studies with physiological assessments of GH release. Indeed, the description of these children as 'short normal' referred more to their response to GH pharmacological testing than their actual growth rate, which was often suboptimal [1–6]. The observation that many of these individuals did have poor GH secretion serves to underline this issue [13]. The term 'neurosecretory dysfunction', although helpful in this respect, only served to create a further subgroup of individuals in whom GH therapy was used [14]. None the less, it does appear that those children who secrete very little GH and grow poorly will have a marked response when given relatively small amounts of exogenous GH. At the other extreme are those children who are small and growing at a normal velocity, whose outcome is described in this chapter. In between are children with a range of reduced pretreatment growth rates and varying degrees of GH insufficiency, who have varying responses to GH treatment.

Physiology of growth hormone secretion

Growth hormone secretion in humans is pulsatile, and the secretory pattern appears to arise from a close interplay between the two hypothalamic peptides growth hormone-releasing hormone (GHRH) and somatostatin. The precise interaction between the two is unclear, but the simplest model available suggests that a GH pulse is generated as a result of a rise in GHRH and a concomitant fall in somatostatin levels in the hypothalamoportal system [15]. The growth rate of an individual in childhood appears to be largely determined by the amplitude of the GH pulses [16,17]. Figure 5.1 illustrates the dose–response nature of this relationship.

The asymptotic nature of this relationship has implications for the therapeutic interventions to be discussed. Children producing very little GH, which leads to a reduced growth rate, may be expected to have the best response to GH therapy, since a small increment (movement to the right in Fig. 5.1) will lead to a marked increase in growth velocity. Children already secreting sufficient GH to maintain a near-normal growth velocity (height veloc-

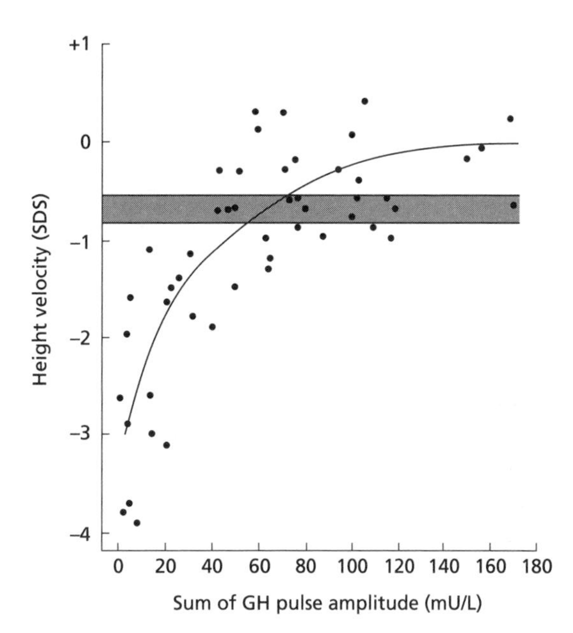

Fig. 5.1 The asymptotic relationship between height velocity, expressed as a standard deviation score, and the sum of GH pulse amplitude measured over a 24-hour period in 50 short prepubertal children. The shaded zone shows the lower boundary of height velocity for short children.

ity SDS 0 to −0.8) might also be expected to grow faster if given extra GH, although the magnitude of their response would be less than their worse-off peers [18].

More importantly, on a dose-for-dose basis the response in the severely GH-deficient individuals will always be greater than that in their more replete peers. This observation is important, and needs to be borne in mind when interpreting the studies outlined below. Very few of the original studies used dosages that were greater than the then-standard GH replacement dose.

Clinical trial problems

There is a large literature on the statistical theory behind clinical trials, and a number of issues are relevant to GH trials in children in general and to the short normal population in particular. In contrast to studies on the use of GH therapy in adults, very few of the paediatric studies have used the randomized controlled trial (RCT) format. Where this format has been used, it has only been over a short period of time. There are clearly problems associated with the RCT approach, not the least of which is the need to administer a placebo by injection for long periods of time. The studies therefore tend to involve treatment versus observation (whether randomized or not), or comparisons with historical controls. The latter data are extremely difficult to interpret, not least due to the secular trend in height that operates in most populations [19]. The former approach is subject to all the problems of bias in determining which patients wish to participate and who is to be randomized—and if not, why not? Even the admirable community approach taken by the Southampton group [20] enlisted only those patients who wanted to be part of the study.

Very few of the studies were designed on an intention-to-treat basis, and this makes it difficult to determine the true effect observed, since there is a tendency over the course of the study to select those individuals who respond well and to continue treatment in that particular group. These problems, coupled with the heterogeneity of the studies in terms of the definition of the condition, age at the start of therapy, management at puberty and dosage used, make a 'meta-analysis' of the available data difficult. Such an analysis would be extremely helpful if it could be done using individuals' data, but it is not clear how easy it would be to overcome the heterogeneity effect. Even if it were possible to remove that obstacle, the absence of the RCT approach limits the deductions that can be made and effectively excludes the possibility of determining clinically useful concepts such as the 'number needed to treat' [21]. In retrospect, it is interesting to contemplate why the RCT method has not been used. The injection issue aside, it does seem that the investigators were sufficiently unsure of the effect that they might expect, which if they were in equipoise would have justified the RCT approach. The RCT approach would not, however, have overcome one of the major deficiencies of all these studies, namely the small sample size. As many of the studies were constructed to determine the short-term effect, the statistical power of sample size was based on that aim and not on the ultimate goal—the final height. Even if it had been, there are good reasons to be wary of the conclusions derived from such small studies.

Consider the following. Suppose that GH therapy is being considered for short normal individuals and the effect is to be compared with placebo or possibly historical controls. It is known that in short normal individuals over a period of one year, 30% of the individuals will display acceleration in growth of 2 cm per year. The trial is set up to determine whether GH intervention is better than this or not. Imagine that the truth is that GH therapy produces this type of response in 50%, but only 20% of the time. If there are 200 000 participants world-wide; if the probability of a false-positive conclusion if the null hypothesis of no treatment difference is true is kept at 0.05; and if there are studies containing 400, 100, 50 or 25 participants—then Table 5.1 shows what begins to happen as the trial size decreases.

The first thing to note is that, as the number of individuals in the trial decreases (and, by the way, the number of studies increases), the number of true positives detected declines in such a way that small trials are less likely to identify effective therapy. The second thing that starts to happen is that, because we have kept the probability fixed, a large number of small trials

Table 5.1 Effects of trial size on results detected.

Size of trial	Number of studies possible	True positives detected (%)	Ratio of false positives to expected true positives
400	50	99	0.2
50	400	43	0.5
25	800	27	0.7

increases the likelihood that a positive conclusion is a false conclusion. These observations explain, in part, why a beneficial effect or no effect can be seen in the studies reported. Both conclusions are equally valid and entirely predictable from statistical theory! There are other problems that arise, however. The limited number of subjects is an important fact and is inadvertently overlooked, e.g. in relation to bone age delay. Randomization should ensure that the effect is roughly similar in all groups. Since the effectiveness of randomization depends on sample size, a large number of small sample studies will not be sufficient to compensate.

There are a number of problems that are also inherent in studies in short normal individuals. The use of historical controls has already been discussed, and the definition of the end point has been alluded to. The end point of any study should be unambiguous and easily measurable, with minimum error attached to it. While final height can be defined and measured, it is in determining effect that problems arise. If large studies had been conducted and the groups were well matched for age, height, bone age, parental height, etc., then a simple difference between the groups could be tested for. This is rarely the case, and what has happened is a comparison between the final height achieved and that predicted at entry to the study or intervention. There are several problems with this approach. First, the height prediction equations make certain assumptions regarding the growth process, particularly with respect to the timing and magnitude of the pubertal growth spurt and the degree of bone age delay allowed for [11]. Secondly, the height prediction equations are only models, and they carry with them an inherent imprecision, which is outlined with the Tanner–Whitehouse system in Table 5.2. Thirdly, and allied to this, there is the issue of regression to the mean operating within the short-statured population. Although many of these factors are quantifiable, very few have been included in the assessment of treatment effects.

Short-term studies

The initial studies reporting treatment effects in short 'normal' individuals

Accuracy of prediction (residual standard deviation)		
Age (years)	Boys (cm)	Girls (cm)
6.0	4.7	3.5
9.0	4.1	3.6
12.0	3.8	3.0

Table 5.2 Accuracy of prediction of final height using Tanner–Whitehouse system.

using pituitary GH are extremely difficult to interpret. Up until the studies of rhGH treatment in short normal individuals, of the 91 children reported in the world literature only 12 had height velocity standard deviation scores between 0 and –0.8 [1–6]. These studies were also complicated by the fact that many of the children did not enter puberty at the correct time, while in others the responses to exogenous GH could easily have been ascribed to the patient's own puberty growth spurt. The studies described an increase in growth rate, but did not consider whether this might have been achieved at the expense of an inappropriate advance in skeletal maturity. The studies reported in the late 1980s using rhGH showed an acceleration in growth rate with the administration of rhGH doses that were higher than those used at the time for GH replacement. The studies can be broadly divided into two groups: those in which an observation group was present, and those in which the patients served as their own controls. Two studies conducted a little later included attempts at randomization [22,23] both using higher doses of growth hormone than hitherto (30–40 units/m^2 body surface area/week). Overall, the changes in growth rate were less than in more severely GH-deficient individuals, confirming the original hypothesis regarding intervention with GH in this group. In contrast to the pituitary-derived studies, many of the short-term studies included an assessment of skeletal maturation, and from this it was possible to obtain an estimate of how final height might be changed by the estimation of height for bone age. Over a period of two or three years, several groups [22,24,25] reported a positive effect on final height using the surrogate marker of height for bone age.

Long-term effects

The effect of rhGH therapy over the longer term has not been borne out. Figure 5.2 shows the changes in predicted final height expressed as a standard deviation score over a period of nine years in subjects from our first study. During the first three or four years, there appears to be a beneficial effect of intervention. Thereafter, the effect appears to begin to wane, so that at the end of therapy the net gain in final height amounted to some

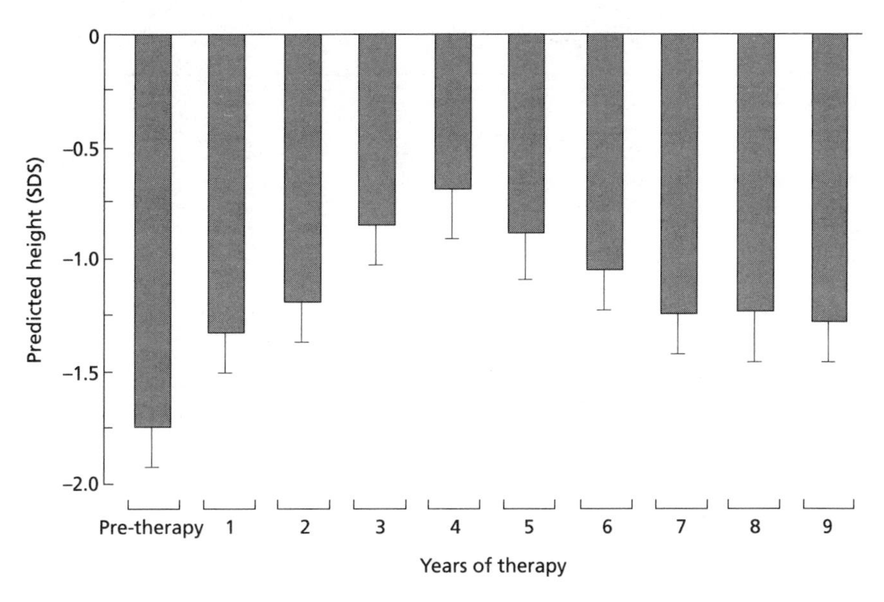

Fig. 5.2 Changes in height prediction expressed as a standard deviation score over a period of nine years in 16 short prepubertal children treated with biosynthetic human growth hormone. Data are shown as mean ± SEM. Reproduced with permission from Hindmarsh and Brook [29].

3 cm. This study also included the end results from the observation group, which minimized further the actual gain in stature. Table 5.3 summarizes the data that are available from a number of studies reporting final height [14,26–33]. To place these results in context, it is important to recall the previous discussion of outcome in untreated individuals [9]. For example, in the FSS group, males are likely to gain + 0.74 SDS and females + 0.71, and those in the CDGP group + 1.68 SDS (males) and + 1.66 SDS (females). As predicted from the discussion of clinical trials above, there has been very little effect of therapy overall, although a number of papers suggested quite impressive improvements. What these data point to is the need to conduct a meta-analysis of all the data sets, including actual individual data and dosage schedules.

Further insight into the differences also comes from analysis of the dosage schedules used. Some of the more impressive effects appear to come from studies in which a higher dosage schedule was used, and further data will be available in the next two to four years from similar studies using higher doses compared to lower doses.

Issues regarding the waning effect

Both in patients with GH deficiency and probably also in the short normal

Table 5.3 Final height in short normal children.

Growth hormone treatment

Study	Number and sex distribution	Age at start (years)	Height SDS at start	Final height SDS
Loche [26]	4 M; 3 F	10.5	–2.5	–1.6
Wit [27]	7 M; 5 F	11.2	–3.5	–2.4
Guyda [28]	60 M; 39 F	11.5	–2.81	–1.93
Hindmarsh [29]	10 M; 6 F	8.4	–2.17	–1.33
Bernasconi [30]	54 M; 17 F	12.0	–2.84	–1.69
Coste [31]	42 M; 39 F	10.5	–3.4	–2.4
Zadik [14]	10 M; 6 F	11.1	–3.32	–1.17
McCaughey [32]	8 F	6.2	–2.52	–1.14*
Buchlis [33]	30 M; 6 F	11.9	–2.9	–1.5

Observation group

Study	Number and sex distribution	Age at start (years)	Height SDS at start	Final height SDS
Wit [27]	16 M; 11 F	10.5	–3.0	–2.4
Hindmarsh [29]	6 M; 1 F	7.6	–2.34	–1.88
McCaughey [32]	20 F	6.2	–2.32	–2.13
Buchlis [33]	41 M; 17 F	12.5	–2.9	–2.1

*Higher dosing schedule than other studies.

population, there appears to be an effect of GH therapy itself on shortening the pubertal process [34,35]. Although there is considerable debate about this particular observation, many groups have considered the possibility of holding up the whole of the pubertal process. The rationale for this type of intervention is that the outcome for patients with Turner syndrome, craniopharyngioma and those with GH and gonadotrophin-releasing hormone (GnRH) deficiency appears to be better than in those individuals who have GH deficiency in whom puberty takes place in the normal fashion [36,37]. A number of studies have been undertaken in children with GH deficiency to try to address this particular issue, and the current status of these studies is summarized in Table 5.4 [38–41]. Again, a number have shown some impressive effects, while others have found very little effect. Note the discrepancy between GH-sufficient plus GnRH individuals in the Adan and Cara studies [40,41]. The situation appears to be the same if cyproterone is used instead of GnRH to hold up puberty, with virtually no effect observed in the study by Kawai *et al.* [42]. The arguments outlined in the clinical trials section above apply here as elsewhere in this chapter, and again an overview is required. Finally, consideration needs to be given to the dosage schedule, and this problem is beginning to be addressed [43].

Table 5.4 Response to a combination of GH and GnRH agonist therapy.

First author (reference)	Number and sex ratio	Age at start (years)	Bone age at start	Duration GnRH + GH (years)	Change in predicted height	
					Subjects	Controls
Toublanc [38]	8 M; 3 F	12.6	11.6	2	+4.0 cm	
Saggese [39]	6 M; 4 F	11.5	9.5	1	+2.1 cm	
Adan [40]	9 M; 15 F	9.7	10.1	3	+1.3 cm	+2.0 cm (GH only) +7.3 cm (GnRHa, GH sufficient)
Cara [41]	2 M; 3 F	9.5	12.0	2–3	+10.0 cm	+2.8 cm (GnRHa, GH sufficient)

Safety issues

The safety issues that have been examined have concerned the metabolic changes associated with GH excess (acromegaly), namely carbohydrate intolerance, lipid abnormalities and cardiovascular changes. The general conclusion is that short-term and long-term GH administration is safe in short normal children. Glucose intolerance has not been reported, although euglycaemia has been achieved at the expense of an elevated serum insulin concentration. The effect is transient, with a maximum rise seen in the first six months of therapy and a gradual return to normal levels during the next two to four years [44,45]. Changes in lipid profiles have not been as well documented in these individuals as those in adult GH deficiency studies, but where tests have been conducted no adverse changes have been observed.

The effects on blood pressure and cardiac dimensions and function have only been examined in two or three studies [46,47]. Blood pressure, as assessed by sphygmomanometric studies, appears to be unchanged [47]. No adverse effect on cardiac dimensions has been discerned. All of the changes detected were appropriate given the change in body size of the individuals. An important aspect is the absence of side-effects over a wide range of rhGH doses (20–40 units/m^2 body surface area/week).

Psychological effects

One of the premises underlying the use of GH intervention in these children was the received wisdom that short children were disadvantaged from the psychological standpoint both during childhood and adulthood. Although this view was prevalent in the mid- to late 1980s, it had never been tested formally in either the clinic population or in the general population. A number of studies then suggested that adults who had suffered CGDP were not significantly disadvantaged in adulthood, even though many of the individuals remained short for the general population. The problem that these individuals wanted most help with was the delay in puberty rather than the actual statural problem per se [48].

Two detailed studies testing the psychometric performance of short children in the community [20] and in the clinical setting [49] were not able to identify any particular difference in the psychometric profile of short children compared to profiles obtained from their average-sized peers in the United Kingdom. These studies were important, because the individuals were carefully matched for age, sex and social class, and the studies concentrated in particular on family and social attitudes to the short child during the childhood years. The only group that showed a difference which might be attributable to stature were those individuals with severe intrauterine growth restriction, where the more dominant factor in determining attitudes to health and social interactions was the general intelligence quotient of the child [49].

This lack of any perceived psychological effect of short stature during childhood or adulthood led many investigators to begin to question the ethics of any form of intervention in these individuals with short stature. Certainly, given the limited effect of intervention of an average of 3–5 cm and at best 8 cm, it could not be seen as a universal therapy for improving the lot of the short child. Indeed, if the actual issue was psychological, one might feel that it would be better addressed by appropriate psychological intervention rather than by resorting to an expensive medication that would probably not address the underlying issues.

Conclusion

Despite the fact that growth in childhood and adolescence is GH-dependent and is modulated by the GH pulse amplitude, there appears to be little evidence to support the contention that treating short normal children with GH will significantly improve their final height. Indeed, it is not entirely clear from the more recent psychological studies precisely what the physician would be treating with the growth-promoting intervention. Given the fact that the majority of individuals are not distressed by their particular condition, and that ultimately their adult adjustment is normal, the case for intervention seems to be rather thin. Even if intervention is contemplated, it is extremely difficult to provide parents with an informed outcome for intervention with rhGH, given the limitations of the studies that have been conducted to date.

This review of studies of GH therapy in short normal children has highlighted the inadequacies of clinical trials in paediatric endocrinology in general and the difficulties associated with this type of study in the short normal population. Future studies of GH therapy will require more clearly defined end points, and large studies need to be conducted to avoid some of the statistical problems described in the clinical trials section above.

References

* 1 Rudman D, Kutner MN, Blackston RD *et al.* Children with normal variant short stature: treatment with human growth hormone for six months. *J Clin Endocrinol Metab* 1982; **35**: 665–70.
2 Frazer T, Gavin JR, Daughaday WH, Hillman RE, Weldon VV. Growth hormone dependent growth failure. *J Pediatr* 1982; **101**: 168–74.
3 Plotnick LP, Van Meter QL, Kowarski AA. Human growth hormone treatment of children with growth failure and normal growth hormone levels by immunoassay: lack of correlation with somatomedin generation. *Pediatrics* 1983; **71**: 324–7.
* 4 Van Vliet G, Styne DM, Kaplan SL, Grumbach MM. Growth hormone treatment for short stature. *N Engl J Med* 1983; **309**: 1016–22.
* 5 Grunt JA, Howard CP, Daughaday WH. Comparison of growth and somatomedin C responses following growth hormone treatment in children with small-for-date short stature, significant idiopathic short stature and hypopituitarism. *Acta Endocrinol* 1984; **106**: 168–74.
* 6 Gertner JM, Genel M, Gianfredi SP. Perspective clinical trial of human growth hormone in short children without growth hormone deficiency. *J Pediatr* 1984; **104**: 172–6.
* 7 Buchanan CR, Law CM, Milner RDG. Growth hormone in short slowly growing children and those with Turner's syndrome. *Arch Dis Child* 1987; **62**: 912–16.
8 Freeman JV, Cole TJ, Chinn S *et al.* Cross-sectional stature and weight reference curves for the UK, 1990. *Arch Dis Child* 1995; **73**: 17–24.
9 Ranke MB, Grauer ML, Kistner K, Blum WF, Wollmann HA. Spontaneous adult height in idiopathic short stature. *Horm Res* 1995; **44**: 152–7.
10 Rekers-Mombarg LT, Wit JM, Massa GG *et al.* Spontaneous growth in idiopathic short stature (European Study Group). *Arch Dis Child* 1996; **75**: 175–80.
11 Tanner JM, Whitehouse RH, Cameron N *et al. Assessment of Skeletal Maturity and Prediction of Adult Height (TW2 Method)*. London: Academic Press, 1983.
* 12 Greco L, Power C, Peckham C. Adult outcome of normal children who are short or underweight at age 7 years. *Br Med J* 1995; **310**: 696–700.
13 Bercu BB, Shulman D, Root AW, Spiliotis BE. Growth hormone provocative testing frequently does not reflect endogenous GH secretion. *J Clin Endocrinol Metab* 1986; **63**: 709–16.
* 14 Zadik Z, Chalew S, Zung A *et al.* Effect of long term growth hormone therapy on bone age and pubertal maturation in boys with and without classic growth hormone deficiency. *J Pediatr* 1994; **125**: 189–95.
15 Tannenbaum GS, Ling N. The interrelationship of growth hormone (GH)–releasing factor and somatostatin in generation of the ultradian rhythm of GH secretion. *Endocrinology* 1984; **115**: 1952–7.
* 16 Hindmarsh PC, Smith PJ, Brook CGD, Matthews DR. The relationship between growth velocity and growth hormone secretion in short prepubertal children. *Clin Endocrinol* 1987; **27**: 581–91.
17 Mauras N, Blizzard RM, Link Johnson ML, Rogol AD, Veldhuis JD. Augmentation of growth hormone secretion during puberty: evidence for a pulse amplitude–modulated phenomenon. *J Clin Endocrinol Metab* 1987; **64**: 596–601.
* 18 Hindmarsh PC, Smith PJ, Pringle PJ,

Brook CGD. The relationship between the response to growth hormone therapy and pretreatment growth hormone secretory status. *Clin Endocrinol* 1988; **28**: 559–63.

19 Chinn S, Hughes JM, Rona RJ. Trends in growth and obesity in ethnic groups in Britain. *Arch Dis Child* 1998; **78**: 513–17.

*20 Downie AB, Mulligan J, Statford RJ, Betts PR, Voss LD. Are short normal children at a disadvantage? The Wessex Growth Study. *Br Med J* 1997; **314**: 97–100.

21 Sackett DL, Haynes RB, Guyatt GH, Tugwell P, eds. *Clinical Epidemiology*. 2nd edn. Boston: Little, Brown, 1991.

*22 McCaughey ES, Mulligan J, Voss LD, Betts PR. Growth and metabolic consequences of growth hormone treatment in prepubertal short normal children. *Arch Dis Child* 1994; **71**: 201–6.

*23 Barton JS, Gardiner HM, Cullen S *et al.* The growth and cardiovascular effect of high dose growth hormone therapy in idiopathic short stature. *Clin Endocrinol* 1995; **42**: 619–26.

*24 Cowell CT, Dietsch S, Greenacre P. Growth hormone therapy for 3 years: the OZGROW experience. *J Paediatr Child Health* 1996; **32**: 86–93.

25 Hopwood NJ, Hintz RL, Gertner JM *et al.* Growth response of children with non-growth-hormone deficiency and marked short stature during three years of growth hormone therapy. *J Pediatr* 1993; **123**: 215–22.

26 Loche S, Cambiaso P, Setzu S *et al.* Final height after growth hormone therapy in non-growth-hormone-deficient children with short stature. *J Pediatr* 1994; **125**: 196–200.

*27 Wit JM, Boersma B, de Muinck Keizer-Schrama SM *et al.* Long-term results of growth hormone therapy in children with short stature, subnormal growth rate and normal growth hormone response to secretagogues (Dutch Growth Hormone Working Group). *Clin Endocrinol (Oxf)* 1995; **42**: 365–72.

28 Guyda HJ. Growth hormone treatment of non-growth hormone deficient subjects: the International Task Force Report. *Clin Paediatr Endocrinol* 1996; **5**: 11–18.

*29 Hindmarsh PC, Brook CG. Final height of short normal children treated with growth hormone. *Lancet* 1996; **348**: 13–16.

*30 Bernasconi S, Street ME, Volta C, Mazzardo G, Italian Multicentre Study Group. Final height in non-growth hormone deficient children treated with growth hormone. *Clin Endocrinol (Oxf)* 1997; **47**: 261–6.

*31 Coste J, Letrait M, Carel JC *et al.* Long-term results of growth hormone treatment in France in children of short stature: population register-based study. *Br Med J* 1997; **315**: 708–13.

*32 McCaughey ES, Mulligan J, Voss LD, Betts PR. Randomised trial of growth hormone in short normal girls. *Lancet* 1998; **351**: 940–4.

*33 Buchlis JG, Irizarry L, Crotzer BC *et al.* Comparison of final heights of growth hormone-treated versus untreated children with idiopathic growth failure. *J Clin Endocrinol Metab* 1998; **83**: 1075–9.

34 Ranke MB, Price DA, Albertsson-Wikland K, Maes M, Lindberg A. Factors determining pubertal growth and final height in growth hormone treatment of idiopathic growth hormone deficiency. *Horm Res* 1997; **48**: 62–71.

35 Darendeliler F, Hindmarsh PC, Preece MA, Brook CGD. Growth hormone increases rate of pubertal maturation. *Acta Endocrinol* 1990; **122**: 414–16.

*36 Burns EC, Tanner JM, Preece MA, Cameron N. Final height and pubertal development in 55 children with idiopathic growth hormone deficiency treated for between 2 and 15 years with human growth hormone. *Eur J Pediatr* 1981; **137**: 155–64.

*37 Hibi I, Tanaka T, Tanae A *et al.* The influence of gonadal function and the effect of gonadal suppression treatment on final height in growth hormone (GH)-treated GH-deficient children. *J Clin Endocrinol Metab* 1989; **69**: 221–6.

38 Toublanc JE, Couprie C, Garnier P, Job JC. The effects of treatment combining an agonist of gonadotropin-releasing hormone with growth hormone in pubertal patients with isolated growth hormone deficiency. *Acta Endocrinol (Copenh)* 1989; **120**: 795–9.

*39 Saggese G, Cesaretti G, Andreani G,

Carlotti C. Combined treatment with growth hormone and gonadotropin-releasing hormone analogues in children with isolated growth hormone deficiency. *Acta Endocrinol* 1992; **127**: 307–12.

*40 Adan L, Souberbielle JC, Zucker JM *et al.* Adult height in 24 patients treated for growth hormone deficiency and early puberty. *J Clin Endocrinol Metab* 1997; **82**: 229–33.

*41 Cara JF, Kreiter ML, Rosenfield RL. Height prognosis of children with true precocious puberty and growth hormone deficiency: effect of combination therapy with gonadotropin-releasing hormone agonist and growth hormone. *J Pediatr* 1992; **120**: 709–15.

42 Kawai M, Momoi T, Yorifuji T *et al.* Combination therapy with GH and cyproterone acetate does not improve final height in boys with non-GH-deficient short stature. *Clin Endocrinol* 1998; **48**: 53–7.

43 Reckers-Momberg LTM, Massa GG, Wit JM *et al.* Growth hormone therapy with three dosage regimens in children with idiopathic short stature. *J Pediatr* 1998; **132**: 455–60.

*44 Saenger P, Attie KM, Di Martino-Naardi J, Fine RN. Carbohydrate metabolism in children receiving growth hormone for 5 years: chronic renal insufficiency compared with growth hormone deficiency, Turner syndrome and idiopathic short stature (Genentech Collaborative Group). *Pediatr Nephrol* 1996; **10**: 261–3.

45 Al-Shoumer KAS, Gray R, Anyaoku V *et al.* Effects of four years' treatment with biosynthetic human growth hormone (GH) on glucose homeostasis, insulin secretion and lipid metabolism in GH-deficient adults. *Clin Endocrinol* 1998; **48**: 795–802.

46 Daubney PE, McCaughey ES, Chase C *et al.* Cardiac effects of growth hormone in short normal children: results after four years of treatment. *Arch Dis Child* 1995; **72**: 337–9.

47 Barton JS, Hindmarsh PC, Preece MA, Brook CGD. Blood pressure and the renin–angiotensin–aldosterone system in children receiving recombinant human growth hormone. *Clin Endocrinol* 1993; **38**: 245–51.

48 Crowne EC, Shalet SM, Wallace WH, Eminson DM, Price DA. Final height in boys with untreated constitutional delay of growth and puberty. *Arch Dis Child* 1990; **65**: 1109–12.

49 Skuse D, Gilmour J, Tian CS, Hindmarsh PC. Psychosocial assessment of children with short stature: a preliminary report. *Acta Paediatr* 1994; **406** (Suppl): 11–16.

6: What is the role of GH therapy in children born small for gestational age?

Anita C.S. Hokken-Koelega

Introduction

Children born small for gestational age (SGA) comprise a heterogeneous group of new-borns [1]. Infants may be born either full-term or prematurely. Conditions interfering with fetal growth have been differentiated into fetal, placental, maternal and environmental factors. In most cases, however, the aetiology of intrauterine growth retardation is not known. The term 'intrauterine growth retardation' (IUGR) refers to the fetal growth pattern, and implies that at least two intrauterine growth assessments have been performed that have indicated a low growth velocity in the fetus. The term 'small for gestational age' (SGA) does not refer to fetal growth, but to the size of the infant at birth. Infants with SGA have a low weight and/or length for their gestational age at birth, e.g. below the third percentile or –2 standard deviation scores (SDS) of the reference data. This does not necessarily imply that the SGA child has suffered from IUGR. Conversely, infants born after a short period of IUGR are not necessarily SGA.

Spontaneous catch-up growth after birth

The majority of SGA infants show sufficient catch-up growth to a height at or above the third height percentile within the first two years of life, independently of whether they were born prematurely or at term (Fig. 6.1) [2]. In a Swedish population-based study, 8–10% of children born SGA were below –2 SDS in height at the age of 18 years in comparison with the total population [3,4].

To date, no auxological, biochemical, or endocrine finding, either at birth or within the first year of life, has been identified to predict the catch-up growth of an infant born SGA. Gestational age, multiple birth and sex were not significantly associated with catch-up in height above the third percentile at the age of two years, while a negative correlation was found between the birth length SDS and the degree of catch-up growth—e.g. the

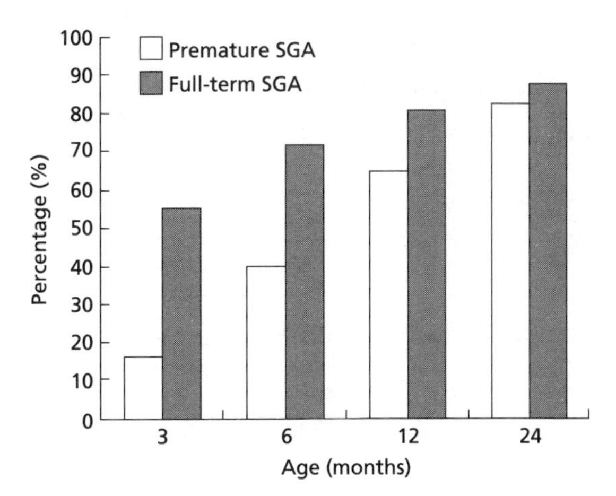

Fig.6.1 Percentage of small-for-gestational-age (SGA) infants with postnatal catch-up growth to a height at or above the third percentile at various ages. Data are given for premature infants (white bars) and full-term infants (shaded bars) in the total study group. Adapted with permission from Hokken-Koelega *et al.* [2].

smaller at birth, the lower the catch-up growth [2]. A positive association was found between the weight gain during the first three and six months and the catch-up growth to a height above the third percentile at the age of two years [2]. It is well known that many short children born SGA have serious feeding problems due to a very low appetite, particularly during the first three years of life. Recent data have shown that six-year-old short children born SGA had a daily food intake according to safe levels for age as advised by the World Health Organization, but significantly less food intake and significantly lower muscle and fat mass than their normal-statured peers [5]. These data suggest that the degree of appetite and food intake may play a role in the process of catch-up in children born SGA. The elucidation of the precise relationships requires further research.

Endogenous growth hormone secretion in short children born SGA

The pathophysiological mechanism underlying the failure to achieve sufficient catch-up in children born SGA is not completely understood. Growth hormone (GH) secretion has only been investigated in prepubertal SGA children with short stature. GH insufficiency may consist of either classical GH deficiency, as diagnosed by two GH stimulation tests, or abnormalities in the GH secretory pattern, as detected by GH profile studies. Approximately 25% of these children show low GH peaks during two GH stimulation tests [6–8]. A reduction in physiological growth hormone (GH) secretion and low levels of insulin-like growth factor-I (IGF-I) and its binding protein-3 (IGFBP-3) have been reported in 50–60% of children with short stature born SGA [8–10]. It is not known why many of these children show abnormal GH patterns. Somatotrope unresponsiveness to growth hormone-releasing hormone (GHRH) does not seem to be implicated [11]. Deficiencies of endogenous GH secretagogues, such as GHRH, galanin, and abnormalities in the diurnal rhythms of GHRH/GH secretion remain to be studied. In addition, it is not yet known why some of these children born SGA experience severe growth retardation without any known disturbance in the GH–IGF axis or the binding proteins.

Growth hormone therapy

Several randomized, controlled GH trials have been performed [11–20]. Most studies included prepubertal children aged two to eight years, born with a birth length and/or weight less than –2 SDS for gestational age, showing persistent short stature with a height less than –2 SDS, and a GH peak over 20 mU/L during either a GH stimulation test or a 24-hour GH profile. Children with a chromosomal disorder or syndrome have been excluded from the trials, with the exception of children with signs of Silver–Russell syndrome. Different GH doses and GH treatment regimens, such as discontinuous versus continuous use of GH, have been investigated during the last 10 years.

Short-term growth results

Most controlled trials have investigated the use of GH therapy over a period of two to three years in short prepubertal children born SGA, in comparison with a randomized untreated group for one to two years [12–19]. Table 6.1 summarizes the two-year growth results, expressed as the gain in height SDS, in the various European GH trials in short prepubertal children born SGA. These trials established the ability of GH therapy to normalize height in these children when GH doses of 3, 6 or 9 IU/m^2/day (0.1, 0.2, 0.3 IU/kg/day) are given for two years. The randomly assigned, fully parallel, untreated control groups confirmed that prepubertal short SGA children do not show any catch-up growth during the one-year to two-year study periods, thus indicating that these children are destined to remain short, at least throughout childhood.

The growth variables (height velocity, gain in height SDS, weight gain) have been assessed in relation to GH dose regimen. Figure 6.2 shows the gain in height SDS after two years of GH therapy at a dose of 3 vs. 6 IU GH/m^2/day in a group of 32 prepubertal short children born SGA, in comparison with 10 untreated children. The children receiving 6 IU GH/m^2/day reached their target height SDS after three years of GH therapy. Meta-analysis of the combined data from three independent randomized GH trials in short prepubertal SGA children in Scandinavia, Germany and Belgium clearly

Table 6.1 Data at start and after 2 years of growth hormone therapy in short prepubertal SGA children: summary of European randomized GH trials. Values given as means ± SD.

First author (country) [ref.]	At start		Gain in height SDS after 2 years			
			Dose of GH IU/m²/day			Untreated
	Age	Height SDS	3	6	9	
Chaussain (France) [13]	5.4 ± 1.4	−2.8 ± 0.6	1.0 ± 0.8[a] (n = 55)	1.4 ± 0.8[b] (n = 48)	–	0.1 ± 0.2 (n = 27)
De Zegher (Belgium, Germany, Scandinavia) [17]	4.9 ± 0.1	−3.6 ± 0.1	1.13 ± 0.1[a] (n = 38)	2.11 ± 0.1[b] (n = 54)	2.64 ± 0.2[c] (n = 16)	0.12 ± 0.1 (n = 31)
Fjellestad-Paulsen (France) [19]	4.5 ± 1.8	−3.4 ± 0.6	–	1.7 ± 0.6[a] (n = 38)	–	0.2 ± 0.3* (n = 31)
Boguszewski et.al. (Scandinavia) [16]	4.7 ± 1.6	−3.1 ± 0.7	1.53 ± 0.3[a] (n = 14)	2.13 ± 0.3[b] (n = 18)	–	0.04 ± 0.1 (n = 12)
Sas (Netherlands) [20]	5.8 ± 1.7	−2.9 ± 0.7	1.52 ± 0.5[d] (n = 23)	2.12 ± 0.5[b] (n = 20)	–	–

[a] $p < 0.005$ compared to untreated group.
[b] $p < 0.005$ compared to GH dose 3 IU/m²/day.
[c] $p < 0.005$ compared to GH dose 6 IU/m²/day.
[d] $p < 0.001$ compared to pretreated values.
*Untreated group for 1 year.
SDS: standard deviation score.

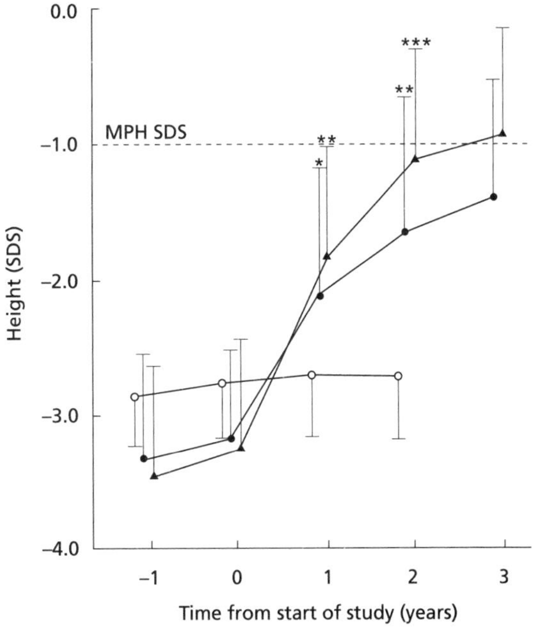

Fig.6.2 Height expressed as standard deviation score (SDS), and the mid-parental height SDS during two to three years of study in untreated children (○) and children given GH at 3 IU/m²/day (▲) or 6 IU/m²/day (●). Values are given as means (SD). *$p < 0.05$, **$p < 0.01$, ***$p < 0.001$ compared to the untreated group. Adapted with permission from Boguszewski *et al.* [16].

established the GH dose-dependency of the two-year growth response [17,18]. This is illustrated in Figure 6.3, showing the gain in height SDS for chronological age and height SDS for bone age during GH therapy at various doses.

The effect of discontinuous treatment with high-dose GH for a few years versus continuous lower-dose GH therapy for a longer period on the gain in height SDS after four years of GH therapy has been analysed in a group of 100 short prepubertal SGA children from independent GH trials in four countries [18]. In France, Germany and Scandinavia, GH at a dose of 3 or 6 IU/m²/day was administered for four years (continuous treatment regimen). In Belgium, high-dose therapy at a dose of 6 or 9 IU GH/m²/day was administered for two years, followed by a two-year period of no GH therapy (discontinuous treatment regimen). The growth results with the two regimens are depicted in Figure 6.4. The continuous and discontinuous treatment regimens, respectively, were found to elicit persistent catch-up growth and catch-up growth followed by catch-down growth, respectively. However, the two treatment regimens ultimately resulted in similar growth responses over time. These data suggest that it is the cumulative GH dose received, and not the design of the treatment regimen, that determines the growth response.

Bone maturation

Bone maturation was determined centrally in many trials [12–20]. However, most trials have used different methods to estimate the bone maturation,

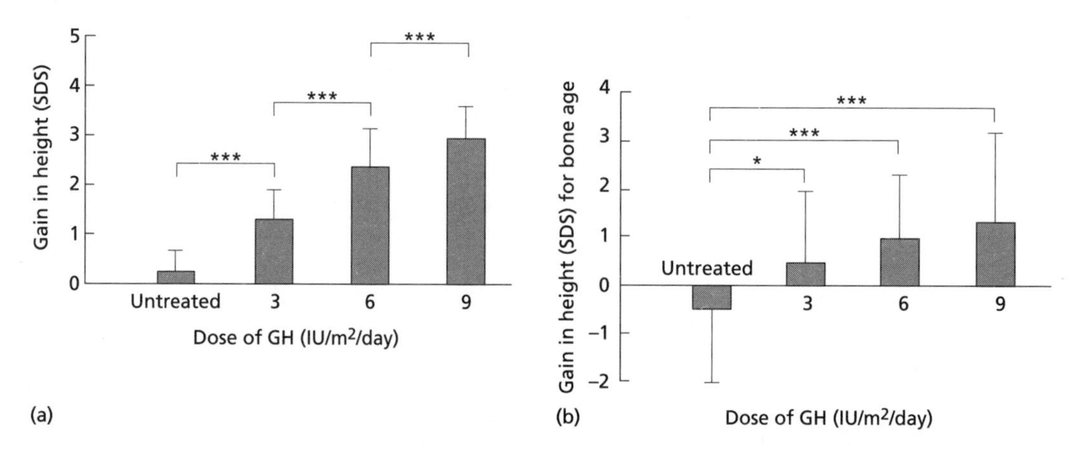

Fig.6.3 Gain in height SDS (a) and gain in height SDS for bone age (b) over two years in children born SGA and treated with different doses of GH. Values are given as means (SD) of combined data for children from Belgium, Germany and Scandinavia. $*p < 0.03$, $**p < 0.005$. Adapted with permission from De Zegher *et al.* [17].

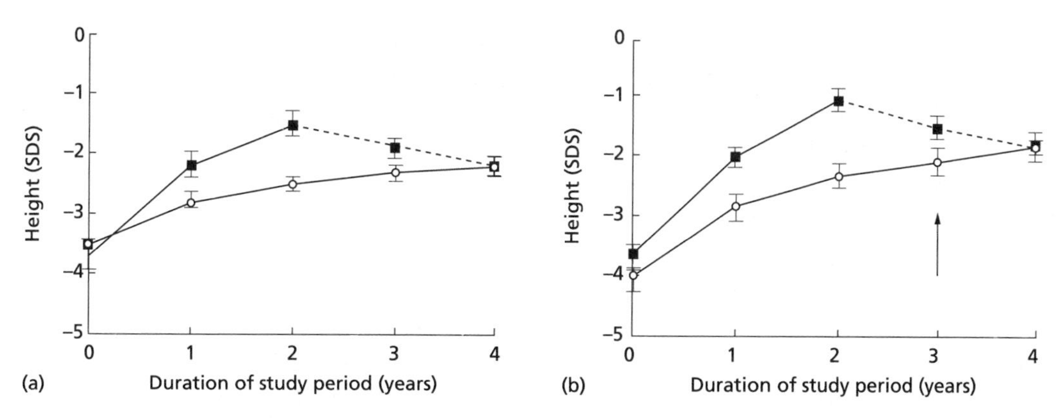

Fig. 6.4 Height SDS of short prepubertal children born SGA who received GH therapy at a mean dose of 3 IU/m²/day over four years (a), administered either as 3 IU/m²/day for four years (continuous, $n = 42$; ○), or as 6 IU/m²/day for two years, followed by two years without GH therapy (discontinuous, $n = 14$; ■). (b) This shows the effect when GH therapy at a mean dose of 6 IU/m²/day was administered over three years, either as 6 IU/m²/day for three years (continuous, $n = 30$; ○), or as 9 IU/m²/day for two years, followed by one year without GH therapy (discontinuous, $n = 14$; ■). Adapted with permission from De Zegher *et al.* [18].

and for this reason it is impossible to compare the data on bone maturation between the trials.

During GH therapy at a dose of 3, 6 or 9 IU/m²/day, the mean annual increase in bone age determined using the Tanner–Whitehouse II (20 bones) method was significantly greater in the treated groups, being 1.0, 1.20 and 1.41 years, respectively, than the annual increase of 0.85 years in the untreated groups [17]. However, as the height velocity increased considerably in the GH-treated groups, a significant gain in height SDS for bone age was observed in the GH-treated children in contrast to the untreated group (Fig. 6.3b). The significant increase in mean height SDS for bone age proved to be virtually dose-independent, suggesting that the different GH doses within the range studied may have quite different short-term but less different long-term effects. Thus, the short-term growth and bone acceleration induced by GH therapy does not appear to occur at the expense of long-term growth.

Studies describing the natural history of persistently short children born SGA have shown a spontaneous acceleration of bone maturation and a decrease in height SDS for bone age in untreated children from the age of six to eight years [21,22]. This phenomenon indicates that, as a group, SGA children have an abnormal rhythm of bone maturation throughout childhood. It may also explain why the final height prediction is so difficult to perform in these children, resulting in an over-prediction of final height in most of the untreated short SGA children. Several GH trials have shown that the rate of bone maturation is slower in children aged 3.0–5.9 years (young) than in those aged 6.0–8.9 years (older), both on GH therapy and in untreated controls [18]. During GH therapy, this age difference is more marked in children receiving the higher dose of 6–9 IU GH/m^2/day (ΔBA/ΔCA 1.13 and 1.36 years in young and older children, respectively), than in those treated with 3 IU/m^2/day (ΔBA/ΔCA 0.94 and 1.12 years in young and older children, respectively). Thus, in this group of children, the age is important as a determinant not only of the growth response, but also of the progression of bone maturation during GH therapy, and consequently, of the effect on final height prognosis. In estimating the capacity of GH therapy to increase final height, both the GH dose and the child's age should be taken into account.

Long-term growth results and final height

Most randomized GH trials show a significant gain in height SDS for bone age, suggesting final height improvement. However, as randomized GH trials in prepubertal short children born SGA have only started fairly recently, final height data are not yet available. Long-term data on the effect of lower-dose GH therapy in a continuous regimen are limited at the moment and will be in the future, as many of the randomized trials had a discontinuous GH regimen, evaluating the effect of high-dose GH therapy during a few years only. Final height data for randomized untreated patients will not be available in the future, as all control groups have started GH therapy after one or two years.

In a long-term, randomized, double-blind, dose–response GH trial, 79 short Dutch children born SGA were treated continuously for six years with either 3 or 6 IU GH/m^2/day. GH was started at a mean age of 7.5 years. Figure 6.5 presents the individual heights of boys before and after six years of GH therapy. It clearly illustrates that long-term GH therapy results in a normalization of height, enabling all children to grow within the normal range for height during their childhood. The five-year data showed that continuous GH therapy at a dose of 3 vs. 6 IU/m^2/day resulted in a significant improvement in the mean (SD) height SDS of 2.12 (0.52) vs. 2.57 (0.72), respectively, and a significant increase in mean (SD) height SDS for bone age of 1.06 (0.78) vs. 1.40 (1.37), respectively [20]. The mean gain in height SDS after five years was significantly greater in children treated with the higher GH dose of 6 IU/m^2/day. However, both dosage groups reached their target height SDS within five years of GH therapy, regardless of the GH dose administered, indicating that longer-term therapy with a lower GH dose of 3 IU/m^2/day is also able to normalize the height of short SGA chil-

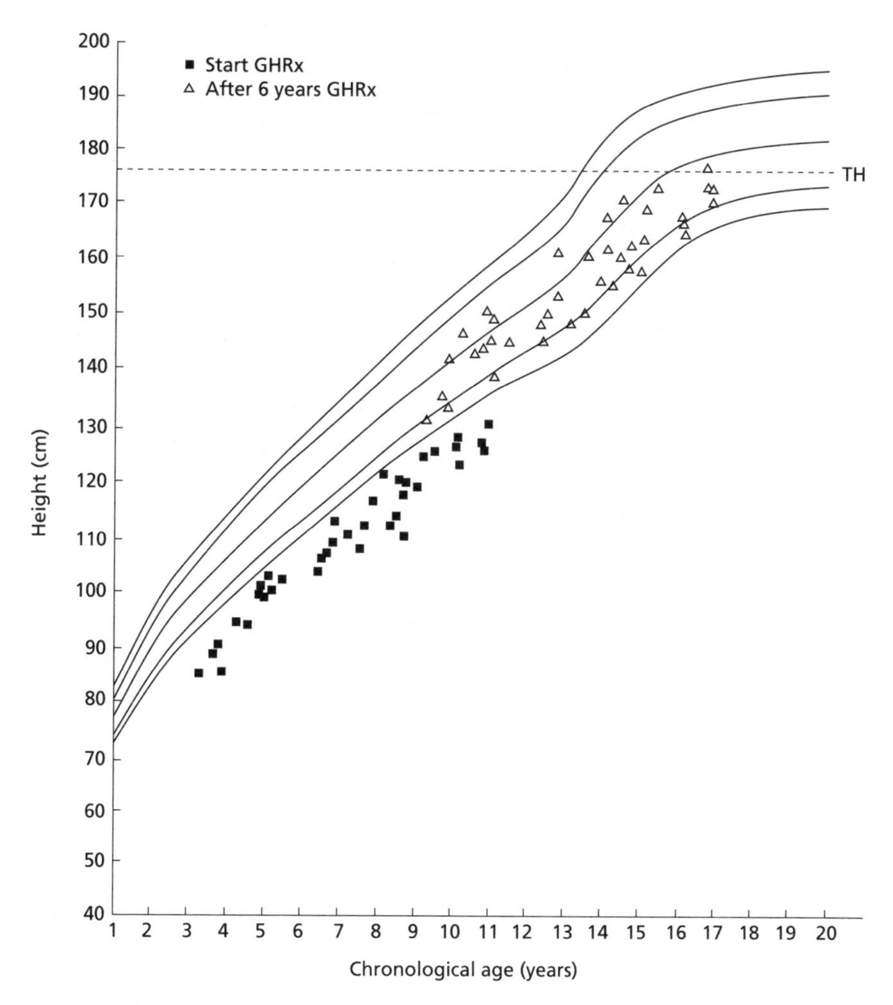

Fig. 6.5 Individual heights in boys at the start of the study (■) and after six years of GH therapy at 3–6 IU/m²/day (△), in comparison with the reference height-for-chronological age curves for healthy Dutch boys. The lines represent the 3rd, 10th, 50th, 75th and 97th percentiles. TH: mean target height for the male study group.

dren during childhood. Final height prediction increased significantly after five years of GH therapy at a dose of 3 IU/m²/day and even more so with 6 IU GH/m²/day, indicating that final height will improve. However, only true final height data can substantiate these positive findings.

Final height results have been reported in one uncontrolled trial in 12 short SGA children after continuous treatment with ≈ 4 IU GH/m²/day from a median age of 7.6 years [23]. Seven children had signs of Silver–Russell syndrome. The mean duration of treatment was 8.5 years. At final height (– 1.5 SDS) a significant gain in height SDS was attained in comparison with the height and the final height prediction at the start of GH therapy.

Predictors of the growth response

In view of the variability of the growth response to GH therapy observed in short children born SGA, methods of identifying predictors of the growth response have been developed. The major determinants of the short-term and longer-term growth response were the GH dose, followed by the age at the start of treatment (the younger the child, the better the growth response) and the parental-adjusted individual height deficit (the greater the deficit, the better the growth response) [12–20]. A fairly broad range of individual gains in height SDS during GH was observed. In some SGA children, GH therapy resulted in an excellent growth response, similar to that obtained in severely GH-deficient patients, whereas in others, the growth response was less pronounced.

Interestingly, the growth response was not significantly different between GH-deficient (GHD) and non-GHD patients, indicating that the magnitude of the growth response to GH therapy in short children born SGA is determined by the condition of SGA rather than by the GH-secretory capacity [24,25]. It is likely that the less pronounced growth response in some children is the result of a lower responsiveness to GH, which might require higher-dose GH therapy. No difference in the growth response was found between children with the classical characteristics of Silver–Russell syndrome and those without any sign of the syndrome [23,24].

To optimize GH therapy for short SGA children, further studies are needed to elucidate the pathophysiological mechanisms underlying the short stature and to identify predictors for the long-term effect of GH therapy.

Safety

All trials have concluded that GH therapy was well tolerated, regardless of dosage and regimen. No adverse events were detected that were GH-related [12–20]. GH therapy apparently does not increase the incidence of precocious puberty in these short children born SGA.

GH therapy resulted in a dose-dependent significant rise in insulin levels with normal glucose and glycosylated haemoglobin levels, without a further increase during longer-term therapy [16,24]. As insulin resistance has been reported in short children born SGA [26], fasting insulin levels as well as glycosylated haemoglobin need careful monitoring during and after GH therapy. Further research should focus on the mechanisms and the long-term consequences of the insulin resistance observed in SGA children.

Most studies report a significant dose-dependent increase in weight for age in GH-treated children. However, the body mass index SDS did not change in any of the treatment groups. One study used magnetic resonance imaging to estimate the muscle and fat mass in short children before and during GH therapy [27]; the muscle mass and fat mass of short children born SGA were smaller than in other short children. The fat mass decreased during GH therapy, whereas the muscle mass increased.

GH therapy induced a significant and sustained rise in IGF-I and IGFBP-3 levels, which were between 1 and 2 SDS during 3 IU GH/m^2/day treatment and above 2 SDS during 6 IU GH/m^2/day treatment [16,24]. Since the long-term effects of persistently high IGF-I levels above 2 SDS are not yet known, and continuous therapy with the lower GH dose appears to be effective in achieving satisfactory longer-term growth, one may decide to give a lower GH dose of 3 instead of 6 IU/m^2/day when children are continuously treated for many years.

Psychological effects

The psychological effects of short stature in 63 children born SGA and the effect of two years of GH therapy at a dose of 3 or 6 IU/m^2/day on the psychological development of these children were studied in the Dutch randomized double-blind trial [28]. Intelligence quotients (IQs) were measured using the Dutch version of the revised Wechsler Intelligence Scale for Children (WISC-RN). Before the trial, the mean total IQ, verbal IQ and performance IQ were significantly lower than those of normative Dutch children. Head circumference and height SDS were significantly associated with the total IQ; the larger the head circumference and the taller the child, the higher the intelligence scores. These results suggest a relation between the degree of intrauterine growth retardation and IQ in these children. A significant increment in mean total and mean performance IQ scores was found after two years of GH therapy, independently of the gain in height. The mean (SD) total IQ increased from 91 (16) to 93 (17) ($p < 0.001$). Attention was assessed using the Sonneville Visual Attention Tasks (SVAT) test. SGA children had significantly greater deficits in all measures of attention than normal children, and they performed less accurately, more impulsively and more slowly than their peers. Head circumference and artificial delivery methods were significantly correlated with the attention measures. Two years of GH therapy had only a very limited effect on the SVAT scores, suggesting that the attention deficits are related to the condition of SGA, and not to GH levels and/or current height. GH therapy had a beneficial effect on behavioural and emotional problems and on the self-concept of the children.

Conclusions and directions for future research

GH therapy is established as an effective and well-tolerated therapy for the normalization of short stature in prepubertal children born SGA. Various GH doses, ranging from 3 to 6 to even 9 IU/m^2/day, have been evaluated in short-term studies, indicating that the short-term growth response is dose-dependent and age-dependent. After long-term GH therapy at doses of 3 or 6 IU/m^2/day for five to six years, many children grow within the height range for healthy children, according to their target height SDS. The acceleration of bone maturation does not appear to occur at the expense of long-term growth, as the final height prognosis improved significantly in all GH-treated groups. The growth-promoting effect of GH appears to be independent of the GH–IGF-I levels at baseline, and is similar in children with and without signs of Silver–Russell syndrome.

Various treatment regimens have been evaluated, ranging from discontinuous high-dose GH therapy at a dose of 6–9 IU/m^2/day for a few years to long-term continuous low-dose GH at a dose of 3–6 IU/m^2/day. The first regimen with high-dose discontinuous GH therapy has the advantage of normalizing stature more rapidly with fewer injections; however, it also has the disadvantage of inducing very high IGF-I levels, with unknown consequences and catch-down growth in many children after the discontinuation of GH. Long-term low-dose GH therapy induces a gradual, more physiological increase in height and IGF-I levels within the normal range, with the disadvantage of involving a higher number of subcutaneous injections. The results of ongoing and new trials will determine the optimal GH regimen for the group, and probably also for the individual children.

No GH-related adverse events have been reported, but GH therapy resulted in a dose-dependent rise in insulin levels, in the presence of normal glucose values. As insulin resistance has been reported in untreated short children born SGA, fasting insulin levels and glycosylated haemoglobin levels need careful monitoring during and after GH therapy. GH therapy has a beneficial effect on the psychosocial development of short SGA children.

Further studies should be directed at optimizing GH modalities and establishing final height results and long-term safety data, and at finding the predictors indicating which SGA children will benefit most from GH therapy.

References

1 Wennergren M, Wennergren G, Vilbergsson G. Obstetric characteristics and neonatal performance in a four-year small for gestational age population. *Obstet Gynecol* 1988; **72**: 615–20.

* 2 Hokken-Koelega ACS, De Ridder MAJ, Lemmen RJ *et al.* Children born small for gestational age: do they catch up? *Pediatr Res* 1995; **38**: 267–71.

* 3 Albertsson-Wikland K, Karlberg J. Natural catch-up growth in children born small for gestational age with and without catch-up growth. *Acta Paediatr Scand* 1994; **399** (Suppl): 64–70.

* 4 Karlberg J, Albertsson-Wikland K. Growth in full-term small-for-gestational-age infants: from birth to final height. *Pediatr Res* 1995; **38**: 733–9.

5 Arends NJT, Hokken-Koelega ACS. Body composition and daily food intake in children with short stature after intrauterine growth retardation (IUGR). *Horm Res* 1998; **50**: 47.

6 Albertsson-Wikland K. Growth hormone secretion and growth hormone treatment in children with intrauterine growth retardation. *Acta Paediatr Scand* 1989; **349** (Suppl): 35–41.

7 Stanhope R, Ackland F, Hamill G *et al.* Physiological growth hormone secretion and response to growth hormone treatment in children with short stature and intra-uterine growth retardation. *Acta Paediatr Scand* 1989; **349** (Suppl): 47–52.

* 8 De Waal WJ, Hokken-Koelega ACS, Stijnen T *et al.* Endogenous and stimulated GH secretion, urinary GH excretion, and plasma IGF-I and IGF-II levels in prepubertal children with short stature after intrauterine growth retardation. *Clin Endocrinol* 1994; **41**: 621–30.

* 9 Boguszewski M, Rosberg S, Albertsson-Wikland K. Spontaneous 24-hour growth hormone profiles in prepubertal small for gestational age children. *J Clin Endocrinol Metab* 1995; **80**: 2599–606.

*10 Boguszewski M, Jansson C, Rosberg S, Carlsson LMS, Albertsson-Wikland K. Changes in serum insulin-like growth factor I and IGF-binding protein-3 levels during growth hormone treatment in prepubertal short children born small for gestational age. *J Clin Endocrinol Metab* 1996; **81**: 3902–8.

11 Job JC, Chatelain P, Rochiccioli P *et al.* Growth hormone response to a bolus injection of 1–44 growth hormone-releasing hormone in very short children with intrauterine onset of growth failure. *Horm Res* 1990; **33**: 161–5.

*12 Chatelain P, Job JC, Blanchard J *et al.* Dose-dependent catch-up growth after 2 years of growth hormone treatment in intrauterine growth retarded children. *J Clin Endocrinol Metab* 1994; **78**: 1454–60.

13 Chaussain JL, Colle M, Landier F. Effects of growth hormone therapy in prepubertal children with short stature secondary to intrauterine growth retardation. *Acta Paediatr* 1994; **399** (Suppl): 74–5.

*14 De Zegher F, Maes M, Gargoskiy SE *et al.* High-dose growth hormone treatment of short children born small for gestational age. *J Clin Endocrinol Metab* 1996; **81**: 1887–92.

15 Job JC, Chaussain JL, Job B *et al.* Follow-up of three years off treatment with growth hormone and of one post-treatment year in children with severe growth retardation of intrauterine onset. *Pediatr Res* 1996; **39**: 354–9.

*16 Boguszewski M, Albertsson-Wikland K, Aronsen S *et al.* Growth hormone treatment of short children born small-for-gestational-age: the Nordic Multicentre Trial. *Acta Paediatr* 1998; **87**: 257–63.

*17 De Zegher F, Albertsson-Wikland K, Wilton P *et al.* Growth hormone treatment of short children born small for gestational age: a meta-analysis of four independent, randomized, controlled, multicentre studies. *Acta Paediatr* 1996; **417** (Suppl): 27–31.

18 De Zegher F, Butenandt O, Chatelain PG *et al.* Growth hormone treatment of short children born small for gestational age: reappraisal of the rate of bone maturation over 2 years and meta-analysis of height gain over 4 years. *Acta Paediatr Scand* 1997; **423** (Suppl): 207–12.

*19 Fjellestad-Paulsen A, Czernichow P, Brauner R *et al.* Three-year data from a comparative study with recombinant human growth hormone in the treatment of short stature in young children with intrauterine growth retardation. *Acta Paediatr Scand* 1998; **87**: 511–17.

*20 Sas TCJ, Mulder P, Houdijk EC *et al.* Five-year results of a randomised double-blind dose–response study of growth hormone (GH) treatment in short stature after intrauterine growth retardation (IUGR). *Horm Res* 1998; **50** (Suppl): 52.

21 Tanner JM, Lejarriaga H, Cameron N. The natural history of the Silver–Russell syndrome: a longitudinal study of thirty-nine cases. *Pediatr Res* 1975; **9**: 611–23.

22 Job JC, Rolland A. Histoire naturelle des retards de croissance à debut intra-uterin. *Arch Fr Pediatr* 1986; **43**: 301–6.

23 Albanese A, Azcona C, Stanhope R. Final height in children with IUGR receiving GH treatment. *Horm Res*

1998; **50** (Suppl): 46.

*24 De Waal W. Influencing the extremes of growth: too tall—too small [PhD dissertation]. Rotterdam: Erasmus University, 1996.

25 De Zegher F, Francois I, Van Helvoirt M *et al.* Growth hormone treatment of short children born small for gestational age. *Trends Endocrinol Metab* 1998; **9**: 233–7.

*26 Hofman L, Cutfield WS, Robinson EM *et al.* Insulin resistance in short children with intrauterine growth retardation. *J Clin Endocrinol Metab* 1997; **82**: 402–6.

27 Leger J, Carel C, Legand I *et al.* Magnetic resonance imaging evaluation of adipose tissue and muscle mass in children with growth hormone deficiency, Turner's syndrome, and intrauterine growth retardation during the first year of treatment with GH. *J Clin Endocrinol* 1994; **78**: 904–9.

*28 Van der Reijden-Lakeman EA. Psychological evaluation of children with short stature after intrauterine growth retardation, before and after two years of growth hormone treatment [PhD dissertation]. Rotterdam: Erasmus University, 1996.

Part 2: GH Replacement in Hypopituitary Adults

7: Is cardiovascular mortality increased in hypopituitary adults?

Salem A. Beshyah

Introduction

The outcome in conventionally treated hypopituitarism, and the possibility of increased cardiovascular disease and risk factors in adults with hypopituitarism who have growth hormone deficiency (GHD), have attracted increasing interest over the past nine years. In the past, it was thought that conventionally treated hypopituitarism is associated with a normal life expectancy, and some authors even suggested that GHD deficiency may impart a protective effect against vascular disease, despite the earlier description of abnormal lipid profiles. Large amounts of data on the overall mortality and cardiovascular mortality, cardiovascular morbidity and prevalence of classical vascular risk factors have accumulated recently. In addition, with the increased availability of biosynthetic human GH, it has become possible to study the effects of GH replacement on cardiac structure and function and on vascular risk factors in much more detail. However, data on the possible effect of GH therapy on the long-term survival are not yet available.

In this chapter, the issue of cardiovascular disease in hypopituitarism will be addressed by examining the available epidemiological and experimental data on cardiovascular morbidity and mortality in adult hypopituitarism, and an attempt will be made to look for potential causes and mechanisms for the cardiovascular risk.

Outcome in growth hormone-deficient hypopituitary adults

Is life expectancy reduced in adults with hypopituitarism?

In 1990, Rosén and Bengtsson [1] were the first to demonstrate that life expectancy is reduced in hypopituitary adults (Table 7.1). They examined

Table 7.1 Life expectancy in hypopituitary adults.

Study origin	Gothenburg, Sweden	North Staffordshire, UK	Lund, Sweden	Birmingham, UK
Reference	1	3	2	4
Study period	1956–87	1967–94	1952–92	1968–92
Total number	333	172	344	349
Deaths (n)	104	60	188	82
Overall mortality	O/E 1.8	O/E 1.7	SMR 2.17	O/E 1.3
p value	$p < 0.001$	$p < 0.01$	–	$p < 0.05$

O/E: Observed to expected ratio of mortality in hypopituitary adults compared with the background population. SMR: standard mortality risk.

the records of 333 patients who attended the endocrine clinic in Gothenburg with a diagnosis of hypopituitarism between 1956 and 1987. During the study period, 104 patients died, and the mortality rate was significantly higher than that in the Swedish population (observed/expected; 1.8, $p < 0.001$). Another report from Sweden [2] on 344 patients diagnosed in Lund between 1952 and 1992 demonstrated increased overall mortality (standardized mortality ratio [SMR] 1.75; 95% CI, 1.4–2.19). Bates *et al.* [3] reviewed the case notes of all patients who were investigated for hypopituitarism between 1967 and 1994 at the North Staffordshire Hospital in England. Of 172 patients who had at least one biochemically demonstrable hormone deficiency, 60 died during the study period. The overall mortality was increased (observed/expected; 1.73, $p < 0.01$). A preliminary report from Birmingham on the mortality in a cohort of 349 patients (200 men and 149 women) with pituitary disease diagnosed between 1968 and 1992 has also been presented by Bates *et al.* [4]. In this report, 53 men and 29 women died during the study period, at a median age of 64 years. The overall observed/expected ratio was 1.3 (95% CI, 0.99–1.6; $p < 0.05$).

Evidence therefore exists for significantly reduced life expectancy in adult-onset hypopituitarism in all major series published so far (Table 7.1).

What are the causes of death in hypopituitarism?

In the seminal study by Rosén and Bengtsson [1], the excess mortality reflected deaths from vascular disease (58%). Myocardial infarction, ischaemic heart disease with congestive cardiac failure and cerebrovascular disease were the most frequent causes of vascular death. The other study from Sweden [2] confirmed the increase in deaths, which in this study was particularly due to cerebrovascular disease (SMR 3.39; 95% CI, 1.41–1.88). The increase in cardiovascular mortality was greater in women than in men

(SMR. 2.39 vs. 1.54). Wüster *et al.* [5] examined vascular disease in 122 patients with hypopituitarism in Heidelberg, Germany. The frequency of arteriosclerotic events (myocardial infarction or stroke) was also higher than that anticipated for the German population in the PROCAM study.

Excess vascular mortality has not been universally observed. Bates *et al.* [3] reported that vascular disease accounted for 42% of deaths in the study from North Staffordshire, but no significant excess vascular mortality was observed (observed/expected 1.35; $p = 0.11$) (Table 7.2). However, in the preliminary report on the larger study from the English Midlands [4], the vascular deaths have so far been relatively low, and there was no significant excess vascular mortality in either sex (Table 7.2).

Assessment of adults with GH deficiency since childhood showed no evidence of increased vascular disease [6–8]. The morbidity and mortality in two other series of patients who had received GH in the United Kingdom [9] and Canada [10] were mostly related to recurrence of the underlying pituitary disease or the metabolic complications of hypopituitarism.

The reduced life expectancy in adult-onset hypopituitarism seems mainly to be due to excess vascular disease in two of the four studies. The differences in the findings of the studies from the UK versus those from Sweden may be related to the greater background vascular risk in the former in comparison with the latter. Studies from other parts of the world may help resolve this dispute. An excess vascular risk does not appear to be evident in those who develop GH deficiency during childhood. Possible explanations include the shorter duration of such studies, the increased contribution of deaths due to the underlying disease such as recurrence of tumours and their local complications and deaths related to metabolic complications of hypopituitarism, such as hypoglycaemia and adrenal crises, during childhood and adolescence.

Table 7.2 Causes of death in hypopituitary adults.

Study origin	Gothenburg, Sweden	North Staffordshire, UK	Lund, Sweden	Birmingham, UK
References	1	3	2	4
Vascular	58%*	42%	44%*	23%
Malignancy	7%	24%	20%	–
Respiratory	4%	10%	6%	–
Other causes	31%	24%	31%	–

*Vascular mortality is significantly increased.

What is the evidence for increased cardiovascular morbidity in asymptomatic hypopituitary patients?

Data on heart size and function

In experimental animals, hypophysectomy causes a decrease in the size of the heart as well as of several other organs [11], and this is reversed with GH administration. Stimulated by the Rosén and Bengtsson study [1], several research groups have examined cardiac structure and function in adults with hypopituitarism and GH deficiency (Table 7.3). All of these patients were free of cardiac symptoms, and the aim of the studies was therefore to look for subtle rather than major changes. The methods employed have included two-dimensional echocardiography, Doppler techniques, radionuclide angiography and exercise tolerance test protocols.

The left ventricular posterior wall thickness (LVPWT) and interventricular septal thickness (IVST), left ventricular internal diameter (LVID) and fractional shortening (FS%) have been examined in two studies from Italy [12,13]. In our own patients at St. Mary's Hospital, London [14,15], LVPWT and IVST were lower in patients than in controls, but the left ventricular mass index (LVMI), LVID and ejection fraction (EF%) did not differ. In two other studies [16,17] there was no significant difference between patients and controls in the echocardiographic measurements of cardiac structure and systolic function. In 20 GH-deficient patients studied by Nass *et al.* [18], the measurements were within the normal range for their laboratory, although controls were not included in the study. A positive relation-

Table 7.3 Summary of findings suggestive of increased early cardiovascular morbidity in symptom-free hypopituitary adults compared with matched normal controls.

Cardiac structure and function
Reduced interventricular septal thickness
Reduced left ventricular posterior wall thickness
Decreased left ventricular mass index
Reduced left ventricular internal diameter
Decreased resting ejection fraction/fractional shortening
Decreased exercise tolerance
Abnormal exercise-induced electrocardiographic changes
Abnormal systolic function response to exercise (decreased ejection fraction)
Abnormal resting diastolic function
Peripheral arteries
Increased carotid intima–media thickness
Increased frequency of carotid and femoral plaques
Increased carotid wall stiffness
Decreased thoracoabdominal aortic distensibility

ship between left ventricular mass, total body fat-free mass and serum IGF-I concentrations was demonstrated in hypopituitary patients [14,15], suggesting that GH does influence cardiac size and components of body composition in humans.

Left ventricular diastolic function, a newer concept in cardiology, has also been assessed in several studies [15–17,19,20]. Prolonged isovolumic relaxation time (IVRT) and decreased early/atrial peak velocity ratio (E/A ratio) were observed in eight of 39 patients, and the patients differed statistically from controls (IVRT: 94.1 ± 22.1 vs. 83.3 ± 15.7 ms, $p < 0.02$; E/A: 1.25 ± 0.38 vs. 1.46 ± 0.47, $p < 0.04$, in patients and controls, respectively) in our own studies [15]. Caidhal *et al.* [19] reported that left ventricular filling was abnormal in three out of 10 patients, but Thuesen *et al.* [16] found no significant difference between patients and controls in the E/A ratio. E wave deceleration time, another marker of left ventricular filling, was increased in 20 patients compared with 20 controls ($p < 0.01$) in another study [17], although the E/A ratio was not abnormal in this study. Using radionuclide angiography, Cittadini *et al.* [20] demonstrated a reduced peak filling rate in hypopituitary patients, again indicating impaired left ventricular diastolic function. The clinical significance of abnormal diastolic function is not clearly understood. In other groups of patients, abnormal diastolic function indicates early myocardial disease or ischaemic damage, and may predate systolic abnormalities and heart failure.

Cardiac function during exercise has been investigated in a smaller number of studies. It has been shown that hypopituitary patients were able to exercise for a shorter time and tolerated a smaller workload [15,21]. In the former study, there was a higher frequency of exercise-induced S–T segment depression greater than 1 mV in the patients (74%) than in normal controls (28%) [15]. The abnormal ischaemic-like exercise-induced S–T segment depression was associated with left ventricular diastolic dysfunction on echocardiography [14,15]. In asymptomatic subjects, such electrocardiographic changes during exercise are associated with an increased risk of developing ischaemic heart disease during the long-term follow-up. Abnormal left ventricular systolic function (cardiac index and EF%) during peak exercise has been reported in hypopituitary patients compared with controls [20]. In the hypopituitary patients, EF% demonstrated an abnormal response (decrease rather than expected increase) during exercise [20,21], and the abnormality correlated with the duration of hypopituitarism [20]. The cardiac dysfunction was present in both childhood and adult-onset hypopituitarism when the two groups were analysed separately [21].

The positive effects of GH replacement on the various indices of cardiac function and exercise tolerance described above provide surrogate evidence for the specific aetiological role of GH deficiency.

Data on arterial disease

The status of peripheral arteries has been assessed using three different methods in four independent centres, and all comparisons were made with normal controls matched for age and sex. The carotid and femoral arteries of hypopituitary patients and matched controls were studied using high-resolution ultrasonography [22]. Carotid intima–media thickness was greater in 34 patients (mainly adult-onset hypopituitarism) than in 39 controls. The percentage of patients with one or more atheromatous plaques and the percentage of individual arteries with a plaque were both higher in the patient group. Increased intima–media thickness has also been reproduced in another study of 14 childhood-onset GHD adults [23]. Increased intima–media thickness is indicative of early atheroma in epidemiological studies, and abnormalities in carotid arteries reflect disease in other arteries, particularly the coronary arteries.

The arterial stiffness index (β), calculated from changes in arterial diameter derived from ultrasound measurements and the brachial blood pressure, has been reported to be significantly greater in hypopituitary patients than in controls [24]. Other investigators have demonstrated that the blood pressure-corrected thoracoabdominal aortic compliance index, measured by non-invasive pulsed-wave Doppler ultrasound, is significantly reduced, suggesting that the aorta also is less distensible [25]. Aortic compliance measured using this technique is reduced in conditions with an established increased risk of vascular disease, such as diabetes and familial hypercholesterolaemia. On the other hand, Dunne *et al.* [26] reported that forearm blood flow, assessed by strain gauge plethysmography, did not differ in controls and hypopituitary patients on conventional replacement. In this last study, both systolic and diastolic BP responded appropriately, and by a similar degree to that seen in controls, in response to several cardiovascular tests, suggesting that cardiovascular reflexes were intact [26].

The bulk of the available evidence suggests that hypopituitarism is associated with abnormalities of the arterial walls compatible with the changes of early atherosclerosis. The impact on arterial reactivity is limited. However, the effect may be more dependent on the integrity of the autonomic nervous system than on the presence or absence of early atherosclerotic changes.

Is the frequency of hypertension increased in hypopituitarism?

An increased frequency (18%) of treated hypertension was observed in one large series of hypopituitary patients in Germany, and was significantly greater than that in the population at large [5]. Rosén *et al.* [27] found that treated hypertension increased in frequency in hypopituitary men compared to the Swedish population (11.3 vs. 8.0%, $p < 0.05$). In the earlier studies, Merimee [7] found no significant difference in the prevalence of treated hypertension in 31 GH-deficient adults with childhood onset, compared with 662 normal controls (19% vs. 24%, N.S.). In other smaller studies, resting blood pressure (BP) has been either similar in patients and controls [13,22] or lower in the patient group [14,16,21]. However, as these measurements were reported as baseline investigations for GH replacement studies, patients with known hypertension may have been excluded beforehand. In longitudinal studies, resting BP did not change significantly after six weeks to 10 months in six patients on routine replacement following hypophysectomy [28]. Using ambulatory BP measurement methods, the mean 24-hour BP during the day and night-time BP (both systolic and diastolic) tended to be lower in hypopituitary patients than in normal controls, but the differences were not significant [26].

An increased prevalence of hypertension compared with normal populations has been observed, although this may reflect the fact that the medical attention received by hypopituitary patients provides an increased opportunity for diagnosis. The excess vascular disease cannot be attributed to hypertension with any certainty.

What are the possible causes of increased vascular disease in hypopituitary adults?

Stimulated by the findings discussed above, several studies have examined the classical factors associated with increased cardiovascular risk [27,29–36]. The findings of such studies are summarized in Table 7.4. Many of the studies attempted to link the findings to GH deficiency and, in support of this notion, observed beneficial changes during GH replacement studies. However, hypopituitarism is a complex condition with variable, though fairly predictable, combinations of hormone deficiency states. Although increased vascular mortality has been attributed to GH deficiency [1,5,27,29], there does not appear to be a simple explanation for all the abnormalities forming the final clinical picture of hypopituitarism in adults. Other pituitary hormones may contribute to the vascular risk (Table 7.5). In particular, inappropriate replacement with sex hormones, thyroxine, or glucocorticoids could induce adverse changes, as the present replacement regimens are unphysiological. All the pituitary-dependent hormones are known to influence the classical risk factors, particularly carbohydrate and lipid metabolic disturbances, that are demonstrable in hypopituitarism (Table 7.4). The possible mechanisms underlying the contribution of these hormones are discussed below.

Table 7.4 Summary of the anthropometric, body compositional and metabolic mechanisms associated with increased cardiovascular risk documented to be abnormal in hypopituitary adults on conventional replacement therapy [27, 29–36].

Anthropometric
Increased overall adiposity: high body mass index, increased observed vs. ideal body weight
Increased skin fold thickness
Increased central body fat deposition reflected in increased waist to hip ratio
Body-compositional
Increased total body fat or percentage body fat
Decreased lean body mass derived from total body water, total body potassium methods etc.
Increased central/truncal/abdominal fat deposition measured by CT, MRI or DEXA scanning
Metabolic
Increased frequency of impaired glucose tolerance and undiagnosed diabetes during oral glucose tolerance tests
Decreased insulin sensitivity
Increased beta cell function
Increased circulating products of pancreatic cell (insulin, proinsulin and split proinsulin)
Elevated plasma total and low-density lipoprotein cholesterol levels
Elevated plasma triglyceride concentrations*
Decreased plasma high-density lipoprotein cholesterol levels*
Abnormal plasma lipoprotein and hepatic lipase activities

*Indicates lack of uniformity of findings between different studies.

Table 7.5 Possible causes and clinical scenarios leading to increased cardiovascular risk factors in hypopituitary adults.

Underlying original disease process such as acromegaly and Cushing's disease*
Growth hormone deficiency
Delayed or inadequate replacement with sex hormones
Under- or over-replacement with thyroxine
Under-, over- or unphysiological replacement with glucocorticoids

*The excess mortality was demonstrable independent of the original disease [1].

The role of GH deficiency

GH deficiency may be the primary mechanism underlying the abnormal body composition and central fat distribution, and may also contribute to the hypercholesterolaemia. Impaired glucose tolerance, hyperinsulinaemia and hypertriglyceridaemia cannot be related directly to GH deficiency. It may in theory contribute to these indirectly via the abnormal body fat distribution, although abnormalities such as hypertriglyceridaemia persist with

GH replacement, despite improvement in body fat distribution. With long-term GH administration, the cardiovascular risk will reflect the balance between its different actions. The anti-insulin effect of GH may be opposed by the beneficial effect for glucose homeostasis of the increase in muscle bulk and decrease in total and central adipose tissue mass. From the available data, it is not yet clear which mode of action will prevail, as the effects on carbohydrate tolerance and insulin sensitivity have tended to recover in some of the trials, but not in others.

Inappropriate glucocorticoid therapy

Glucocorticoid excess is associated with increased total and central body fat. It is also associated with glucose intolerance and hyperinsulinaemia [37,38]. Triglyceride concentrations are elevated. Chronic glucocorticoid excess, if untreated, also causes death from vascular disease [39]. Chronic over-replacement in hypopituitary adults, if it has occurred, is therefore a potential contributor to vascular risk.

Current recommended glucocorticoid replacement regimens (hydrocortisone both twice and three times daily) do not mimic the normal diurnal cortisol profile, in which peak values occur between 6.00 a.m. and 8.00 a.m. [40–42]. Predictably, cortisol values overnight with the twice-daily regimen are low or even unmeasurable. Circulating cortisol profiles remain unphysiological, with low values overnight. Hydrocortisone is the preferred agent in the majority of centres. There is no uniformity of opinion as to the optimum method of monitoring hydrocortisone therapy [43,44]. Some advocate that the hydrocortisone dose should be gauged by clinical assessment only, others measure diurnal cortisol profiles and in other centres 24-hour urinary cortisol estimations are performed. Where biochemical assessment is performed, most data suggest over-replacement during the day and under-replacement at night [45,46]. The importance of this for the long-term vascular risk is unknown. However, there is increasing concern that over-replacement with glucocorticoids may be an important contributor to the observed excess cardiovascular disease. Furthermore, it is now recognized that GH inhibits 11β-hydroxysteroid dehydrogenase type 1 [47,48]; the effect of this phenomenon is a reduction in the net conversion of cortisol to inactive cortisone in untreated GH deficiency.

Glucocorticoid deficiency could also contribute to apparent vascular deaths. It is associated with abnormal cardiovascular reflexes which might have adverse effects in patients with established vascular disease. It is evident that both cortisol deficiency and cortisol excess occur with conventional replacement regimens over a 24-hour period. The long-term implications for vascular disease are unknown.

Because GH reduces serum cortisol binding globulin levels [47,49], serum cortisol concentrations should be interpreted with caution in hypopituitary patients both on and off GH replacement therapy.

Inappropriate thyroid hormone replacement

Primary hypothyroidism is associated with a two-fold increased risk of coronary artery disease. Thyroxine deficiency potentiates the development of atheroma in the hyperlipidaemic rat. Hypothyroidism in man causes an elevation in serum cholesterol and triglyceride concentrations. Overt hypothyroidism is therefore associated with excess vascular risk and with established risk factors. Subclinical hypothyroidism, defined as a normal circulating thyroid hormone concentration with an elevated thyroid-stimulating hormone level (in the absence of pituitary disease) causes dyslipidaemia and an increased cardiovascular risk [50]. The lipid status improves with replacement treatment, at least in some studies. In patients with hypopituitarism, the thyroid-stimulating hormone (TSH) level is not a suitable guide to the adequacy of thyroid hormone replacement. Thyroid hormone concentrations themselves (free or total T_4, total T_3) have therefore been employed. For all the measures of thyroid hormone concentration, normal values extend over a two-fold to three-fold range. It is therefore difficult to be certain whether patients are receiving adequate replacement, even though circulating thyroid hormone levels may be in the normal range. Chronic mild under-replacement may also contribute to vascular risk and risk factors. There is also a risk of dysrhythmias developing with mild over-replacement, particularly in elderly patients. This is suggested by studies in patients with an intact pituitary gland demonstrating that atrial fibrillation was associated with borderline hyperthyroidism.

Inadequate sex hormone replacement

Sex hormones have important and complex actions on the cardiovascular system and cardiovascular risk factors in both men and women [51]. Sex hormone replacement may be initiated and monitored on a symptomatic basis, and late or inadequate replacement is therefore possible, particularly in females. The mortality in hypopituitary patients was higher in women than in men in the three published series (observed/expected: 2.8 vs. 1.5 [1]; observed/expected: 2.3 vs. 1.5 [3]; and standardized mortality ratio 2.39 vs. 1.54 [2] in women vs. men). Hypopituitary women had more marked abnormalities of carotid wall structure and function [24]. Body compositional and metabolic abnormalities, such as glucose intolerance and hyperlipidaemia, were also more marked in women [33,34]. An excess central body fat disposition has been demonstrated in women at the menopause [52], and has been shown to be prevented by oestrogen replacement therapy [53]. These observations suggest a possible contribution from inadequate sex hormone replacement in these groups of patients.

Conclusions

There is a substantial body of evidence for reduced life expectancy in adults with hypopituitarism who are receiving conventional replacement therapy. The increased mortality was attributable to vascular disease at least in two major series. Studies of asymptomatic patients support this hypothesis, showing an increased frequency of early markers of atherosclerosis in the heart and peripheral arteries, as well as an excess of the classical risk factors of vascular disease. GH deficiency is likely to be an aetiological factor, but although this has been claimed as the sole mechanism for excess vascular disease, there is the potential for some bias given the large volume of published work focusing on the role of GH. An additional contribution from unphysiological replacement by other pituitary hormones is possible, and this topic deserves further study.

References

** 1 Rosén T, Bengtsson BÅ. Premature mortality due to cardiovascular disease in hypopituitarism. *Lancet* 1990; **336**: 285–8.

** 2 Bulow B, Hagmar L, Mikoczy Z, Nordstrom CH, Erfurth EM. Increased cerebrovascular mortality in patients with hypopituitarism. *Clin Endocrinol (Oxf)* 1997; **46**: 75–81.

** 3 Bates AS, Bullivant B, Clayton RN, Sheppard MC, Stewart PM. Increased mortality in hypopituitarism is not due to an increase in vascular mortality. *J Endocrinol* 1997; **152** (Suppl): OC9.

** 4 Bates AS, Van't Hoff W, Jones PJ, Clayton RN. The effect of hypopituitarism on life expectancy. *J Clin Endocrinol Metab* 1996; **81**: 1169–72.

* 5 Wüster C, Slenczka E, Ziegler R. Erhöhte Prävalenz von Osteoporose und Arteriosklerose bei konventionell substituierter Hypophysenvorderlappeninsuffizienz: Bedarf einer zusätzlichen Wachstumshormonsubstitution? *Klin Wochenschr* 1991; **69**: 769–73.

6 Merimee TJ, Fineberg S, Hollander W. Vascular disease in the chronic HGH-deficient state. *Diabetes* 1973; **22**: 813–19.

7 Merimee TJ. A follow-up study of vascular disease in growth hormone-deficient dwarfs with diabetes. *N Engl J Med* 1978; **298**: 1217–22.

* 8 Libber SM, Plotnick LP, Johanson AJ, *et al.* Long-term follow-up of hypopituitary patients treated with human growth hormone. *Medicine* 1990; **69**: 46–55.

* 9 Buchanan CR, Preece MA, Milner RDG. Mortality, neoplasia and Creutzfeldt–Jakob disease in patients treated with human pituitary growth hormone in the United Kingdom. *Br Med J* 1991; **302**: 824–8.

* 10 Taback SP, Dean HJ, Canadian Growth Hormone Advisory Committee. Mortality in children with growth hormone (GH) deficiency receiving GH therapy, 1967–92. *J Clin Endocrinol Metab* 1996; **81**: 1693–6.

11 Hjalmarson A, Isaksson O, Ahren K. Effects of growth hormone and insulin on amino acid transport in perfused rat heart. *Am J Physiol* 1969; **217**: 1795–802.

*12 Merola B, Cittadini A, Colao A *et al.* Cardiac structural and functional abnormalities in adult patients with growth hormone deficiency. *J Clin Endocrinol Metab* 1993; 77: 1658–61.

*13 Amato G, Carella C, Fazio S *et al.* Body composition, bone metabolism and heart structure and function in growth hormone (GH) deficient adults before and after GH replacement therapy. *J Clin Endocrinol Metab* 1993; 77: 1671–6.

*14 Shahi M, Beshyah SA, Hackett D *et al.* Myocardial dysfunction in treated adult hypopituitarism: a possible explanation for increased cardiovascular mortality. *Br Heart J* 1992; **67**: 92–6.

*15 Beshyah SA, Shahi M, Mayet J, Foale R, Johnston DG. Growth hormone and the cardiovascular system. In: Ranke M, Christiansen JS, eds. *The Complexity of Endocrine Systems.* Mannheim: J & J-Verlag, 1996: 131–55.

16 Thuesen L, Jorgensen JOL, Müller JR *et al.* Short- and long-term cardiovascular effects of growth hormone therapy in growth hormone deficient adults. *Clin Endocrinol (Oxf)* 1994; **41**: 615–20.

17 Valcavi R, Gaddi O, Zini M *et al.* Cardiac performance and mass in adults with hypopituitarism: effects of one year of growth hormone treatment. *J Clin Endocrinol Metab* 1995; **80**: 659–66.

18 Nass R, Huber RM, Klauss V *et al.* Effect of growth hormone (hGH) replacement therapy on physical work capacity and cardiac and pulmonary function in patients with hGH deficiency acquired in adulthood. *J Clin Endocrinol Metab* 1995; **80**: 552–7.

19 Caidhal K, Éden S, Bengtsson BÅ. Cardiovascular and renal effects of growth hormone. *Clin Endocrinol (Oxf)* 1994; **40**: 393–400.

20 Cittadini A, Cuocolo A, Merola B *et al.* Impaired cardiac performance in GH-deficient adults and its improvement after GH replacement. *Am J Physiol* 1994; **267**: E219–E225.

21 Longobardi S, Cuocolo A, Merola B *et al.* Left ventricular function in young adults with childhood and adulthood onset growth hormone deficiency. *Clin Endocrinol (Oxf)* 1989; **48**: 137–44.

22 Markussis V, Beshyah SA, Fisher C *et al.* Abnormal carotid wall arterial wall dynamics in symptom-free hypopituitary adults. *Eur J Endocrinol* 1997; **136: 157–64.

*23 Capaldo B, Patti L, Oliviero U *et al.* Increased arterial intima–media thickness in childhood-onset growth hormone deficiency. *J Clin Endocrinol Metab* 1997; **82**: 1378–81.

24 Markussis V, Beshyah SA, Fisher C *et al.* Detection of premature atherosclerosis by high-resolution ultrasonography in symptom-free hypopituitary adults. *Lancet* 1992; **340: 1188–92.

25 Lehmann ED, Hopkins KD, Weissberger AJ *et al.* Blood pressure corrected aortic distensibility in growth hormone deficient adults. *Clin Sci* 1994; **83 (Suppl 30): 64.

26 Dunne FP, Elliot P, Gammage MD *et al.* Cardiovascular function and glucocorticoid replacement in patients with hypopituitarism. *Clin Endocrinol (Oxf)* 1995; **43**: 623–9.

*27 Rosén T, Éden S, Larson G, Wilhemsen L, Bengtsson BÅ. Cardiovascular risk factors in growth hormone deficient adults. *Acta Endocrinol* 1993; **129**: 195–200.

28 Bojs G, Falkheden T, Sjögren B, Varnauskas E. Haemodynamic studies in man before and after hypophysectomy. *Acta Endocrinol* 1965; **39**: 308–22.

29 Cuneo RC, Salomon F, MacGauley GA, Sonksen PH. The growth hormone deficiency syndrome in adults. *Clin Endocrinol (Oxf)* 1992; **37: 387–97.

*30 Salomon F, Cuneo RC, Hesp R, Sönksen PH. The effects of treatment with recombinant human growth hormone on body composition and metabolism in adults with growth hormone deficiency. *N Engl J Med* 1989; **321**: 1797–803.

31 Rosén T, Bosaeus I, Tolli J, Lindstedt G, Bengtsson BÅ. Increased body fat mass and decreased extracellular fluid volume in adults with growth hormone deficiency. *Clin Endocrinol (Oxf)* 1993; **38**: 63–71.

32 Cuneo RC, Salomon F, Watts GF, Hesp R, Sönksen PH. Growth hormone treatment improves serum lipids and lipoproteins in adults with growth hormone deficiency. *Metabolism* 1993; **42**: 1519–23.

33 Beshyah SA, Freemantle C, Thomas E *et al.* Abnormal body composition and reduced bone mass in growth hormone

deficient hypopituitary adults. *Clin Endocrinol (Oxf)* 1995; **42**: 178–89.

34 Beshyah SA, Henderson A, Niththyananthan R *et al.* Metabolic abnormalities in growth hormone deficient adults, 2: carbohydrate tolerance and lipid metabolism. *Endocrinol Metab* 1994; **1**: 173–80.

35 Beshyah SA, Rutherford OM, Thomas E, Murphy M, Johnston DG. Assessment of regional body composition by dual energy X-ray absorptiometry in hypopituitary adults before and during growth hormone treatment. *Endocrinol Metab* 1995; **2**: 147–55.

36 Weaver JU, Monson JP, Noonan K *et al.* The effect of low-dose recombinant human growth hormone replacement on regional fat distribution, insulin sensitivity, and cardiovascular risk factors in hypopituitary adults. *J Clin Endocrinol Metab* 1995; **80**: 153–9.

37 Pupo AA, Wajchenberg BL, Schnaider J. Carbohydrate metabolism in hyperadrenocortisolism. *Diabetes* 1966; **15**: 24–9.

38 Howlett TA, Rees LH, Besser GM. Cushing's syndrome. *Clin Endocrinol Metab* 1985; **14**: 911–45.

39 Plotz M, Kowlton AI, Ragan C. Natural history of Cushing's syndrome. *Am J Med* 1952; **13**: 597–614.

40 Kehlet H, Binder C, Blichert-Toft M. Glucocorticoid maintenance therapy following adrenalectomy: assessment of dosage and preparation. *Clin Endocrinol (Oxf)* 1976; **5**: 37–41.

41 Scott RS, Donald RA, Espiner EA. Plasma ACTH and cortisol profiles in Addisonian patients receiving conventional substitution therapy. *Clin Endocrinol (Oxf)* 1978; **9**: 571–6.

42 Feek CM, Ratcliffe JG, Seth J *et al.* Patterns of plasma cortisol and ACTH concentrations in patients with Addison's disease treated with conventional corticosteroid replacement. *Clin Endocrinol (Oxf)* 1981; **14**: 451–8.

43 Besser GM, Jeffcoate WJ. Adrenal diseases. *Br Med J* 1976; **1**: 448–51.

44 Burke CW. Primary adrenocortical failure. In: Grossman A, ed. *Clinical Endocrinology*. Oxford: Blackwell Scientific, 1992: 393–404.

*45 Peacey SR, Guo CY, Robinson AM *et al.* Glucocorticoid replacement therapy: are patients over-treated and does it matter? *Clin Endocrinol (Oxf)* 1997; **46**: 255–62.

*46 Howlett TA. An assessment of optimal hydrocortisone replacement therapy. *Clin Endocrinol (Oxf)* 1997; **46**: 263–8.

47 Weaver JU, Thaventhiran L, Noonan K *et al.* The effect of growth hormone replacement on cortisol metabolism and glucocorticoid sensitivity in hypopituitary adults. *Clin Endocrinol* 1994; **41**: 639–48.

48 Gelding SV, Taylor NF, Wood PJ *et al.* The effect of growth hormone replacement therapy on cortisol–cortisone interconversion in hypopituitary adults: evidence for growth hormone modulation of extrarenal 11-β-hydroxysteroid dehydrogenase activity. *Clin Endocrinol* 1998; **48**: 153–62.

49 Rodriguez-Arnao J, Perry L, Besser GM, Ross RJM. Growth hormone treatment in hypopituitary GH deficient adults reduces circulating cortisol levels during hydrocortisone replacement therapy. *Clin Endocrinol* 1996; **45**: 33–7.

50 Bastenie PA, Vanhaelst L, Bonnyns M, Neve P, Staquet E. Preclinical hypothyroidism: a risk factor for coronary heart disease. *Lancet* 1971; **i**: 203–4.

51 Lip GYH, Beevers G, Zarifis J. Hormone replacement therapy and cardiovascular risk: the cardiovascular physicians' viewpoint. *J Intern Med* 1995; **238**: 389–99.

52 Ley CJ, Lees B, Stevenson JC. Sex- and menopause-associated changes in body-fat distribution. *Am J Clin Nutr* 1992; **55**: 950–4.

53 Haarbo J, Marslew U, Gotfredsen A, Christiansen C. Postmenopausal hormone replacement therapy prevents central distribution of body fat after menopause. *Metabolism* 1991; **40**: 1323–6.

8: How should we diagnose GH deficiency in hypopituitary adults?

Asad Rahim and Stephen M. Shalet

Introduction

With the availability of a potentially limitless supply of synthetically produced growth hormone (GH) since the mid-1980s, and the demonstration that patients with GH deficiency (GHD) have benefited from GH replacement [1], there has been a great deal of interest in the use of GH in adults with either adult-onset or childhood-onset GHD. In childhood, the commonest aetiology is isolated idiopathic GHD. In contrast, isolated idiopathic GHD has never been described in adults, and GHD, acquired in adulthood, occurs most frequently as a consequence of a pituitary adenoma and/or treatment with surgery and/or radiotherapy [2].

Clinical aspects

An increase in fat mass, particularly truncal, reduced lean mass, osteopenia, glucose intolerance, altered cardiac structure and function, reduced exercise capacity, reduced quality of life, insulin resistance and impaired fibrinolysis are all associated with GHD in adulthood. In childhood, poor growth offers a useful biological marker that raises the suspicion of GHD. No similar specific markers occur in patients with adult-onset GHD, and indeed the clinical features are non-specific; however, the history of pituitary disease or previous cranial irradiation alerts the endocrinologist to the possibility of GHD.

With expansion of pituitary tumours or following radiotherapy, GH is usually the first of the pituitary hormones to be affected [3]. It therefore follows that, in those patients in whom additional pituitary hormone deficits are present, there is an extremely high probability of GHD. This, and the issue of severity of GHD, were addressed in the study by Toogood *et al.* [2], in which 190 non-acromegalic patients (96 male) with pituitary disease, whose ages ranged from 16 to 72 (mean 39.4) years, were divided into four groups according to the number of anterior pituitary hormone deficiencies demonstrated; isolated GHD (GHD0), or GHD plus an additional one, two or three pituitary hormone deficits (GHD1, GHD2, GHD3) (Fig. 8.1). GH status was determined using the insulin tolerance test (ITT). Ninety-one per cent of patients in the combined groups GHD2 and GHD3, 55% in GHD1 and 24% in GHD0 had a peak GH of less than 2 ng/mL, demonstrating that the degree of GHD varies according to the number of additional pituitary hormone deficits—i.e. the greater the number of pituitary hormone deficits, the more severe the GHD. Assessment of GH status with 24-hour GH profiles [4] or urinary GH measurement [5] confirmed the relationship between GHD and additional pituitary hormone deficits.

GH provocative tests

Historically, knowledge of GH status in adults with pituitary disease was gained coincidentally from an ITT, performed to assess the pituitary–adrenal axis and the need for hydrocortisone replacement therapy. The vast experience of many investigators in using the ITT over the years has led to it being accepted as the gold standard test to assess GH status. The ITT has several advantages in that it not only allows simultaneous assessment of the hypothalamic–pituitary–adrenal axis, but also provides a very potent stimulus

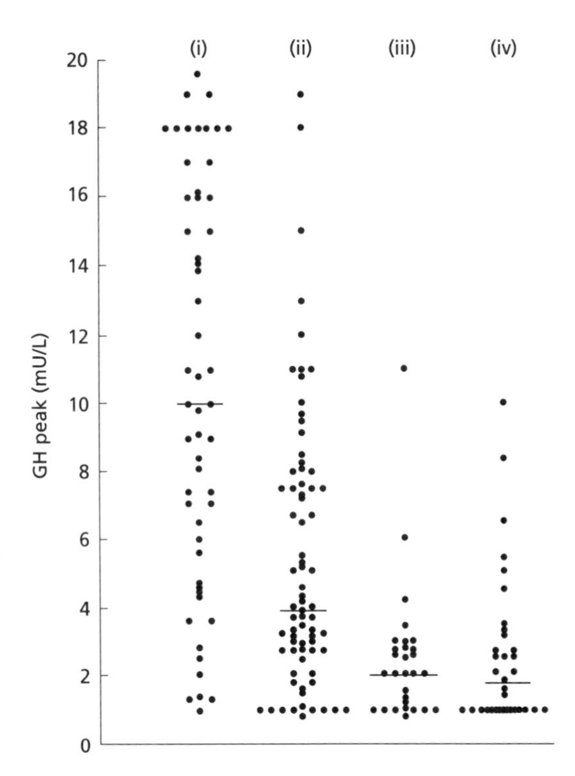

Fig. 8.1 The distribution of the peak serum growth hormone (GH) levels in response to an insulin tolerance test (ITT) in 190 patients divided into groups according to the degree of hypopituitarism present, i.e. the number of anterior pituitary hormone deficiencies, in each patient. Horizontal bars represent medians. (i) Isolated growth hormone deficiency (GHD0); (ii) GHD plus an additional one pituitary hormone deficit (GHD1); (iii) GHD plus an additional two pituitary hormone deficits (GHD2); (iv) GHD plus an additional three pituitary hormone deficits (GHD3). 1 ng/mL = 2.6 mU/L. Reproduced with permission from Toogood *et al.* [2].

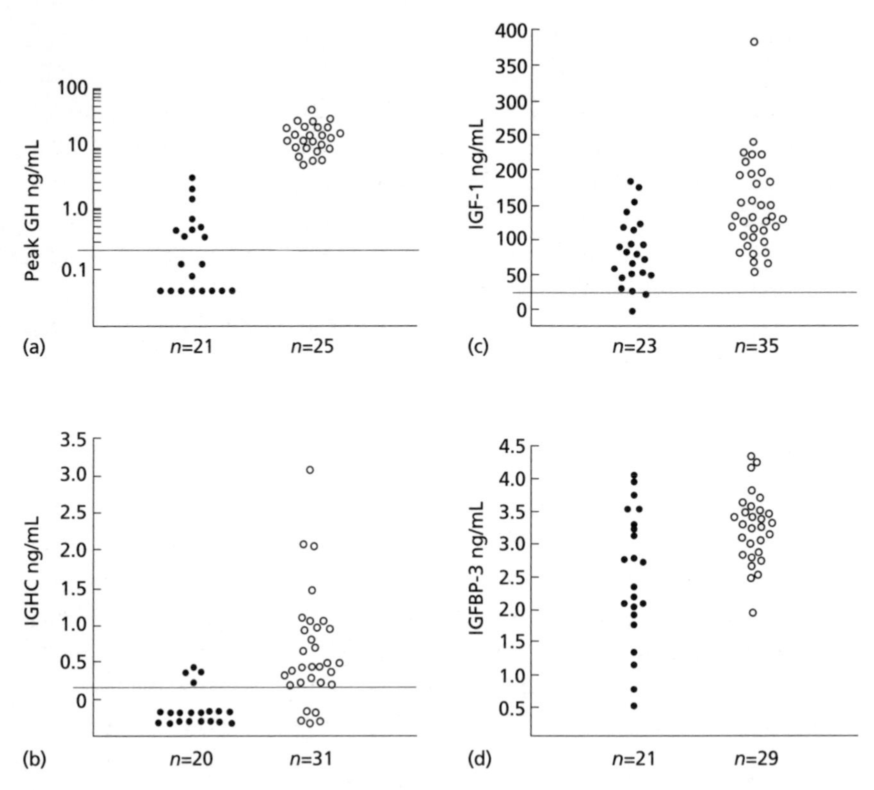

Fig. 8.2 Results of tests of growth hormone (GH) deficiency in normal individuals (○) and hypopituitary subjects (●): (a) peak GH response to insulin tolerance test (ITT); (b) mean 24-hour integrated GH concentration (IGHC); (c) insulin-like growth factor-I (IGF-I) concentration; (d) insulin-like growth factor binding protein-3 (IGFBP-3) concentration. S = assay sensitivity for GH (0.2 ng/mL) and IGF-I (25 ng/mL). Reproduced with permission from Hoffman *et al.* [8].

for GH release [6]. At a recent consensus meeting on adult GHD, the recommended criterion for diagnosing severe GHD was a peak GH response of less than 3 ng/mL to an ITT [7]. In particular, data from the study by Hoffman *et al.* [8] support the use of the ITT to diagnose severe GHD. Using several potential markers of GH status—peak GH response during an ITT, integrated GH concentration (IGHC) from 20-minute sampling over 24 hours, serum insulin-like growth factor-I (IGF-I) and insulin-like growth factor binding protein-3 (IGFBP-3) estimations—in 23 hypopituitary patients and 35 normal age-matched, sex-matched and body mass index (BMI)-matched subjects, Hoffman *et al.* [8] observed that only the peak GH response during the ITT allowed complete separation of subjects from patients (Fig. 8.2). Despite a significantly lower IGHC, IGF-I and IGFBP-3 in the hypopituitary patients, overlap for each of these markers of GH status prevented complete discrimination between normal subjects and patients with hypopituitarism. In addition, all the control subjects had a peak GH response during an ITT equal to or greater than 5 ng/mL, whilst all those with hypopituitarism had a peak GH response of less than 3 ng/mL.

Although it is the gold standard test, the ITT does have several disadvantages. It is unpleasant, and in inexperienced hands frankly dangerous [9]. In experienced hands, however, it is safe, but it should not be used in patients with a history of ischaemic heart disease or epilepsy [10]. The ITT is also contraindicated in the elderly, i.e. those over 65 years, because of the potential risk of occult ischaemic heart disease. Hence, in the light of these disadvantages, an alternative provocative agent may need to be used to assess GH status in adults. Although plentiful comparative data exist in children, relatively few studies have compared the ITT with other provocative agents in adults. Glucagon, arginine, L-dopa, nocturnal sleep, clonidine and exercise have all been used in various studies to assess GH status. Cain *et al.* [11] assessed the GH status of 20 normal young subjects (14 men) aged between 21 and 34 years using glucagon, arginine, ITT and tolbutamide-induced hypoglycaemia. They observed a significant GH increment from baseline to peak after each agent, and concluded that glucagon was at least as good, if not better than, the ITT at stimulating GH release. However, this study reported the mean change in serum GH from baseline to peak and the mean peak GH response to each agent, but did not compare one agent with another. Only 10 subjects had an ITT, and although the mean serum GH peak was not significantly different from that in response to glucagon (21.5 ± 5.2 ng/mL vs. 21.8 ± 3.9 ng/mL), none of the subjects had what Cain *et al.* [11] considered to be a subnormal response to the ITT—whereas the 'failure rates' following arginine and glucagon were three subjects and one subject, respectively. The subjects whom Cain *et al.* [11] considered as having 'failed' a provocative test were those in whom a peak GH response or rise in serum GH from baseline to peak was less than 5 ng/mL. However, no information was provided about the sex of the 10 subjects who underwent the ITT. This may be relevant, as Hoeck *et al.* [12] have demonstrated that the peak GH response to an ITT in female subjects was lower than in males.

Lin and Tucci [13] studied multiple different provocative stimuli to assess the peak GH response in normal, non-obese subjects. In this study, the ITT was the most effective stimulus for GH release, with all 19 individuals undergoing the ITT achieving a peak GH response of greater than 5 ng/mL. Although glucagon and L-dopa were not as effective as the ITT, 17 of 21 (81%) and 21 of 24 (87%) subjects, respectively, produced a GH response of greater than 5 ng/mL; there was no significant difference between the peak GH response to the ITT, glucagon or L-dopa. In a similar study, Eddy *et al.* [14] assessed the GH response to L-dopa, arginine, glucagon, vasopressin and the ITT in 20 normal, non-obese adults (12 women), and demonstrated that the administration of L-dopa resulted in 'normal' findings, i.e. a GH increment of > 5 ng/mL, in 95% of the subjects. The responses after the ITT, arginine, vasopressin and glucagon were 90%, 80%, 60% and 50%, respectively. Furthermore, retesting using the same agents in nine subjects elicited a 'normal' GH response to L-dopa and the ITT in 100% and 89%, respectively.

Comparing physiological stimuli, i.e. sleep and exercise, with pharmacological agents, in nine male subjects Sutton and Lazarus [15] demonstrated that the GH response during exercise was similar to that following the ITT, and both were markedly superior to arginine at stimulating GH release.

The ITT consistently appears to be more effective at stimulating GH release, and more recent studies have added support to these findings. In a prospective, randomized, placebo-controlled study in 18 normal male subjects, Rahim *et al.* [16] demonstrated that the ITT was the most potent stimulus of GH release, followed by glucagon and then arginine. All but one subject (10.7 ng/mL) achieved a peak GH response of greater than 15 ng/mL during the ITT. Intramuscular glucagon was not as impressive as the ITT, but 16 of 18 subjects achieved a peak GH response of greater than 7.7 ng/mL with the remaining two between 3.8 and 7.7 ng/mL. The GH response during an arginine infusion was less impressive than after glucagon, with two of 18 subjects achieving a peak GH response of less than 1.9 ng/mL. In addition, clonidine, which is widely accepted as a useful agent for assessing GH release in paediatric practice [17], when administered orally at doses of 100–200 μg was no better than placebo at stimulating GH release. Oral clonidine should therefore not be used to assess GH status in adults.

Most studies mentioned thus far have included individuals of both sexes and have not explored the influence of sex on stimulated GH release. Hoeck *et al.* [12] observed that the GH response during an ITT was greater in male subjects compared with females. Arginine, in contrast, appears to be a more potent stimulus of GH release in females [18,19]. We have observed similar findings (unpublished data) in 35 subjects (18 male), with the peak GH response during an ITT being significantly greater in male subjects, whilst the GH response during an arginine infusion was significantly greater in female subjects. In 31 subjects (13 females), we observed no significant difference in the peak GH response between the sexes following administration of glucagon. In the 13 female subjects who underwent all three provocative tests, the peak GH response was not significantly different between the tests.

Physiological assessment of GH secretion

Twenty-four-hour profiles with estimation of serum GH at 20-minute intervals, or measurement of 24-hour urinary GH excretion, allow assessment of physiological GH secretion. Unfortunately, the integrated GH concentration (IGHC) derived from a 24-hour profile does not allow satisfactory discrimination between GH-deficient and normal adults [8] (Fig. 8.2). Thus, estimation of IGHC failed to discriminate satisfactorily between GHD and normality, and at least part of the failure was due to the number of normal and hypopituitary subjects with values that were below the limit of detection of the GH radioimmunoassay [8].

Using a newer-generation enzyme-linked immunosorbent assay (ELISA) to enhance the sensitivity (1 ng/L), the same group reassessed the discriminatory capacity of IGHC [20]. Twenty-six per cent of hypopituitary subjects had IGHC values within the normal range [20]. Age stratification improved the separation, but an overlap between normal and GHD remained in those adults aged less than, or over 50 years of age. All samples, including nadirs, from GHD subjects were well within the GH assay detection limit (1 ng/L) [20]. Peak 24-hour GH levels in GHD subjects were lower, and did not overlap with those in the normal subjects [21]. Nadir GH concentrations were significantly lower in GHD subjects, but the range overlapped that of normal subjects [21]. Total daily GH production in GHD was ≈ 5% of the production in matched normal subjects [21]. This difference resulted from a greater reduction in the pulsatile component (by 96%) than in the tonic component (by 47%), so that the fractional daily contribution by tonic GH release in GHD subjects was markedly greater [21].

In summary, the peak 24-hour GH level was useful at discriminating between GH-deficient and normal adults. The nadir GH level and IGHC did not allow satisfactory discrimination between GH-deficient subjects and controls. The practical demands in collecting samples throughout 24 hours for a GH profile, as well as those of time and finance, mean that there is little future for this method of estimating GH status in day-to-day clinical endocrine practice.

Estimation of urinary GH may allow assessment of physiological GH secretion. Bates *et al.* [5] demonstrated reasonable separation of normal controls and GH-deficient patients with known pituitary disease and a peak GH response to an ITT of less than 5 ng/mL. Age stratification demonstrated that the younger the patient, the more useful urinary GH was at allowing discrimination. In those aged less than 40 years, with a sensitivity of 90%, the specificity for diagnosing GHD was 79%. The specificity fell to 67% in those aged 40–60 years, and was only 36% in those aged over 60 years.

IGF-I and IGFBP-3

The changes in IGF-I levels throughout life are similar to those of GH. With the onset of puberty, there is a two-fold to three-fold rise in serum IGF-I concentrations followed by a decline, so that average adult levels are reached by the early twenties [22]. There follows a gradual decline with advancing age [22]. Like IGF-I but less age-dependent, serum IGFBP-3 levels rise to a peak during the pubertal years and then slowly decline in adulthood [23]. Thus, in the adult patient with potential GHD, serum IGF-I and IGFBP-3 measurements can only be interpreted if decade-based normative data are available.

The usefulness of an IGF-I estimation in the diagnosis of adult GHD is contentious, although it has become clear that at least some of the disparity between studies is explained by the timing of the onset of GHD. Hoffman *et al.* [8] found that 70% of IGF-I and 72% of IGFBP-3 values in adult-onset GHD patients, mean age 45 years, were within the range of normal subjects, even allowing for the effects of age.

De Boer *et al.* [24] focused on childhood-onset GHD, and re-evaluated GH status in 89 young adult males who had previously received GH replacement in childhood. Approximately 93% of the patients had an IGF-I level below the normal range, with a similar number of subnormal IGFBP-3 levels amongst the patients. Attanasio *et al.* [25] pursued the same theme in a large study of 74 childhood-onset and 99 adult-onset GHD patients. They concluded that there are profound differences between these two forms of GHD [25]. With group data, they observed that the serum IGF-I levels were below normal in both groups of GHD patients, but were significantly lower in childhood-onset than in adult-onset GHD patients [25]. Other authors have made similar observations [26,27]. In these studies, however, it is not clear if sufficient decade-based normative data were available, nor whether the severity of GHD was equal in the childhood-onset and adult-onset groups. Thus, apart from the potential impact of the timing of onset of GHD, it is possible that the severity of GHD influences the interpretation of IGF-I results.

Obesity

Obesity is one pathophysiological state that may be difficult to distinguish from organic GHD in the adult. There is substantial evidence that morbid obesity is accompanied by suppression of GH release, and that substantial weight loss may restore spontaneous [28,29] and stimulated GH secretion [29]. Veldhuis *et al.* [30] observed mean 24-hour GH levels in obese men, body mass index (BMI) 41–58 kg/m^2, to be 25% that of an age-matched control group, BMI 23–33 kg/m^2. Even in subjects of near-normal weight, relative obesity as determined by BMI is negatively correlated with GH secretion [19,31,32] due both to reduced GH production and increased clearance [30]. Each unit increase in BMI, at a given age, reduced daily GH secretion by 6% [31]. Furthermore, in clinically non-obese healthy adults, relative adiposity in the abdominal region is a major negative determinant of stimulated GH secretion [19].

Changes in IGF-I levels in obesity are disputed. There are data showing decreased [31,33–35] or normal [30] IGF-I levels in adults, and even elevated [36] IGF-I levels in obese children. In the largest population-based study of IGF-I levels in adults, an apparent decline in IGF-I levels attributable to adiposity disappeared when age was accounted for in a multivariate analysis [37].

Exactly how obesity reduces GH secretion has not been clarified. However, there is increasing evidence that free fatty acids (FFAs) play a significant role [38,39]. Recent studies with acipimox, a nicotinic acid analogue that blocks lipolysis and is devoid of side-effects, indicate that FFA reduction enhanced the GH response to a variety of GH secretagogues but did not increase spontaneous GH secretion in obese individuals [40–42]. The possible site of action of FFA is at the pituitary somatotroph [43].

A recent study in age-matched and sex-matched subjects suggests that there is a much more profound reduction in total GH secretion in a group of individuals with organic GHD compared with a group of obese subjects [44]. However, in the obese individual with pituitary disease and no other pituitary hormone deficit, a reduced GH response to any of the standard provocative tests may reflect organic GHD or obesity itself. In this clinical situation, distinction between the two possibilities is not possible with any degree of certainty at present, although the generally held view is that a low IGF-I level would favour the diagnosis of genuine GHD.

Elderly (see also Chapter 12)

GH release gradually declines with increasing age [45,46]. Iranmanesh and colleagues [31] demonstrated that for every decade of life between ages 21 and 71 years, the GH secretion rate decreases by ≈ 14%. Twenty-four-hour integrated GH concentrations are lower in elderly subjects compared with young healthy adults, and are similar to those in patients with known organic GHD [46–48]. A reduction in the mean pulse amplitude and duration has been demonstrated [47], and this led to the conclusion that the reduction in total GH secretion is due predominantly to a reduction in pulse amplitude. Ultrasensitive GH assays used in more recent studies [49] have confirmed these findings. The trend of decreasing GH secretion with increasing age constitutes the somatopause. In addition, ageing itself is associated with a series of biological changes [50,51], e.g. increased body fat, reduced lean tissue and reduced bone mineral density, which are also seen in adults with organic GH deficiency.

The prevalence of non-functioning pituitary adenomas increases with age [52], and thus pituitary disease is common in the elderly. Toogood *et al.* [48] have now established that GH secretion is significantly reduced in the elderly with pituitary disease compared with normal controls of similar age. They studied GH secretion in 24 patients (age 61–85 years) with hypothalamic–pituitary disease and in 24 controls (age 60–87 years) matched for BMI. The median (range) area under the curve of the 24-hour GH profile, < 9.6 ng/mL (< 9.6–20 ng/mL) vs. 18.5 ng/mL (10.7–74.4 ng/mL); the median stimulated peak GH response to arginine, < 0.4 ng/mL (< 0.4–7.7 ng/mL) vs. 8.0 ng/mL (1.6–37.0 ng/mL); and the median serum IGF-I concentration, 102 ng/mL (< 14–162 ng/mL) vs. 147 ng/mL (65–255 ng/mL), were significantly lower in the patients than in the controls [48]. For each of the three latter indices of GH status, there was overlap between the two groups. Fifteen of the 24 patients showed no evidence of spontaneous or stimulated GH secretion (using a standard immunoradiometric assay with a sensitivity of 0.4 ng/mL), whereas all controls showed evidence of both.

Under circumstances in which GH secretion is known to be reduced, however, most values during a 24-hour GH profile cannot be estimated using conventional radioimmunoassays. Thus, a limitation of the original study by Toogood *et al.* [48] was that 93% and 61% of the GH estimations in the patients' and controls' 24-hour profiles (72 samples), respectively, fell below the sensitivity of the conventional assay. Subsequently, the latter GH profiles were re-examined using an ultrasensitive chemoluminescent assay (sensitivity 2 ng/L). GH secretion was detectable throughout the 24-hour profile in all patients and controls [4]. Furthermore, the pattern of GH secretion remained pulsatile in all but one of the patients, despite previous pituitary pathology, surgery and irradiation. GH pulse frequency was unaffected in the patients studied by Toogood *et al.* [4], and as in the findings of Reutens *et al.* [21] in middle-aged adults with organic GHD, absolute peak GH levels were reduced to a greater extent than absolute nadir GH levels. None the less, the overlap between patients' and controls' values remained unless the analysis was restricted to those patients with the most severe degree of GHD, i.e. those with two or more additional pituitary hormone deficits, in whom the absolute peak GH concentration was distinctly separate from control values [4].

In addition, using the ultrasensitive GH assay on the samples obtained during the provocative GH test revealed a highly significant correlation between the peak GH response to arginine and the area under the 24-hour GH profile, which was more pronounced in the patients ($r = 0.9$) than in the controls ($r = 0.5$) [4].

IGF-I estimation appears to offer relatively little to the diagnosis of GHD in the elderly; despite the group differences, overlap between patients with organic GHD and controls is considerable, with only 21% of elderly GHD patients having a serum IGF-I level below the range found in the elderly controls [48]. Serum IGFBP-3 measurement proved even less rewarding, and there is absolutely no difference between the values in the GHD patients and controls [53].

Thus, despite the fact that GH secretion in the elderly with pituitary disease may be reduced to about 13% of that of age-matched controls [4], confirming GH deficiency in an individual patient remains a problem. Twenty-four-hour GH profiles, as well as IGF-I and IGFBP-3 measurements, are unsatisfactory either because of impracticality or lack of discriminatory power. The excellent correlation seen between the peak GH response to arginine and spontaneous GH secretion supports the view that the arginine stimulation test is a reasonable choice for assessing GH status in patients with GHD2 or GHD3 disease, particularly in an age group in whom an ITT may carry an increased risk of morbidity or mortality. In an elderly patient with pituitary disease and either one or no additional pituitary hormone deficits, the distinction between GHD and normality remains a challenge.

Childhood-onset GHD (see also Chapter 14)

Currently, GH replacement in childhood is continued until growth is completed or there is an agreement between the family and the endocrinologist that a satisfactory height has been achieved. Over the last 10 years, there have been a number of studies in which the GH status of children and young adults who had received GH replacement during childhood was reassessed after completion of growth and puberty [54–59].

The purpose of the studies was at least two-fold: firstly, to establish how many individuals no longer appeared GH-deficient at re-evaluation and secondly, to determine how many individuals still had severe GHD requiring consideration of GH replacement during adult life. All individuals were documented biochemically to be GHD in childhood, but at reassessment their GH status was considered to be normal in 20–87% [55–60]. The aetiological classification of the childhood diagnosis in the vast majority of the latter subjects was isolated idiopathic GHD. This raises interesting questions regarding the nature of their defect in GH secretion during childhood. The diagnostic threshold for GHD is arbitrarily defined, and the reproducibility of the GH response to provocative testing within individuals is not high. On these grounds alone, it would be anticipated that a percentage of those considered GHD at one point in time might be considered normal at re-evaluation. Furthermore, it is likely that in a proportion of these patients, the childhood diagnosis was CDGP and not isolated idiopathic GHD, with the initial GH provocative tests carried out in the 'unprimed' state failing to make the distinction. Finally, it remains possible that transient GHD in childhood is a real entity, although longitudinally obtained proof is lacking.

In contrast to the findings in the isolated idiopathic GHD population, those young adults diagnosed as having organic GHD in childhood as a consequence of either a mass lesion, pituitary surgery, or irradiation damage to the hypothalamic–pituitary axis rarely revert to normal GH status [58].

In addition to influencing the incidence of reversal of GH status, the aetiology of the childhood diagnosis of GHD also affects the strategy of how to re-evaluate. In our centre [58], two provocative tests were carried out in two-thirds of 88 young adults undergoing reassessment of GH status, and one GH provocative test in the remainder. By far the most commonly used GH provocative tests were the ITT and the arginine stimulation test. Of those undergoing two tests, 55 patients had a peak GH of less than 3.5 ng/mL to one test, of whom 26 had a peak GH greater than 3.5 ng/mL to the second test; similarly, of the 41 patients with a GH peak less than 1.9 ng/mL to one test, 26 had a GH peak greater than 1.9 ng/mL to the second test [58]. Although there is a considerable degree of disagreement between the results of the two tests in individuals, the discordance was not explained by weighting associated with a persistently greater or lesser GH response to one particular test [58]. In addition, the discordance between test results is not distributed randomly. Of the 58 patients undergoing two tests at reassessment, 15 had additional anterior pituitary hormone deficits; all of them had a GH peak response below 3.5 ng/mL to both tests. At a diagnostic threshold of 3.5 ng/mL, discordance between the two GH test results only occurred amongst those patients with a diagnosis of isolated GHD [58].

The results from these studies emphasize that all children who have received GH replacement therapy in childhood should undergo reassessment of GH status in young adult life. The percentage of such patients who merit consideration for GH replacement in adult life will vary, depending on the definition of severe GHD in use and the aetiological mixture of GHD patient populations from different centres. Patients with isolated GHD should undergo two tests of GH status, but those with additional anterior pituitary hormone deficiencies require only one test at reassessment.

Newer strategies for assessment of GH status

The use of a combination of GHRH with pyridostigmine as a diagnostic test of childhood GHD is not recommended [61], since the test is considered to be less discriminating if the GHD is hypothalamic in origin and since in most children with GHD, the site of abnormal pathophysiology is believed to be hypothalamic rather than pituitary. In adults, the different epidemiology of the pathophysiology of GHD makes the combined GHRH–pyridostigmine test more attractive. The most common cause of adult-onset GHD is a pituitary adenoma, or treatment with pituitary surgery and/or irradiation. It is likely, but not proven, that the majority of patients with a pituitary mass lesion who are GH-deficient before or after pituitary surgery have sustained a pituitary insult; in contrast, GHD as a consequence of irradiation to the hypothalamic–pituitary axis is more likely to be hypothalamic in origin [62]. Thus, provided that the endocrinologist is comfortable in the knowledge that the patient has a pituitary defect rather than a hypothalamic one, the GHRH–pyridostigmine test can be used to assess GH status.

Anderson *et al.* [63] carried out a GHRH–pyridostigmine dose–response study to determine the optimal dose of pyridostigmine, and found it to be 120 mg combined with GHRH 1 μg/kg. They observed a lower reference limit of 21 ng/mL for the peak GH response to the combined test in 40 normal adults aged between 20 and 60 years [63]; there was no sex difference.

Using the threshold of 21 ng/mL for the peak GH response to the GHRH–pyridostigmine test, and an arbitrarily defined level of 10 ng/mL for the ITT, they found consistent classification of normal and subnormal GH responses in 44 of 47 patients with pituitary disease [63].

Perhaps even more promising than GHRH plus pyridostigmine is the combination of GHRH and arginine. Unlike the former combination, the GH response to GHRH–arginine is independent of age; in addition, the GH response is pronounced, and shows much less interindividual and intraindividual variability than the more conventional GH provocative tests [64]. No overlap was found between GH peak responses in 24 adults with organic GHD and normal subjects [65].

In the future, it is likely that the strategy for evaluation of GH status may take into account the degree of GHD, the aetiology of the hypothalamic–pituitary disease and the probable site of the defect within the hypothalamic–pituitary region.

Approach to the diagnosis of GHD

Up to the present, the main suggestion in the literature has been that the peak GH response to an ITT should be the gold standard for the biochemical diagnosis of severe GHD. The ITT does provoke a pronounced GH response in normal subjects, it allows the pituitary–adrenal axis to be tested at the same time, and the morbidity associated with the performance of the test is low in experienced endocrine units. A diagnostic cut-off of either 3 ng/mL or 5 ng/mL has been evaluated by pooling the ITT data available from the literature, albeit with the necessary assumptions required when GH values from different centres are considered together [8]. None the less, based on a cut-off of 5 ng/mL, the ITT provides a specificity of 97%, a sensitivity of 100%, a positive predictive value of 99% and a negative predictive value of 100% [8].

It is not possible, however, to rely on the ITT alone. Under specific circumstances, the ITT is contraindicated, and on other occasions two provocative tests are required. In addition, the lack of standardization of GH assays means that each laboratory has to establish its own diagnostic threshold values, rather than simply accepting the recommended cut-off level of 3 ng/mL or 5 ng/mL.

With the knowledge that they need to gather their own data, endocrine centres should concentrate on establishing the credentials of the test that they favour by acquiring GH data in normal individuals and in patients with hypothalamic–pituitary disease. The patients with additional pituitary hormone deficiencies could provide the gold standard for the diagnosis of severe GHD. Thus, in our centre, ≈90% of patients in groups GHD2 and GHD3 had a peak GH response of < 2 ng/mL to an ITT and in 100% the peak GH response was less than 4.2 ng/mL [2]. This model can be applied to any single test of GH status. The performance requirements for the test in normal individuals are a pronounced GH response, with very few individual 'failures'. Provided that these criteria are satisfied, glucagon, arginine, or even urinary GH estimation may be suitable. Clonidine is unsuitable [16], and the GHRH–pyridostigmine test would be more attractive if the physician could be certain that he or she was dealing with a pituitary rather than a hypothalamic defect. The combined GHRH–arginine test appears promising, but as yet the numbers of adult patients with hypothalamic–pituitary disease who have been studied are low.

To discriminate between GHD and normality, an IGF-I SDS estimation is extremely useful for retesting the young adult with a diagnosis of childhood-onset GHD, moderately helpful (≈30–50% positive predictive value) in the middle-aged adult (25–55 years) and rarely helpful in the elderly (over 60 years).

Within the limitations of the tests described above, it would be reasonable to perform only one provocative test of GH release in patients with two or three additional pituitary hormone deficits. In the patient with pituitary disease and a possible diagnosis of adult-onset isolated GHD or GHD plus one additional pituitary hormone deficit, two provocative tests of GH release would be appropriate.

The same strategy can be applied for reassessing GH status in young adults who received GH replacement for childhood GHD, as has been recommended for establishing the diagnosis of adult-onset GHD, but IGF-I SDS estimation should be considered suitable for the role of the only test in those with multiple pituitary hormone deficiencies and one of the two tests of GH status in the much larger cohort of individuals with a diagnosis of isolated GHD, all of whom require to be retested.

Finally, in patients with pituitary disease who are either morbidly obese or elderly, establishing the diagnosis of isolated GHD remains a challenge.

References

** 1 Cuneo RC, Salomon F, McGauley GA, Sonksen PH. The growth hormone deficiency syndrome in adults. *Clin Endocrinol* 1992; **37**: 387–97.

* 2 Toogood AA, Beardwell CG, Shalet SM. The severity of growth hormone deficiency in adults with pituitary disease is related to the degree of hypopituitarism. *Clin Endocrinol* 1994; **41**: 511–16.

3 Littley M, Shalet S, Beardwell C *et al.* Hypopituitarism following external radiotherapy for pituitary tumours in adults. *Q J Med* 1989; **70**: 145–60.

4 Toogood A, Nass R, Pezzoli S *et al.* Preservation of growth hormone pulsatility despite pituitary pathology, surgery and irradiation. *J Clin Endocrinol Metab* 1997; **82**: 2215–21.

5 Bates A, Evans A, Jones P, Clayton R. Assessment of GH status in adults with GH deficiency using serum growth hormone, serum insulin-like growth factor-I and urinary growth hormone. *Clin Endocrinol* 1995; **42**: 425–30.

6 Gale E, Bennett T, MacDonald I, Holst J, Matthews J. The physiological effects of insulin-induced hypoglycaemia in man: responses at differing levels of blood glucose. *Clin Sci* 1983; **65**: 263–71.

** 7 Thorner M, Bengtsson BÅ, Ho K *et al.* The diagnosis of growth hormone deficiency (GHD) in adults. *J Clin Endocrinol Metab* 1995; **80**: 3097–8.

** 8 Hoffman DM, Aj OS, Baxter RC, Ho KK. Diagnosis of growth-hormone deficiency in adults. *Lancet* 1994; **343**: 1064–8.

9 Shah A, Stanhope R, Matthew D. Hazards of pharmacological tests of growth hormone secretion in childhood. *Br Med J* 1992; **304**: 173–4.

10 Jones S, Trainer P, Perry L *et al.* An audit of the insulin tolerance test in adult subjects in an acute investigation unit over one year. *Clin Endocrinol* 1994; **41**: 123–8.

11 Cain JP, Williams GH, Dluhy RG. Glucagon-initiated human growth hormone release: a comparative study. *Can Med Assoc J* 1972; **107**: 617–22.

12 Hoeck HC, Vestergaard P, Jakobsen PE, Laurberg P. Test of growth hormone secretion in adults: poor reproducibility of the insulin tolerance test. *Eur J Endocrinol* 1995; **133**: 305–12.

* 13 Lin T, Tucci JR. Provocative tests of growth-hormone release: a comparison of results with seven stimuli. *Ann Intern Med* 1974; **80**: 464–9.

14 Eddy RL, Gilliland PF, Ibarra JD Jr, McMurry JF Jr, Thompson JQ. Human growth hormone release: comparison of provocative test procedures. *Am J Med* 1974; **56**: 179–85.

15 Sutton J, Lazarus L. Growth hormone in exercise: comparison of physiological and pharmacological stimuli. *J Appl Physiol* 1976; **41**: 523–7.

16 Rahim A, Toogood AA, Shalet SM. The assessment of growth hormone status in normal young adult males using a variety of provocative agents. *Clin Endocrinol* 1996; **45**: 557–62.

17 Health Services Human Growth Hormone Committee. Comparison of the intravenous insulin and oral clonidine tolerance tests for growth hormone secretion. *Arch Dis Child* 1981; **56**: 852–4.

18 Merimee TJ, Rabinowtitz D, Fineberg SE. Arginine-initiated release of human growth hormone: factors modifying the response in normal man. *N Engl J Med* 1969; **280**: 1434–8.

19 Vahl N, Jorgensen JO, Jurik AG, Christiansen JS. Abdominal adiposity and physical fitness are major determinants of the age associated decline in stimulated GH secretion in healthy adults. *J Clin Endocrinol Metab* 1996; **81**: 2209–15.

20 Reutens AT, Hoffman DM, Leung KC, Ho KK. Evaluation and application of a highly sensitive assay for serum growth hormone (GH) in the study of adult GH deficiency. *J Clin Endocrinol Metab* 1995; **80**: 480–5.

21 Reutens AT, Veldhuis JD, Hoffman DM, Leung KC, Ho KKY. A highly sensitive growth hormone (GH) enzyme-linked immunosorbent assay uncovers increased contribution of a tonic mode of GH secretion in adults with organic GH deficiency. *J Clin Endocrinol Metab* 1996; **81**: 1591–7.

*22 Juul A, Bang P, Hertel NT *et al.* Serum insulin-like growth factor-I in 1030 healthy children, adolescents, and adults: relation to age, sex, stage of puberty, testicular size, and body mass index. *J Clin Endocrinol Metab* 1994; **78**: 744–52.

23 Blum WF, Ranke MB, Kietzmann K *et al.* A specific radioimmunoassay for the growth hormone-dependent somatomedin-binding protein: its use for diagnosis of GH deficiency. *J Clin Endocrinol Metab* 1990; **70**: 1292–8.

24 De Boer H, Blok G, Van der Veen E. Clinical aspects of growth hormone deficiency in adults. *Endocrinol Rev* 1995; **16: 63–86.

25 Attanasio AF, Lamberts SW, Matranga AM *et al.* Adult growth hormone (GH)-deficient patients demonstrate heterogeneity between childhood onset and adult onset before and during human GH treatment (Adult Growth Hormone Deficiency Study Group). *J Clin Endocrinol Metab* 1997; **82: 82–8.

26 Janssen YJ, Frolich M, Roelfsema F. A low starting dose of genotropin in growth hormone-deficient adults. *J Clin Endocrinol Metab* 1997; **82**: 129–35.

27 Beshyah SA, Anyaoku V, Newton P, Johnston DG. Metabolic abnormalities in growth hormone deficient adults, 1: serum insulin-like growth factor-I. *Endocrinol Metab* 1994; **1**: 167–72.

28 Rasmussen MH, Hvidberg A, Juul A *et al.* Massive weight loss restores 24 hour growth hormone release profiles and serum insulin-like growth factor-I levels in obese subjects. *J Clin Endocrinol Metab* 1995; **80**: 1407–15.

29 Williams T, Berelowitz M, Joffe SN *et al.* Impaired growth hormone response to growth hormone-releasing factor in obesity: a pituitary defect reversed with weight reduction. *N Engl J Med* 1984; **311**: 1403–7.

*30 Veldhuis JD, Iranmanesh A, Ho KK *et al.* Dual defects in pulsatile growth hormone secretion and clearance subserve the hyposomatotropism of obesity in man. *J Clin Endocrinol Metab* 1991; **72**: 51–9.

*31 Iranmanesh A, Lizarralde G, Veldhuis JD. Age and relative adiposity are specific negative determinants of the frequency and amplitude of growth hormone (GH) secretory bursts and the half-life of endogenous GH in healthy men. *J Clin Endocrinol Metab* 1991; **73**: 1081–8.

32 Weltman A, Weltman JY, Hartman ML *et al.* Relationship between age, percentage body fat, fitness, and 24-hour growth hormone release in healthy young adults: effects of gender. *J Clin Endocrinol Metab* 1994; **78**: 543–8.

33 Poehlman ET, Copeland KC. Influence of physical activity on insulin-like growth factor-I in healthy younger and older men. *J Clin Endocrinol Metab* 1990; **71**: 1468–73.

34 Rudman D, Kutner MH, Rogers CM *et al.* Impaired growth hormone secretion in the adult population. *J Clin Invest* 1981; **67**: 1361–9.

35 Gama R, Teele JD, Marks V. The effect of synthetic very low calorie diets on the GH–IGF-I axis in obese subjects. *Clin Chim Acta* 1990; **188**: 31–8.

36 Loche S, Pintor C, Cappa M *et al.* Pyridostigmine counteracts the blunted growth hormone response to growth hormone-releasing hormone in obese children. *Acta Endocrinol* 1989; **120**: 624–8.

37 Landin-Wilhelmsen K, Wilhelmsen L, Lappas G *et al.* Serum insulin-like growth factor-I in a random population sample of men and women: relation to age, sex, smoking habits, coffee consumption and physical activity, blood pressure, and concentrations of plasma lipids, fibrinogen, parathyroid hormone and osteocalcin. *Clin Endocrinol* 1994; **41**: 351–7.

38 Blackard W, Hull E, Lopez A. Effect of

lipids on growth hormone secretion in humans. *J Clin Invest* 1971; **50**: 1439–43.

39 Imaki T, Shibasaki T, Shizume K *et al.* The effect of free fatty acids on growth hormone (GH)-releasing hormone-mediated GH secretion in man. *J Clin Endocrinol Metab* 1985; **60**: 290–3.

40 Cordido F, Peino R, Penalva A *et al.* Impaired growth hormone secretion in obese subjects is partially reversed by acipimox-mediated plasma free fatty acid depression. *J Clin Endocrinol Metab* 1996; **81**: 914–18.

41 Lee EJ, Nam SY, Kim KR *et al.* Acipimox potentiates growth hormone (GH) response to GH-releasing hormone with or without pyridostigmine by lowering serum free fatty acid in normal and obese subjects. *J Clin Endocrinol Metab* 1995; **80**: 2495–8.

42 Maccario M, Procopio M, Grottoli S *et al.* Effects of acipimox, an antilipolytic drug, on the growth hormone (GH) response to GH-releasing hormone alone or combined with arginine in obesity. *Metabolism* 1996; **45**: 342–6.

43 Alvarez C, Mallo F, Burguera B *et al.* Evidence for a direct pituitary inhibition by free fatty acids of in-vivo growth hormone responses to growth hormone-releasing hormone in the rat. *Neuroendocrinology* 1991; **53**: 185–9.

44 Ho K, Rasmussen M, Hoffman D, Hilstead J, Leung K. How growth hormone deficient are the obese [abstract]? Paper presented at the 78th Annual Meeting of the Endocrine Society, San Francisco, 1996: 115.

45 Finkelstein J, Roffwarg H, Boyar R *et al.* Age-related change in the twenty-four-hour spontaneous secretion of growth hormone. *J Clin Endocrinol Metab* 1972; **35**: 665–70.

*46 Zadik Z, Chalew S, McCarter RJ, Meistas M, Kowarski A. The influence of age on the 24-hour integrated concentration of growth hormone in normal individuals. *J Clin Endocrinol Metab* 1985; **60**: 513–16.

47 Vermeulen A. Nyctohemeral growth hormone profiles in young and aged men: correlation with somatomedin-C levels. *J Clin Endocrinol Metab* 1987; **64**: 884–8.

*48 Toogood AA, Pa ON, Shalet SM. Beyond the somatopause: growth hormone deficiency in adults over the age of 60 years. *J Clin Endocrinol Metab* 1996; **81**: 460–5.

*49 Veldhuis JD, Liem AY, South S *et al.* Differential impact of age, sex steroid hormones, and obesity on basal versus pulsatile growth hormone secretion in men as assessed in an ultrasensitive chemiluminescence assay. *J Clin Endocrinol Metab* 1995; **80**: 3209–22.

50 Rudman D. Growth hormone, body composition, and aging. *J Am Geriatr Soc* 1985; **33: 800–7.

51 Forbes G. The adult decline in lean body mass. *Hum Biol* 1976; **48**: 161–73.

52 Mindermann T, Wilson CB. Age-related and gender-related occurrence of pituitary adenomas. *Clin Endocrinol* 1994; **41**: 359–64.

53 Toogood A, O'Neill P, Shalet S. The diagnosis of severe growth hormone deficiency in elderly patients with hypothalamic–pituitary disease. *Clin Endocrinol* 1998; **48**: 569–76.

*54 De Boer H, Blok G, Van Lingen A *et al.* Consequences of childhood-onset growth hormone deficiency on adult bone mass. *J Bone Miner Res* 1994; **9**: 1319–26.

55 Clayton P, Price D, Shalet S. Growth hormone state after completion of treatment with growth hormone. *Arch Dis Child* 1987; **62**: 222–6.

56 Wacharasindhu S, Cotterill AM, Camacho Hubner C, Besser GM, Savage MO. Normal growth hormone secretion in growth hormone insufficient children retested after completion of linear growth. *Clin Endocrinol* 1996; **45**: 553–6.

57 Longobardi S, Merola B, Pivonello R *et al.* Re-evaluation of growth hormone secretion in 69 adults diagnosed as GH deficient patients during childhood. *J Clin Endocrinol Metab* 1996; **81**: 1244–7.

*58 Nicolson A, Toogood AA, Rahim A, Shalet SM. The prevalence of severe growth hormone deficiency in adults who received growth hormone replacement in childhood. *Clin Endocrinol* 1996; **44**: 311–16.

59 Tauber M, Moulin P, Pienkowski C,

Jouret B, Rochiccioli P. Growth hormone (GH) retesting and auxological data in 131 GH-deficient patients after completion of treatment. *J Clin Endocrinol Metab* 1997; **82**: 352–6.

60 Juul A, Kastrup K, Pedersen S, Skakkebaek N. Growth hormone (GH) provocative retesting of 108 young adults with childhood-onset GH deficiency and the diagnostic value of insulin-like growth factor I (IGF-I) and IGF-I-binding protein-3. *J Clin Endocrinol Metab* 1997; **82**: 1195–201.

*61 Ghigo E, Bellone J, Aimaretti G *et al.* Reliability of provocative tests to assess growth hormone secretory status: study in 472 normally growing children. *J Clin Endocrinol Metab* 1996; **81**: 3323–7.

62 Shalet S. Radiation and pituitary dysfunction. *N Engl J Med* 1993; **328: 131–3.

63 Anderson M, Hansen T, Stoving R *et al.* The pyridostigmine–growth hormone-releasing hormone test in adults: the reference interval and a comparison with the insulin tolerance test. *Endocrinol Metab* 1996; **3**: 197–206.

64 Guevara-Aguirre J, Rosenbloom AL, Fieldre PJ, Diamond FB, Rosenfeld RG. Growth hormone receptor deficiency in Ecuador: clinical and biochemical phenotype in two populations. *J Clin Endocrinol Metab* 1993; **76**: 417–23.

65 Ghigo E, Aimaratti G, Gianotti L *et al.* New approach to the diagnosis of growth hormone deficiency in adults. *Eur J Endocrinol* 1996; **134: 352–6.

9: What is the impact of GH deficiency on bone density in childhood-onset and adult-onset GH deficiency, and how effective is GH replacement?

Mark Vandeweghe and Jean-Marc Kaufman

Introduction

A great deal of information has been gathered in recent years concerning the effects of growth hormone on bone and mineral metabolism [1,2]. Stimulation of longitudinal bone growth results from local effects of growth hormone on cartilage, based on a dual mechanism of action: direct growth hormone stimulation of the differentiation of the growth-plate precursor cells, and indirect stimulation of cell proliferation through insulin-like growth factor-I (IGF-I). On the other hand, it is now evident that growth hormone also stimulates the proliferation of osteoblasts, while a recent study shows that IGF-I exerts direct anabolic effects on bone-forming cells *in vivo* [3]. Less is known about the mechanisms involved in the effect of growth hormone on promoting bone resorption, although a study by Nishiyama *et al.* [4] suggests that growth hormone stimulates osteoclastic bone resorption through both direct and indirect actions on osteoclast differentiation and activation.

Apart from a direct action on the skeleton, the effects of growth hormone on bone and mineral metabolism also involve extraskeletal sites of action [5–7]. Of special interest in this respect is the fact that the anabolic effects of growth hormone in increasing muscle mass and strength may represent an important indirect influence on bone physiology.

Thus, it seems clear that growth hormone can affect bone metabolism through both direct and indirect mechanisms of action. It is well established that growth hormone plays an important role in the build-up of bone mass in childhood and adolescence, primarily through the regulation of longitudinal bone growth. Much less is known about the importance of growth hormone in adult bone physiology. However, accumulating evidence from observations during administration of growth hormone to individuals with and without growth hormone deficiency indicates that growth hormone has a significant impact on the regulation of bone turnover, and may therefore also have a role in the preservation of adult bone mass.

In the early 1990s, pioneering work was carried out in relation to bone mineral status in childhood-onset growth hormone-deficient adults, particularly by the groups from Ghent (Kaufman and Vandeweghe), Amsterdam (De Boer and Van der Veen) and Manchester (Shalet and collaborators). The first positive long-term results (18 months or more) of growth hormone treatment on bone mineral mass in a homogeneous population with childhood-onset deficiency were published by Vandeweghe *et al.* [8].

The data concerning the role of growth hormone deficiency *per se* on bone mineral status in adult-onset hypopituitary patients show more conflicting results, which are due in the first instance to confounding factors related to the associated pathology and other hormone-substitution treatments. On the other hand, a positive effect of long-term growth hormone treatment on bone mineral mass in this population appears to be convincingly documented [9–11].

Altered states of activity of the somatotrophic hormonal axis are accompanied by changes in bone size, body composition and bone metabolism that may complicate the interpretation of measurements of bone mineral density and affect the accuracy of measurements in growth hormone deficiency and during growth hormone treatment. These potential sources of methodological bias deserve close scrutiny.

These various controversial topics are examined in this chapter under the following five headings:

- Bone mineral mass in growth hormone deficiency of childhood onset: the role of growth hormone in the build-up of bone mass.
- Should growth hormone be continued for some additional years after the completion of linear growth and attainment of final height, in order to optimize peak bone mass?
- Bone mineral mass in growth hormone deficiency of adult onset: does growth hormone play a role in the maintenance of bone mass?
- The effect of growth hormone treatment on bone mineral mass in adult growth hormone deficiency of childhood and adult onset.
- Factors affecting the measurements of bone mineral density and their interpretation in altered growth hormone status: are possible methodological pitfalls a matter for concern?

Bone mineral mass in growth hormone deficiency of childhood onset: the role of growth hormone in the build-up of bone mass

There is convincing evidence that adults with childhood-onset growth hormone deficiency present with a substantial deficit in bone mass. Kaufman *et al.* [12] measured bone mineral content (BMC) and bone mineral density

(BMD) at the forearm and at the lumbar spine, using single-photon and dual-photon absorptiometry, respectively, in a population of 30 young adult males with growth hormone deficiency of childhood onset, and compared the findings with those for a control group matched for age and height. This cross-sectional study revealed a markedly lower bone mass in the patients as compared to the controls, with the deficit at the forearm being proportionally larger than that at the lumbar spine (Fig. 9.1). The observation

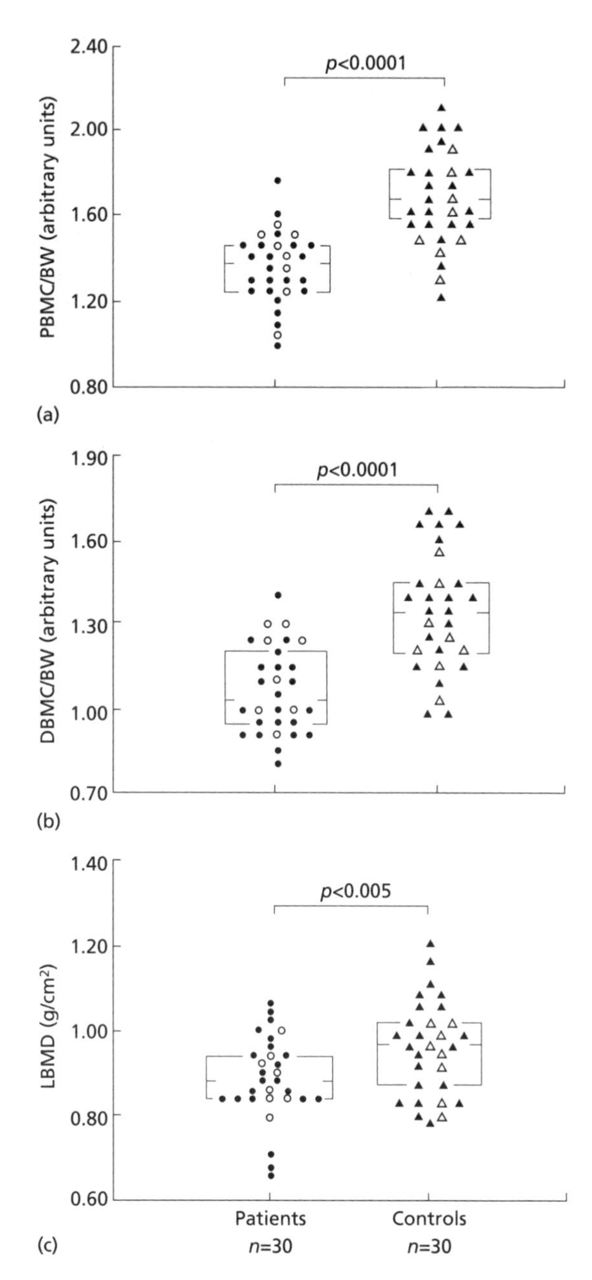

Fig. 9.1 Individual values for bone mineral measurements in 30 men with isolated GH deficiency (○) or multiple pituitary deficiencies (●) of childhood onset, and in 30 age-matched and height-matched control individuals (the open triangles (△) are the controls for patients with isolated GH deficiency; the filled triangles (▲) are the controls for patients with multiple pituitary deficiencies). The figure shows the results for forearm measurements using single-photon absorptiometry of bone mineral content (BMC) normalized for bone width at (a) a proximal site (> 95% cortical bone), (b) a more distal site (30–40% trabecular bone), and (c) measurements of the lumbar spine (L2–L4) bone mineral density (BMD) using dual-photon absorptiometry. The horizontal lines represent the 25th, 50th and 75th percentiles. Reproduced with permission from Kaufman *et al.* [12].

that patients with multiple pituitary deficiencies and those with isolated growth hormone deficiency were both osteopenic in comparison with their respective controls suggests that growth hormone deficiency *per se* is responsible for at least part of the observed bone mass deficit.

These findings were confirmed by De Boer *et al.* [13], who carried out a similar cross-sectional study in 70 men with childhood-onset growth hormone deficiency, using dual energy X-ray absorptiometry (DEXA). This study showed a significant bone mass deficit at the lumbar spine and the proximal femur in patients compared with age-matched controls, and a substantial deficit was still demonstrable after correction of the data for differences in bone size between the patients and the controls, who were not matched for height. In agreement with the observations by Kaufman *et al.* [12], the extent of the bone mass deficit was similar in the patients with isolated growth hormone deficiency and in those with multiple pituitary deficiencies.

A markedly reduced bone mass in comparison with age-matched controls was also observed by O'Halloran and colleagues [14] in 10 males with isolated growth hormone deficiency present since childhood or adolescence. More recently, Saggese *et al.* [15] measured lumbar BMD using DEXA in 11 patients with isolated growth hormone deficiency at the time when they reached their final height. Compared with a control group of 17 individuals with familial short stature who were approximately the same age, a significantly reduced Z-score was observed in the growth hormone-deficient patients. Other authors [16–19] have confirmed these data in partly mixed male/female populations. The view that childhood-onset growth hormone deficiency *per se* results in a low adult bone mass is further supported by the finding by De Boer *et al.* [13] of a positive correlation between plasma IGF-I levels and volumetric BMD in their patients, and by observations of osteopenia and an increased incidence of osteoporotic fractures in adult Laron dwarfs with a functional defect in the growth hormone receptor and consequent growth hormone insensitivity [20].

Several lines of evidence appear to suggest that the low bone mass in adults with childhood-onset growth hormone deficiency primarily results from a deficient build-up of peak bone mass, rather than from subsequent bone loss. Indeed, some data indicate that bone mass is already low in growth hormone-deficient children before and during growth hormone replacement therapy [21–23].

In addition, Kaufman *et al.* [12] carried out longitudinal measurements of forearm and lumbar spine BMC in a cohort of their adult patients with childhood-onset growth hormone deficiency, and were not able to detect any significant bone loss.

Table 9.1 BMC follow-up after cessation of growth hormone treatment.

First author (ref.)	Patients (*n*)	Follow-up	Comments
Vandeweghe [24] (unpublished data)	*n* = 10 Childhood onset	First 12 months	Slight increase in lumbar BMC (DPA) no change in forearm BMC (SPA)
		12–42 months	No changes at lumbar spine or forearm (DEXA)
De Boer (personal communication, 1997)	*n* = 33 Childhood onset	24 months	No change at lumbar spine or femoral neck (DEXA)
Holmes [25]	*n* = 10 Childhood onset	12 months	Increase in forearm BMC (SPA); no change in lumbar BMC (QCT)
Rahim [26]	*n* = 8 Adulthood onset	24 months	Increase in trochanter BMD (DEXA); no changes at other sites

BMC: bone mineral content; DEXA: dual-energy X-ray absorptiometry; DPA: dual-photon absorptiometry; QCT: quantitative computed tomography.

The data available concerning the course of BMC after the cessation of growth hormone treatment in growth hormone-deficient adults point in the same direction (Table 9.1) [24–26]. Finally, the lack of correlation between adult bone mass and either the total duration of growth hormone deficiency or the time elapsed since cessation of growth hormone replacement therapy in patients with childhood-onset growth hormone deficiency [12,13] is also in agreement with the view that the bone mass deficit is acquired before adulthood.

Should growth hormone be continued for some additional years after completion of linear growth and attainment of final height, in order to optimize peak bone mass?

Although the osteopenia observed in young adults with childhood-onset growth hormone deficiency may result from an inadequate dosage of growth hormone treatment, or from the fact that the treatment may have been initiated too late, there is a real possibility that growth hormone treatment is interrupted too early, i.e. at the completion of linear growth, when peak bone mass has not yet been fully acquired.

Until recently, opinion was divided as to whether peak bone mass is achieved in the second decade [27,28] or whether it is not achieved until the third or fourth decade of life [29]. Additional recent data [30,31], however, seem to indicate that most of the bone mass at the femoral neck and lumbar spine is accumulated by late adolescence. On the other hand, it is also clear that in normal individuals the cessation of longitudinal bone growth takes place several years earlier than the cessation of the rapid accumulation of bone mass at various skeletal sites. Probably at least some 20% of peak bone mass has still to be acquired after final height has been reached [32].

In view of the well-known beneficial effect of growth hormone therapy on the increment of bone mass in growth hormone-deficient children [15,33], it seems plausible that growth hormone treatment has an important role to play in the additional bone mass accumulation still to be achieved after the completion of linear growth. As suggested by Saggese *et al.* [15], growth hormone treatment should perhaps be continued until the attainment of peak bone mass, irrespective of the height achieved. However, this has not yet been convincingly documented.

In order to evaluate the effect of continued growth hormone treatment after cessation of linear growth on the build-up of BMC toward final peak bone mass, an ongoing study has been initiated in our department in which, at the attainment of (near) final height, growth hormone-treated growth hormone-deficient patients are randomly assigned to either continuation of daily growth hormone injections for two more years, or discontinuation of growth hormone administration.

Bone mineral mass in growth hormone deficiency of adult onset: does growth hormone play a role in the maintenance of bone mass?

In several, but not all, recent reports, it has been suggested that bone mass may be reduced in patients with adult-onset growth hormone deficiency, and that this reduction is at least partly caused by growth hormone deficiency *per se*. In our opinion, however, the data do not permit straightforward conclusions.

Johansson *et al.* [34] described a significantly reduced BMD measured by DEXA at the lumbar spine, but not for the total skeleton, in 17 growth hormone-deficient adult men, as compared to age-matched controls. In the same study, a similar trend was observed in growth hormone-deficient women, but the comparison was only made with historical reference data obtained elsewhere. Apart from the non-uniformity of the results for spine and total skeleton BMD, the interpretation of the study is further complicated by the fact that most patients presented with multiple pituitary deficiencies, suggesting that at least some patients may have been temporarily hypogonadal before the initiation of sex steroid replacement therapy. Moreover, three of the patients had been treated for Cushing's disease, which is well known to induce osteoporosis.

The data published by Hyer *et al.* [35] on BMD measured by DEXA at the lumbar spine and the proximal femur in 50 growth hormone-deficient adults with adult-onset growth hormone deficiency are inconclusive, as the BMD was not compared with adequately matched controls. Measuring BMC at the third lumbar vertebra using dual-photon absorptiometry in 95 patients with adult-onset growth hormone deficiency, Rosen *et al.* [36] described a reduced BMC in both men and women, which was significant only in the younger age group (under 55 years). Again, several confounding factors (the majority of the patients presented with multiple deficiencies, including 19 with prolactinoma) make correct interpretation of the data very difficult.

In a small group of 14 growth hormone-deficient adults treated for pituitary adenoma, Bing-You *et al.* [37] reported osteopenia measured by DEXA, in comparison with controls matched for age, sex and weight. Since the BMD parameters appeared to be positively correlated with indices of growth hormone secretion, it was suggested that at least part of the bone mass deficit may have resulted from growth hormone deficiency. The interpretation of the findings from a study by Beshyah *et al.* [38], showing a low lumbar spine and total skeleton BMD in 64 growth hormone-deficient adults, is hampered by the inclusion of some patients with childhood-onset growth hormone deficiency and of several patients with a history of pituitary pathology that may have adversely affected bone mass (Cushing's disease and prolactinoma); moreover, the majority of the patients were receiving substitution therapy for other hormone deficiencies.

Perhaps most informative is the study by Holmes *et al.* [39], who measured BMD in 26 patients with adult-onset growth hormone deficiency. The authors observed a significantly reduced BMD in the forearm, lumbar spine and total skeleton, as compared to reference values appropriate for the local population. Although (as in most other studies) a majority of the patients presented with multiple pituitary deficiencies, the authors did carry out an analysis of several subgroups, which suggested the existence of osteopenia independent of associated gonadotrophin deficiency or glucocorticoid substitution. In a more recent study, the same group [40] reported that adults over the age of 60 with adult-onset growth hormone deficiency, who were receiving appropriate replacement therapy for other anterior pituitary hormone deficits but not growth hormone, are not osteopenic when compared with healthy subjects of similar age, sex and BMI. Similarly negative results were reported by Degerblad *et al.* [41]; in men with adult-onset growth hormone deficiency, BMD did not differ at any site in comparison with controls, whereas the slightly reduced BMD in women was probably related to suboptimal oestrogen substitution. Finally, normal findings for spine and mid-radius BMD in subjects with acquired multiple pituitary deficiencies, including growth hormone, were also recently reported by Kaji *et al.* [42].

In conclusion, whereas some of the data suggest that adult-onset growth hormone deficiency may result in osteopenia, none of the available studies allow definitive conclusions on this issue, due to methodological problems in some reports and confounding factors related to multiple pituitary deficiencies in nearly all of them [43] (Table 9.2). Most importantly, so far as we are aware there have been no reports of longitudinal studies showing the occurrence of accelerated bone loss in growth hormone-deficient adults. Clearly, only this type of longitudinal study will be able to clarify the effect of growth hormone deficiency on the maintenance of adult bone mass definitively [44].

The effect of growth hormone treatment on bone mineral mass in adult growth hormone deficiency of childhood and adult onset

The observation that adults with childhood-onset growth hormone deficiency are osteopenic, and the suggestion that adult-onset growth hormone deficiency may also result in osteopenia, have prompted the initiation of a large number of studies, reporting on the potential beneficial effects of growth hormone replacement therapy on bone mass in adult patients with growth hormone deficiency.

The first positive long-term results (18 months or more) of growth hormone treatment on bone mineral mass in a relatively large, homogeneously childhood-onset population were published by Vandeweghe *et al.* [8]. In all, we identified reports of 17 studies published since 1992 that present original data on the effects of at least 12 months of uninterrupted growth hormone treatment on bone mineral status (Table 9.3).

Table 9.2 Osteopenia due to growth hormone deficiency not proven in the adult-onset growth hormone-deficient population. Why?

Doubts about value of control population
Most patients have associated sex steroid deficiency (temporarily unsubstituted)
Doubts about correct substitution with other hormones (thyroid, hydrocortisone)
Influence of other associated pathology (prolactinoma, Cushing syndrome)
No data available on isolated growth hormone deficiency of 'adult-onset'
Bone loss during a longitudinal study has never been demonstrated in a growth hormone-deficient adult population

The findings concerning the effects on BMC and/or BMD of a six-month course of growth hormone treatment have been fairly consistent, showing an unchanged bone mineral status at the end of the treatment period [8,11,41,45–51]. The limited decrease of BMD observed during the initial phase of growth hormone treatment in some studies might be explained by an expansion of the so-called 'remodelling space', resulting from the increased bone turnover [8]. That this is indeed the case was documented recently by Bravenboer *et al.* [52], who investigated the effects of growth hormone on bone structure and turnover by histomorphometry in growth hormone-deficient adults. O'Halloran *et al.* [14] are the only group to have reported an increase in (vertebral) BMD, as estimated by single-energy quantitative computed tomography (QCT), an increment that may possibly have resulted in part from changes in the fat content of the bone marrow during treatment.

Seventeen reports present data on the effect of a more prolonged treatment course of 12–48 months. In all except two [9,53], the long-term data were obtained in an open, uncontrolled manner. Twelve of these studies reported an increase in bone mass [8–11,14,19,26,49–51, 54,55]. The bone mineral status remained unchanged in the five other studies [17,41,48, 53,56]. The results of the study by Vandeweghe *et al.* [8] in childhood-onset growth hormone deficiency are shown in Figure 9.2. Some of the data published by Kann *et al.* [11] concerning adult-onset growth hormone deficiency are provided in Figure 9.3.

More than half of the studies suggesting a beneficial long-term effect of growth hormone treatment on bone mineral status specifically refer to patients with childhood-onset growth hormone deficiency, and include predominantly young growth hormone-deficient adults. Although there are indications that the attainment of peak bone mass in young adults may be achieved earlier than previously suggested [27,28,30,31], it cannot be excluded with certainty that at least part of the increase in bone mass observed during growth hormone treatment in some of these studies represents spontaneous bone accretion towards peak bone mass. Most of the data, however, argue in favour of a consistent and undeniable effect of growth hormone *per se* on bone mineral mass.

Even though the only two placebo-controlled studies have produced divergent results, the findings in six out of a total of seven long-term studies in patients with adult-onset growth hormone deficiency strongly suggest a net increase in bone mass obtained with growth hormone treatment, especially in skeletal sites rich in trabecular bone.

In view of the incompletely established adverse effect of growth hormone deficiency on bone mineral mass in growth hormone deficiency of adult onset, these findings may appear rather confusing. In this respect, it

Table 9.3 Findings on the effects of recombinant growth hormone (GH) administration on bone density (BMD)[*] in GH-deficient adults.[†]

First author (ref.)	Dosage, patients, BMD assessment technique[‡]	Type of study, findings (number of patients)[§]	
		First 6 months' treatment	Prolonged treatment
Degerblad [54]	0.28 to 0.7 IU/kg/week Mainly childhood onset; multiple deficiencies; mainly women; SPA	Open ($n = 9$) Unchanged BMD at forearm	Open; non-significant increase in forearm BMD at 12 months ($n = 6$); increase at 18 months ($n = 5$)
O'Halloran [14]	0.125 → 0.250 IU/kg/week Childhood onset; isolated deficiency; mainly men; SPA, single-energy QCT, dual-energy QCT	Placebo-controlled ($n = 11$) Unchanged BMD at forearm; increased spine BMD by single-energy QCT; unchanged spine BMD by dual-energy QCT	Open; 12 months; increase at forearm ($n = 12$); increase at spine by single-energy QCT ($n = 11$) and by dual-energy QCT ($n = 11$); remark: 6 patients < 20 y of age!
Vandeweghe [8]	0.25 IU/kg/week Childhood onset; isolated/multiple deficiencies; men; SPA, DPA	Placebo-controlled ($n = 10$) Unchanged BMD at forearm and spine (non-significant decrease at 3 months)	Open; 18 months ($n = 15$); significant increase in spine and forearm BMD
Juul [55]	14 IU/m²/week Childhood onset; isolated/multiple deficiencies; men/women; SPA	–	Open; ($n = 13$); increase in forearm BMD between 7th and 14th month of treatment
Degerblad [41]	0.125 to 0.25 IU/kg/week Mainly adult onset; isolated/multiple deficiencies; men/women; DEXA	Placebo-controlled ($n = 68$) Decrease in all BMD parameters	Open; 12 months ($n = 64$) Unchanged at all sites, except for decrease in total BMD
Holmes [48]	0.125 to 0.25 IU/kg/week Adult onset; mainly multiple deficiencies; men/women; SPA, DPA and QCT	Placebo-controlled ($n = 22$) Decrease in forearm, lumbar spine and femoral neck BMD	Open; 12 months ($n = 13$) Decrease in most BMD parameters vs. baseline
Beshyah [56]	0.14 to 0.35 IU/kg/week Mainly adult onset; multiple deficiencies; men/women; DEXA	Placebo-controlled ($n = 12$) Unchanged BMD at spine and whole body	Open; 12 to 18 months ($n = 11$) Unchanged BMD at spine and whole body
Balducci [17]	0.6 IU/kg/week Childhood onset; multiple deficiencies; men/women; DPA	–	Open No change spine BMD at 12 months ($n = 13$) and 24 months ($n = 10$)
Amato [19]	70 μg/kg/week Childhood onset; multiple deficiencies; mainly males; DPA (forearm)	–	Open ($n = 9$) 12 months; increase in BMD

Baum [9]	10 μg/kg/day Adult onset; mainly multiple deficiencies; men; DEXA	–	Placebo controlled; 18 months ($n = 16$ on GH) Increased BMD at sites of trabecular bone (lumbar spine, femoral neck); not at site of cortical bone (proximal radius)
Hansen [53]	2 IU/m²/day Adult onset; mainly multiple deficiencies; men/women; DEXA	–	Placebo-controlled; 12 months ($n = 14$ on GH) Decrease in BMD of whole body and radius
Johansson [10]	0.1 to 0.25 IU/kg/week Adult onset; mainly multiple deficient; men/women; DEXA	–	Open; 24 months ($n = 44$) Increase in all BMD parameters: especially in those patients with the lowest pretreatment values
Finkenstedt [49]	2.4 IU daily Adult onset; multiple deficiencies; men/women; DEXA	Placebo-controlled ($n = 20$) No change in BMD	Open; 12 months ($n = 20$) Increase in lumbar spine and proximal femur, especially in those patients with the lowest pretreatment values
Ter Maaten [50]	1.3 IU/m²/day Childhood onset; isolated/multiple deficiencies; men; DEXA	Decline at lumbar spine and for total mineral content at 6 months	Open; 48 months ($n = 27$) Net increase for all BMD parameters
Kann [11]	0.25 IU/kg/week Adult-onset; mainly multiple deficiencies; men/women; SPA/DPA	Placebo-controlled ($n = 20$) Decrease at all sites	Open 48 months ($n = 20$) Net increase of BMD at lumbar spine and forearm
Rahim [26]	0.125 → 0.250 IU/kg/week Adult onset; mainly multiple deficiencies; men/women; DEXA	–	Open 36 months ($n = 7$) Significant increase of spine and trochanter BMD; no changes at other sites
Kotzmann [51]	0.125 → 0.250 IU/kg/week Adult onset; multiple deficiencies; men/women DEXA	Placebo-controlled ($n = 19$) Unchanged at lumbar spine and femoral neck	Open; 12 months Increase at lumbar spine and femoral neck (only significant for lumbar spine)

*Bone density (BMD) is used here in a broad sense to include the best estimates of density according to the measurement technique (i.e. bone mineral content by cm^3, by cm^2 or by cm of bone width).

†Only those studies with uninterrupted GH treatment for at least 12 months are included.

‡Techniues for assessment of bone density (BMD). SPA: single-photon absorptiometry; DPA: dual-photon absorptiometry; DEXA: dual-energy X-ray absorptiometry; QCT: quantitative computed tomography.

§Number of patients on active treatment.

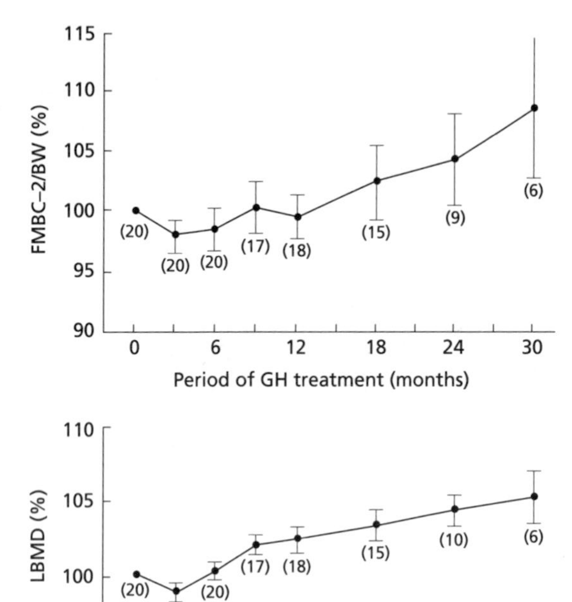

Fig. 9.2 Forearm bone mineral content divided by bone width (FBMC-2/BW) and lumbar spine (L2–L4) bone mineral density (LBMD), expressed as changes from the initial values in adult men with childhood-onset GH deficiency during prolonged treatment with GH (0.25 IU/kg/week). Each individual observation at the time of initiation of GH treatment taken as 100%. The data shown are means ± SEM; the number of observations at each time interval is indicated in parentheses. Reproduced with permission from Vandeweghe *et al.* [8].

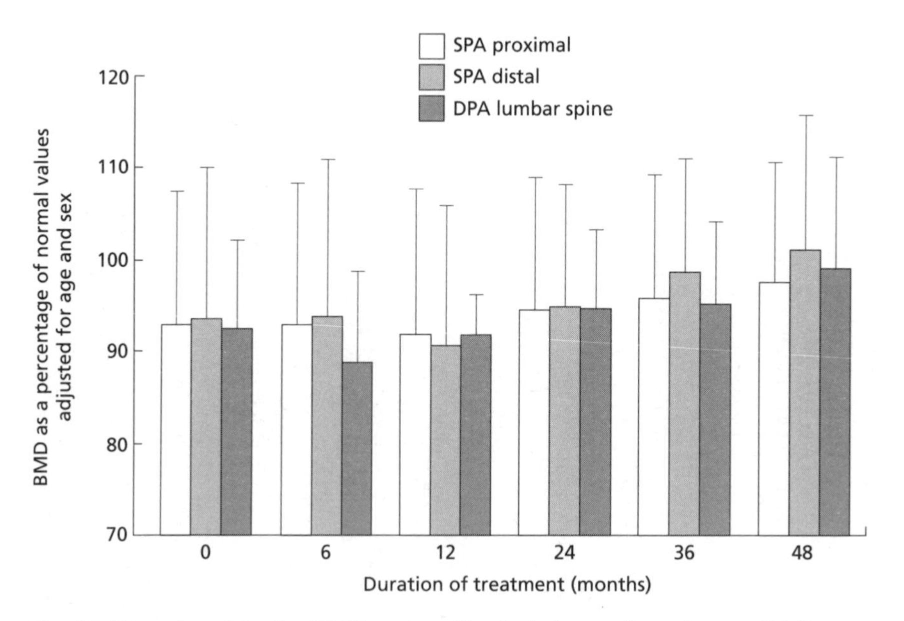

Fig. 9.3 Bone mineral density (BMD) measured by dual-photon absorptiometry (DPA) at the lumbar spine, and by single-photon absorptiometry (SPA) at proximal and distal sites in the nondominant forearm, GH group. The results are shown as mean + 1 SD of age-adjusted and sex-adjusted normal values. Reproduced with permission from Kann *et al.* [11].

cannot be excluded that the increase in bone mineral mass obtained with long-term growth hormone therapy in these patients in fact represents a more general pharmacological effect of growth hormone on the skeleton, rather than a specific consequence of providing substitution treatment for a deficiency.

Factors affecting the measurements of bone mineral density and their interpretation in altered growth hormone states: are possible methodological pitfalls a matter for concern?

Single-photon absorptiometry (SPA), dual-photon absorptiometry (DPA), and more recently mainly dual-energy X-ray absorptiometry (DEXA) and quantitative computed tomography (QCT) are the most commonly applied techniques for non-invasive, *in vivo* assessment of bone mineral density (BMD) in clinical investigations and clinical practice. These techniques are generally reliable, with good to excellent precision, although their accuracy is somewhat less impressive. However, it is important to bear in mind the limitations inherent to these measurement techniques.

An initial methodological problem is related to the fact that neither SPA, DPA nor DEXA allow measurement of a true volumetric BMD. These projectional techniques measure an areal BMD (grams per square centimetre) as a substitute for volumetric BMD (grams per cubic centimetre). Consequently, these techniques fail to correct fully for differences in bone size between individuals, or for changes in bone size that may occur in the same patient during longitudinal follow-up, as the areal BMD values are 'normalized' for only two of the three dimensions of the measured bone [57–59]. This is obviously a potential source of bias when comparing BMD in adults with childhood-onset growth hormone deficiency and healthy controls; the short stature and smaller bone size in the growth hormone-deficient population will result in a systematic underestimation of BMD. Using SPA and DPA in a study of bone mineral status in adults with childhood-onset growth hormone deficiency, Kaufman *et al.* [12], addressed this problem by comparing the patients with height-matched controls; a similar approach was followed by Saggese *et al.* [15] in a study with DEXA. De Boer *et al.* [13] took a different approach, correcting the areal BMD obtained using DEXA for an estimation of bone volume derived from the dimensions of the bone projections [60]. As childhood-onset growth hormone deficiency may differentially affect adult bone size in different parts of the skeleton (e.g. long bones vs. vertebrae) [61], this may in turn contribute to the observed heterogeneity of the extent of apparent BMD deficit for different skeletal sites [12,13].

The failure of DEXA and related techniques to fully normalize BMD values for bone size is even more critical when performing studies in children, particularly in children with growth hormone deficiency. Finally, growth hormone treatment in adults may result in limited increases in bone size at some skeletal sites. Observed increases in areal BMD during prolonged growth hormone treatment may therefore to some extent represent changes in bone size, rather than simply changes in true volumetric BMD.

Several methods have been proposed for estimating volumetric BMD from the areal BMD. Some approaches use only the data derived from the projectional anteroposterior DEXA scan, while others incorporate additional data obtained by performing a lateral-view DEXA or from conventional radiography [57–60]. QCT is the only technique that directly measures true volumetric BMD; however, this method involves a much greater exposure to radiation, has lower precision, and has been extensively validated only for measurements of vertebral BMD. An interesting development is the introduction of peripheral QCT (pQCT) with dedicated equipment, allowing QCT measurements at distal peripheral skeletal sites (forearm, tibia).

In summary, the failure of areal BMD to account fully for differences in bone size is a potential source of methodological bias in studies related to growth hormone deficiency, which should and can be addressed properly. It is also important that clinicians should be aware of this problem when interpreting BMD measurements of individual patients with growth hormone deficiency, especially in children. Unfortunately, the tools and expertise that would allow this problem to be corrected for are not usually readily available in everyday clinical practice.

Other possible methodological pitfalls relate to the influence of the soft tissues surrounding the measured bone and the fat content of the bone marrow. The techniques used for *in vivo* measurement of BMD are based on the principle that attenuation of a photon beam passing through bone tissue is proportional to the bone mineral content in the path of the beam. The accuracy of such methods is critically dependent on the ability to correct adequately for photon absorption by the soft tissues surrounding the measured bone—which is made difficult in the projectional techniques by the nonhomogeneity of the soft tissues (lean mass and fat tissue) surrounding the skeleton and the large interindividual variability in body size and in the composition of the soft tissues. Generally speaking, corrections for the influence of soft tissues are achieved more reliably with techniques such as DPA and DEXA that take advantage of differential attenuation by soft tissue and bone tissue of photons with higher and lower energies, respectively.

Suboptimal correction for soft-tissue interference should not be a source of significant methodological bias in studies comparing BMD values between groups of subjects, except in extreme conditions (very large differences in the body mass index between groups). For this purpose, even the single-energy SPA technique can be regarded as reliable, given that this method is applied only to measurements at the appendicular skeleton surrounded by limited amounts of soft tissue, and the fact that a correction is usually applied for soft-tissue interference. However, an absolute prerequisite for the reliable use of projectional techniques is that all the subjects are measured under exactly the same conditions with the same equipment under strict quality control. It is therefore methodologically unsound to compare BMD values in subjects who have an altered state of growth hormone secretion with values from reference databases provided by the manufacturer of the equipment, or even with historical controls previously measured with the same equipment.

Suboptimal correction for the effects of soft tissues may be of more critical importance in longitudinal studies attempting to demonstrate changes in BMD during growth hormone treatment. Growth hormone treatment induces substantial changes in body composition, with a decrease in the fat mass and an increase in the lean mass. Any failure of the technique used to correct for these changes fully will result in a systematic bias (usually overestimation) in the BMD value, which although limited in absolute terms may still be relevant in view of the small amplitude of expected changes in BMD during growth hormone treatment (not more than a few percentage points per year). For longitudinal studies, preference should therefore be given to the more reliable dual-energy techniques such as DEXA. One should in any case bear in mind the possibility of this type of bias. The performance of the equipment used can be verified, e.g. by performing measurements in volunteers with and without addition of soft-tissue equivalents (e.g. animal fat, water bags). It may also be useful to monitor the changes in body composition occurring during the treatment and to compare the timing of the changes for soft tissues and BMD. In most studies that have shown an increase in BMD during treatment with growth hormone (Table 9.3) gains in BMD were usually only observed after a delay of 6–12 months. It is unlikely that in these cases the observed BMD increases would have resulted from a methodological bias related to changes in soft-tissue composition, as the most important changes in soft tissues occur well within the first six months of growth hormone treatment [43].

A third methodological problem is related to the use of QCT for longitudinal follow-up of vertebral BMD during growth hormone treatment. Indeed, BMD values can be substantially affected by changes in the fat content of the bone marrow, with higher fat contents resulting in systematic underestimation of BMD [62–65]. Apparent increases in vertebral BMD on QCT during anabolic treatment may therefore be artefacts resulting from a treatment-induced decrease in the bone marrow fat content. Single-energy

QCT is critically affected by this problem, and this makes the technique unsuitable for longitudinal studies of BMD during growth hormone treatment. Dual-energy QCT techniques have been reported to reduce substantially the errors resulting from changes in bone marrow fat content, but these techniques have a lower precision [63–65]. In the light of this discussion, it is interesting to note that O'Halloran *et al.* [14], in a study of growth hormone therapy in growth hormone-deficient adults, observed a significant apparent increase in the vertebral BMD measured by single-energy QCT during the first six months of treatment, while they found no significant change with dual-energy QCT.

To conclude this discussion, we would also like to draw attention to a more fundamental conceptual issue related to the clinical interpretation of BMD measurements in growth hormone-deficient adults. Low BMD is the best-documented risk factor for osteoporotic fractures in postmenopausal women, and more limited data indicate a similar relationship between BMD and the fracture risk in men with primary osteoporosis. However, low BMD is not the only determinant of fracture risk; other factors, such as bone size, bone geometry, bone quality, trabecular connectivity and extraskeletal risk factors (risk of falls) are also important. In previous sections of this discussion, we have pointed out that adverse effects of growth hormone deficiency on adult bone mineral status have been documented convincingly only in the childhood-onset population, and that the observed BMD deficit results from a deficient build-up of bone mass rather than from bone loss at adult age. It should be emphasized that this situation differs fundamentally from that in osteoporotic postmenopausal women presenting with low bone mass as a consequence of substantial bone loss at adult age. Bone fragility in postmenopausal women may be greatly enhanced by a marked deterioration in the trabecular microarchitecture resulting from increased osteoclastic activity. Therefore, the gradient of increased fracture risk with decreased BMD documented for postmenopausal women (roughly a doubling of the fracture risk for each decrease of BMD by one standard deviation of peak bone mass) should not be simply extrapolated to the markedly different situation in adults with childhood-onset growth hormone deficiency, as is often done. More work is needed to evaluate the clinical implications of low BMD in the latter group of patients, who may benefit from a better-maintained trabecular architecture, on the one hand, but may have the disadvantage of smaller bone dimensions, on the other.

Conclusions

- Childhood-onset growth hormone deficiency results in substantial osteopenia due to a deficient build-up toward peak bone mass.

• After completion of linear growth and attainment of final height in growth hormone-deficient patients of childhood onset, growth hormone treatment should probably be continued for an as yet undefined period in order to obtain a more satisfactory peak bone mass, but this remains to be established in a prospective randomized trial.
• It has, at present, not been demonstrated with certainty that adult-onset growth hormone deficiency results in osteopenia. The failure of the available studies to allow definite conclusions is due to the existence of methodological problems and confounding factors related to multiple pituitary deficiencies and other associated pathologies in nearly all of the populations studied, and to the fact that there are no longitudinal data available to document the occurrence of accelerated bone loss in patients with adult-onset growth hormone deficiency.
• Most of the data on the effects of long-term treatment (12 months or more) with growth hormone in growth hormone-deficient adults support the view that this type of treatment has a significant beneficial effect on bone mineral mass, both in the childhood-onset and adult-onset patient populations.
• Some limitations of the techniques for BMD measurement deserve particular attention in the design and interpretation of studies in growth hormone-deficient patients, possible pitfalls being related to the failure of areal BMD to account fully for differences in bone size, the possibility of suboptimal correction for soft-tissue interference and possible interference caused by bone marrow fat in QCT measurements. These methodological problems can, however, be suitably addressed, as has been done in many, but not all, of the published studies on growth hormone deficiency.

References

** 1 Inzucchi SE, Robbins RJ. Growth hormone and the maintenance of adult bone mineral density. *Clin Endocrinol* 1996; **45**: 665–73.

** 2 Ohlsson C, Bengtsson BA, Isaksson OGP, Andreassen TT, Slootweg MC. Growth hormone and bone. *Endocr Rev* 1998; **19**: 55–79.

* 3 Bianda T, Glatz Y, Bouillon R, Froesch ER, Schmid C. Effects of short-term insulin-like growth factor-I (IgF-I) or growth hormone (GH) treatment on bone metabolism and on production of 1,25-dihydroxycholecalciferol in GH-deficient adults. *J Clin Endocr Metab* 1998; **83**: 81–7.

* 4 Nishiyama K, Sugimoto T, Kaji H *et al.* Stimulatory effect of growth hormone on bone resorption and osteoclast differentiation. *Endocrinology* 1996; **137**: 35–41.

* 5 Bouillon R. Growth hormone and bone. *Horm Res* 1991; **36** (Suppl 1): 49–55.

* 6 Slootweg MC. Growth hormone and bone. *Horm Metab Res* 1993; **25**: 335–43.

* 7 Inzucchi SE, Robbins RJ. Clinical review: effects of growth hormone on human bone biology. *J Clin Endocr Metab* 1994; **79**: 691–4.

** 8 Vandeweghe M, Taelman P, Kaufman JM. Short- and long-term effects of growth hormone treatment on bone turnover and bone mineral content in adult growth hormone-deficient males. *Clin Endocrinol* 1993; **39**: 409–15.

** 9 Baum HBA, Biller BMK, Finkelstein JS *et al.* Effects of physiologic growth hormone therapy on bone density and body composition in patients with adult-onset growth hormone deficiency. *Ann Intern Med* 1996; **125**: 883–90.

*10 Johansson G, Rosen T, Bosaeus I, Sjöström L, Bengtsson BA. Two years of growth hormone (GH) treatment increased bone mineral content and density in hypopituitary patients with adult-onset GH deficiency. *J Clin Endocr Metab* 1996; **81**: 2865–73.

11 Kann P, Piepkorn B, Schehler B *et al.* Effect of long-term treatment with GH on bone metabolism, bone mineral density and bone elasticity in GH-deficient adults. *Clin Endocrinol* 1998; **48: 561–8.

12 Kaufman JM, Taelman P, Vermeulen A, Vandeweghe M. Bone mineral status in growth hormone-deficient males with isolated and multiple pituitary deficiencies of childhood onset. *J Clin Endocrinol Metab* 1992; **74: 118–23.

13 De Boer H, Blok GJ, Van Lingen A *et al.* Consequences of childhood-onset growth hormone deficiency for adult bone mass. *J Bone Miner Res* 1994; **9: 1319–26.

*14 O'Halloran DJ, Tsatsoulis A, Whitehouse RW *et al.* Increased bone density after recombinant human growth hormone (GH) therapy in adults with isolated GH deficiency. *J Clin Endocrinol Metab* 1993; **76**: 1344–8.

15 Saggese G, Baroncelli GI, Bertelloni S, Barsanti S. The effect of long-term growth hormone treatment on bone mineral density in children with GH-deficiency: role of GH in the attainment of peak bone mass. *J Clin Endocrinol Metab* 1996; **81: 3077–83.

16 Nussey SS, Hyer SL, Brada M, Leiper AD. Bone mineralization after treatment of growth hormone deficiency in survivors of childhood malignancy. *Acta Paediatr* 1994; **399** (Suppl): 9–14.

17 Balducci R, Toscano V, Pasquino AM *et al.* Bone turnover and bone mineral density in young adult patients with panhypopituitarism before and after long-term growth hormone therapy. *Eur J Endocrinol* 1995; **132**: 42–6.

18 Sartorio A, Ortolani S, Conti A *et al.* Effects of recombinant growth hormone (GH) treatment on bone mineral density and body composition in adults with childhood onset growth hormone deficiency. *J Endocrinol Invest* 1996; **19**: 524–9.

19 Amato G, Izzo G, La Montagna G, Bellastella A. Low dose recombinant human growth hormone normalizes bone metabolism and cortical bone density and improves trabecular bone density in growth hormone deficient adults without causing adverse effects. *Clin Endocrinol* 1996; **45**: 27–32.

20 Guevara-Aguirre J, De la Torre W, Rosenbloom AL, Acosta M, Rosenfeld G. Osteopenia in menstruating women with IGF-I deficiency due to growth hormone receptor deficiency [abstract]. Paper presented at the 73rd Annual Meeting of the Endocrine Society, Washington, D.C., 1991.

21 Shore RM, Chesney RW, Mazess RB, Rose PG, Borgman GJ. Bone mineral status in growth hormone deficiency. *J Pediatr* 1980; **96**: 393–6.

22 Zamboni G, Antoniazzi F, Radetti G, Musumeci C, Tato L. Effects of two different regimens of recombinant human growth hormone therapy on the bone mineral density of patients with growth hormone deficiency. *J Pediatr* 1991; **119**: 483–5.

23 Saggese G, Baroncelli GI, Bertelloni S, Cinquanta L, Di Nero G. Effects of long-term treatment with growth hormone on bone and mineral metabolism in children with growth hormone deficiency. *J Pediatr* 1993; **122**: 37–45.

24 Vandeweghe M. Can we predict and prevent adult morbidity in males with childhood-onset growth hormone deficiency? *Acta Paediatr* 1997; **423** (Suppl): 121–3.

25 Holmes SJ, Whitehouse RW, Economou G *et al.* Further increase in forearm cortical bone mineral content after discontinuation of growth hormone replacement. *Clin Endocrinol* 1995; **42**: 3–7.

26 Rahim A, Holmes SJ, Adams JE, Shalet SM. Long-term change in the bone mineral density of adults with adult onset growth hormone (GH) deficiency in response to short or long-term GH replacement therapy. *Clin Endocrinol* 1998; **48**: 463–9.

*27 Bonjour JP, Theintz G, Buchs B, Slosman D, Rizzoli R. Critical years and stages of puberty for spinal and femoral bone mass accumulation during adolescence. *J Clin Endocr Metab* 1991; 73: 555–63.

28 Theintz G, Buchs B, Rizzoli R *et al.* Longitudinal monitoring of bone mass accumulation in healthy adolescents: evidence for a marked reduction after 16 years of age at the levels of lumbar spine and femoral neck in female subjects. *J Clin Endocr Metab* 1992; **75: 1060–5.

*29 Ott SM. Attainment of peak bone mass. *J Clin Endocr Metab* 1990; **71**: 1082A–1082C.

30 Matkovic V, Jelic T, Wardlaw GMI *et al.* Timing of peak bone mass in Caucasian females and its implication for the prevention of osteoporosis: inference from a cross-sectional model. *J Clin Invest* 1994; **93**: 799–808.

31 Takahashi Y, Minamitani K, Kobayashi Y *et al.* Spinal and femoral bone mass accumulation during normal adolescence: comparison with female patients with sexual precocity and with hypogonadism. *J Clin Endocr Metab* 1996; **81**: 1248–53.

*32 Carrascosa A, Gussinye M, Yesta D *et al.* Skeletal mineralization during infancy, childhood and adolescence in the normal population and in populations with nutritional and hormonal disorders: dual X-ray absorptiometry (DEXA) evaluation. *Pediatr Osteol* 1996; **1105**: 93–102.

33 Boot A, Engels M, Boerma G, Krenning E, De Muinck Keizer-Schrama S. Changes in bone mineral density, body composition and lipid metabolism during growth hormone treatment in children with GH deficiency. *J Clin Endocr Metab* 1997; **82**: 2423–8.

34 Johansson AG, Burkman P, Westermark K, Ljunghall S. The bone mineral density in acquired growth hormone deficiency correlates with circulating levels of insulin-like growth factor I. *J Intern Med* 1992; **232**: 447–52.

35 Hyer SL, Rodin DA, Tobias JH, Leiper A, Nussey SS. Growth hormone deficiency during puberty reduces adult bone mineral density. *Arch Dis Child* 1992; **67**: 1472–4.

36 Rosen T, Hansson R, Granhed H, Szucs J, Bengtsson BA. Reduced bone mineral content in adult patients with growth hormone deficiency. *Acta Endocrinol* 1993; **129**: 201–6.

*37 Bing-You RG, Denis MC, Rosen CJ. Low bone mineral density in adults with previous hypothalamic–pituitary tumors: correlations with serum growth hormone responses to GH-releasing hormone, insulin-like growth factor I, and IGF binding protein 3. *Calcif Tissue Int* 1993; **32**: 326–30.

38 Beshyah SA, Freemantle C, Thomas E *et al.* Abnormal body composition and reduced bone mass in growth hormone deficient hypopituitary adults. *Clin Endocrinol* 1995; **42**: 179–89.

39 Holmes SJ, Economou G, Whitehouse RW, Adams JE, Shalet SM. Reduced bone mineral density in patients with adult onset growth hormone deficiency. *J Clin Endocr Metab* 1994; **78: 669–74.

*40 Toogood AA, Adams JE, O'Neill PA, Shalet SM. Elderly patients with adult-onset GH deficiency are not osteopenic. *J Clin Endocr Metab* 1997; **82**: 1462–6.

*41 Degerblad M, Bengtsson BA, Bramnert M *et al.* Reduced bone mineral density in adults with growth hormone (GH) deficiency: increased bone turnover during 12 months of GH substitution therapy. *Eur J Endocrinol* 1995; **133**: 180–8.

*42 Kaji H, Abe H, Fukase M, Chihara K. Normal bone mineral density in patients with adult onset GH deficiency. *Endocr Metab* 1997; **4**: 163–6.

43 De Boer H, Blok GJ, Van der Veen EA. Clinical aspects of growth hormone deficiency in adults. *Endocr Rev* 1995; **16: 63–86.

*44 Johannsson G, Bengtsson BA. Growth hormone and the acquisition of bone mass. *Horm Res* 1997; **48** (Suppl 5): 72–7.

45 Whitehead HM, Boreham C, McIlrath EM *et al.* Growth hormone treatment of adults with growth hormone deficiency: results of a 13-month placebo controlled cross-over study. *Clin Endocrinol (Oxf)* 1993; **36**: 42–5.

46 Binnerts A, Swart GR, Wilson JHP *et al.* The effect of growth hormone administration in growth hormone deficient adults on bone, protein, carbohydrate and lipid homeostasis as well as on body composition. *Clin Endocrinol* 1992; **37**: 79–87.

47 Thoren M, Soop M, Degerblad M, Sääf M. Preliminary study of the effects of

growth hormone substitution therapy on bone mineral density and serum osteocalcin levels in adults with growth hormone deficiency. *Acta Endocrinol* 1993; **128** (Suppl 2): 41–3.

48 Holmes SJ, Whitehouse RW, Swindell R *et al.* Effect of growth hormone replacement on bone mass in adults with adult onset growth hormone deficiency. *Clin Endocrinol* 1995; **42**: 627–33.

49 Finkenstedt G, Gasser RW, Höfle G, Watfah C, Fridrich L. Effects of GH replacement on bone metabolism and mineral density in adult onset GH deficiency: results of a double blind placebo-controlled study with open follow-up. *Eur J Endocrinol* 1997; **136**: 282–9.

*50 Ter Maaten JC, De Boer H, Roos JC, Lips P, Van der Veen EA. Long-term effects of GH treatment on bone density [abstract]. *Endocr Metab* 1997; **4** (Suppl A): 8.

51 Kotzmann H, Riedl M, Bernecker P *et al.* Effect of long-term growth-hormone substitution therapy on bone mineral density and parameters of bone metabolism in adult patients with growth hormone deficiency. *Calcif Tissue Int* 1998; **62**: 40–6.

*52 Bravenboer N, Holzmann P, De Boer H *et al.* The effect of growth hormone (GH) on histomorphometric indices of bone structure and bone turnover in GH-deficient men. *J Clin Endocr Metab* 1997; **82**: 1818–22.

*53 Hansen TB, Brixen K, Vahl N *et al.* Effects of 12 months of GH treatment on calciotropic hormones, calcium homeostasis, and bone metabolism in adults with acquired GH deficiency: a double-blind, randomized, placebo-controlled study. *J Clin Endocr Metab* 1996; **81**: 3352–9.

*54 Degerblad M, Elgindy N, Hall K, Sjöberg HE, Thoren M. Potent effect of recombinant growth hormone on bone mineral density and body composition in adults with panhypopituitarism. *Acta Endocrinol* 1992; **126**: 387–93.

55 Juul A, Pedersen S, Sorensen S *et al.* Growth hormone (GH) treatment increases serum insulin-like growth factor binding protein-3, bone isoenzyme alkaline phosphatase and forearm bone mineral content in young adults with GH deficiency of childhood onset. *Eur J Endocrinol* 1994; **131**: 41–9.

*56 Beshyah SA, Kyd P, Thomas E, Fairney A, Johnston DG. The effects of prolonged growth hormone replacement on bone metabolism and bone mineral density in hypopituitary adults. *Clin Endocrinol* 1995; **42**: 249–54.

57 Jergas M, Breitenseher M, Glüer CC, Yu W, Genant HK. Estimates of volumetric bone density from projectional measurements improve the discriminatory capability of dual X-ray absorptiometry. *J Bone Miner Res* 1995; **10**: 1101–10.

*58 Lu PW, Cowell CT, Lloyd-Jones SA, Briody JN, Howman-Giles R. Volumetric bone mineral density in normal subjects, aged 5–27 years. *J Clin Endocrinol Metab* 1996; **81**: 1586–90.

59 Ott SM, O'Hanlan M, Lipkin EW, Newell-Morris L. Evaluation of vertebral volumetric vs. areal bone mineral density during growth. *Bone* 1997; **20**: 553–6.

*60 Kröger H, Kotaniemi A, Vainio P, Alhava E. Bone densitometry of the spine and femur in children by dual-energy X-ray absorptiometry. *Bone Miner* 1992; **17**: 75–85.

61 Tanner JM, Whitehouse RH, Hughes PCR, Carter BS. Relative importance of growth hormone and sex steroids for the growth at puberty of trunk length, limb length and muscle width in growth hormone deficient children. *J Pediatr* 1976; **89**: 1000–9.

62 Mazess RB. Errors in measuring trabecular bone by computed tomography due to marrow and bone composition. *Calcif Tissue Int* 1973; **35**: 148–52.

*63 Laval-Jeantet AM, Roger B, Bouysse S, Bergot C, Mazess R. Influence of vertebral fat content on quantitative CT density. *Radiology* 1986; **159**: 463–6.

64 Pacifici R, Susman N, Carr PL, Birge SJ, Avioli LV. Single and dual energy tomographic analysis of spinal trabecular bone: a comparative study in normal and osteoporotic women. *J Clin Endocrinol Metab* 1987; **64**: 209–14.

*65 Glüer CC, Genant HK. Impact of marrow fat on accuracy of quantitative CT. *J Comput Assist Tomogr* 1989; **13**: 1023–35.

10: Individual susceptibility and GH dosing regimen in adults

William M. Drake and John P. Monson

Introduction

Although there can be little doubt about the potential beneficial effects of recombinant human growth hormone (rhGH) in the management of adult-onset hypopituitary patients [1–5], the issue of dosage and dose monitoring remains controversial. As with other aspects of endocrine replacement therapy (such as hydrocortisone replacement therapy in primary or secondary hypoadrenalism), reproduction of the complex and subtle patterns of hormone secretion that occur in health are limited by available modes and routes of administration. Furthermore, the lack of a sensitive tissue marker specific for the action of the replaced hormone dictates that the aim of much endocrine replacement therapy (glucocorticoids, sex steroids and GH) is to ensure adequate overall tissue exposure to the actions of that hormone whilst trying to minimize the risks that may result from excess exposure. In contrast to the situation in children, in whom linear growth provides an accurate, reproducible and physiological means of monitoring GH action, there is no ideal marker by which to judge the optimum dosage and dose monitoring regimen for GH replacement in hypopituitary adults.

This chapter will discuss the relative merits and shortfalls of the various schemes that have been used in recent years, and will attempt to draw some conclusions about the optimum method of restoring GH status in adult hypopituitarism. In order to do this, it is first necessary to consider those abnormalities that physicians are aiming to treat in the adult growth hormone deficiency (GHD) syndrome, and what measurable parameters are available by which to judge whether they are being successful.

Potential indices of therapeutic efficacy in GHD: clinical

Much attention has focused on the possible aetiological role of GHD in the reported decreased life expectancy amongst patients with hypopituitarism [6–8] after exclusion of patients treated for acromegaly and Cushing's disease, both of which are known to be associated with increased mortality [9,10]. Such epidemiological studies are limited by their retrospective nature and the fact that there are very few patients with similar pituitary disease and normal GH reserve to serve as controls. However, the finding of approximately twice the mortality during the long-term follow-up in patients with hypopituitarism compared to an age-matched control population has been a relatively consistent figure. The relative contribution of cardiovascular causes to this increased mortality has not been a universal finding [6–8], although it seems likely that GHD does worsen the overall cardiovascular risk profile. Various atherogenic factors such as dyslipidaemia [2,11], insulin resistance [12] and circulating concentrations of fibrinogen and plasminogen-activating inhibitor [13] have all been shown to be adversely affected in patients with GHD. These studies have been paralleled by the observation that the large arteries of patients with hypopituitarism demonstrate an increased intima–media thickness and plaques as judged by ultrasound [14] and, recently, that these changes are reversible during GH replacement therapy [15]. However, it is important to remember that the growth hormone-deficient state is characterized by a wide variety of clinical abnormalities, emphasizing the diverse actions of GH (which should, perhaps, more accurately be referred to as somatotrophin) in health. In theory, each of these might be thought of as a potential candidate for use as a marker of the efficacy of GH replacement therapy, although—as will be seen below—many of them are not practical for use in this way.

Body composition

Patients with GHD have decreased lean body mass, salt and water depletion and increased central adiposity compared to age-matched healthy controls [1,2]. All of these factors have been shown to improve or normalize following treatment with GH, and these parameters have received considerable attention as candidates for use as markers of GH action during replacement therapy. Dual energy X-ray absorptiometry (DEXA) scanning provides a highly accurate and reproducible means of measuring and monitoring total and regional body fat and lean body mass. In routine clinical practice, central fat may be assessed using measurements of the waist and/or waist : hip ratio. Total body water may be measured either by using bioimpedance analysis or by isotope dilution of tritiated water. Both of these measurements are highly reproducible in the context of clinical trials, although their suitability for widespread clinical use is questionable.

A number of studies have monitored the effect of GH on indices of body composition, and some have simultaneously attempted to correlate these changes with alterations of GH-dependent serum markers such as insulin-like growth factor-I (IGF-I), insulin-like growth factor binding protein-3 (IGFBP-3) and acid-labile subunit (ALS) [16,17]. In one study [16], patients were randomly assigned to receive one of two dosage schedules of GH: a high-dose (HD) weight-based regimen of 12 μg/kg/day; and an individualized dose (ID) in which, after a low starting dose, adjustments were made on the basis of serum IGF-I levels and changes in body composition, with relative weighting given to the parameter with the greater deviation from normal. In most patients, dosage changes were made on the basis of serum IGF-I levels below the age-related reference range. However, in those patients with a normal serum IGF-I at baseline, dose adjustments were made on the basis of measurements of body composition. Parameters of body composition were all favourably influenced by GH replacement therapy in both groups (HD and ID), although the extent of improvement varied considerably between individual patients. Furthermore, in an attempt to 'normalize' abnormal body composition, GH doses were increased in many patients in such a way that serum IGF-I levels were above the age-related reference range. This would suggest that attributing the totality of abnormal body composition to GH deficiency may not be appropriate, and that increasing the dosage of GH on the basis of a failure to normalize body composition may, in some patients, result in excess GH exposure, as judged by serum IGF-I levels.

In a separate study [17], changes in body composition during 12 months of GH replacement therapy were monitored and related to changes in serum levels of IGF-I, IGFBP-1 and IGFBP-3, and ALS. Body composition, measured by DEXA scanning and bioimpedance estimation of total body water, improved significantly in the treated group compared to placebo. However, despite dose reductions in seven out of 20 patients due to side-effects of fluid retention, serum IGF-I remained above the age-related reference range in seven patients (35%). Serum ALS and IGFBP-3 were outside the reference range in five patients (25%) and three patients (15%), respectively. This would suggest that the presence or absence of the classical symptoms of GH overtreatment is a relatively crude method of judging excess GH exposure compared to the measurement of GH-dependent biochemical markers.

These studies suggest that selecting and monitoring GH dosage solely on the basis of a clinical parameter such as body composition that has such a wide variation in health is likely to lead to the use of higher than optimal doses of GH, judged by measurements of GH-dependent serum markers. It is clear that using symptoms of GH excess is a relatively crude indicator of GH overexposure, particularly since the incidence of side-effects has fallen markedly with the practice of using lower starting doses of GH before building up to the standard target weight-based dose of 0.25 IU/kg/week. It seems that basing the GH dosage on measurements of body composition will inevitably lead, in some patients, to overtreatment, and we consider this strategy to be inappropriate for determining the GH dose in hypopituitarism.

Indices of well-being

Depression of mood and impaired social function are well-recognized features of the GHD syndrome [18,19]. The early trials of GH replacement in adult hypopituitarism used a variety of generic methods in an attempt to document improvements in psychosocial well-being during treatment. The most commonly used was the Nottingham Health Profile (NHP), which aims to quantify perceived health problems and the extent to which they affect daily activities [20]. The NHP questionnaire contains specific questions, with yes/no responses, about energy levels, sleep, relationships, emotional responses, physical mobility and pain. Other tests, such as the Psychological General Well-Being Schedule (PGWB) [21] are similar in principle, but involve the use of a rating (from 0, worst, to 5, best) for a series of questions about such affective categories as anxiety, depression, positive well-being, general health and vitality. Significant improvements in well-being, judged by NHP scores, have been documented following commencement of GH in double-blind placebo-controlled trials [1,5]. However, the doses used in these studies were higher than would now be considered optimal, and the numbers of patients with supranormal serum IGF-I levels was, in some cases, as high as 56% [22]. An important study was that by Carroll *et al.* [23], who monitored NHP scores during GH replacement therapy in patients randomly assigned to receive placebo or one of two GH dosage regimens: 0.024 mg/kg per day or 0.012 mg/kg per day. Identical improvements in psychosocial well-being, judged by NHP and PGWB scores, were documented in both groups, yet 45% of the higher-dose group had supranormal serum IGF-I levels compared to 24% in the lower-dose group. Furthermore, of those who elected to continue with GH therapy after the trial, 33% had a serum IGF-I above the age-related reference range, compared to 30% who declined to continue with GH on account of a lack of perceived benefit in well-being. In other words, any decision to increase the dose of GH based on a perceived lack of improvement in well-being in the patients studied by Carroll *et al.* would have driven serum IGF-I further into the acromegalic range in a third of the patients and would, in all likelihood, have led to an overall increase in the number of patients with serum IGF-I levels above the age-related reference range. If it is accepted that a serum IGF-I above the age-related reference range should be avoided in hypopituitary patients receiving GH replacement, this would suggest that determining the GH dose on the basis of improvements in well-being may lead to overexposure to GH.

More recently, attempts have been made to use scoring systems for psychosocial morbidity that are more specific to GHD. The adult growth hormone deficiency assessment (AGHDA) score [24] provides a sensitive and highly reproducible method of monitoring improvements in the psychosocial consequences of GHD that may accompany GH replacement therapy. The AGHDA questionnaire consists of 25 questions derived from the symptoms most frequently reported by patients with adult-onset GHD. A score of 25/25 represents the worst possible 'well-being' score, while scores of 4/25 or less have been recorded in a normal control population [25]. Improvements in AGHDA scores in the majority of patients occur within three months of GH replacement therapy and are maintained at six and 12 months [25]. Interestingly, improvements in AGHDA scores may be seen in some patients treated with GH whose dose is insufficient to have caused a significant increment in serum IGF-I, suggesting that improvements in the psychological aspects of GHD may, at least in part, be mediated directly by GH rather than via generation of IGF-I [25]. It is not known whether patients exposed to excess GH (either in the context of acromegaly or by over-treatment with GH in hypopituitarism) have AGHDA scores that are different from control populations. However, an interesting comparison can be made between hypopituitary patients treated initially on weight-based dosing schedules, with subsequent dose adjustment during clinical follow-up; and patients initially commenced on low doses of GH with subsequent careful dose titration on the basis of levels of serum IGF-I [26]. Maintenance doses of GH and serum IGF-I levels were significantly higher in the patients initially treated with weight-based dosing schedules, although well-being, as judged by the AGHDA score, was no different. Again, this would suggest that monitoring of GH dose on the basis of indices of well-being has a significant potential for excess GH exposure.

A further, theoretical, reason why the use of well-being scores is not ideal in the monitoring of GH efficacy is the issue of patient bias. The clinical indications for replacement therapy are varied, although impaired psychosocial well-being is among the most important. It is well recognized that those patients with the most unfavourable indices of well-being are more likely to enter a study of GH replacement [27], more likely to experience an improvement in well-being and therefore more likely to wish to continue with GH replacement therapy [28]. However, in routine clinical practice, those patients who start GH replacement for other reasons (such as osteopaenia or abnormal body composition) and who may have a nearer-normal 'well-being' score, are likely to derive relatively less benefit in terms of improvement in well-being. In such patients, relative lack of improvement in AGHDA, NHP or other well-being scores as indices of GH efficacy may prompt erroneous increments in GH dosage with, consequently, inadvertent overtreatment.

Bone density and markers of bone remodelling

GH secretion throughout life is an important factor in the maintenance of bone mass. Adult hypopituitary patients have decreased bone mineral density (BMD) compared to age-matched healthy controls, despite adequate endocrine replacement therapy with gonadal steroids, thyroxine and glucocorticoids [29,30]. In adult-onset growth hormone deficiency (AO-GHD) this represents genuine bone loss (osteopaenia), whereas in childhood onset disease (CO-GHD) reduced BMD represents a failure to reach peak mass. In neither case are the precise reasons clear, but a significant factor is likely to be a reduction in the activity of the bone remodelling unit. The bone remodelling unit, in turn, consists of osteoclastic and osteoblastic activity, the cells of which are both regulated in part by GH. Osteoclastic activity may be monitored by measuring levels of the pyridinium crosslinks, pyridinoline, deoxypyridinoline and type I collagen carboxyterminal telopeptide (ICTP). The presence of these markers in urine and/or serum is a reflection of the resorption of mature skeletal collagen. Osteoblastic activity may be monitored by measuring levels of osteocalcin, bone-specific alkaline phosphatase and procollagen type I carboxyterminal propeptide (PICP) in serum.

A number of studies have examined the effect of GH replacement in GH-deficient adults on BMD (measured by DEXA scanning) and markers of bone remodelling, all using what would now be regarded as higher than optimum dosing schedules [31–35]. The individual response to treatment is highly variable but, in general, after an initial apparent fall in BMD (probably as a result of an increase in the bone remodelling space) [35] there then follows a sustained rise such that BMD reaches baseline after 12–18 months of treatment with GH, and improves thereafter [31]. Markers of bone remodelling are significantly increased four months after the commencement of GH (probably a reflection of the length of the bone remodelling cycle), although there are comparatively few data documenting the effects of physiological GH replacement earlier than this time point. Markers of bone remodelling remain elevated for at least two years following the commencement of GH replacement, although it is not known for how long BMD continues to increase [31].

In studies examining the effect of GH on bone metabolism, the individual response in terms of markers of bone remodelling was highly variable. Serum IGF-I rose in all patients treated with GH and, in many patients, was outside the age-related reference range [33,34]. In some patients, therefore, a lack of beneficial effect on markers of bone remodelling would have led to increases in GH dosage that may well have caused serum IGF-I levels to rise into the acromegalic range. The long-term effects on bone of supranormal serum IGF-I levels in hypopituitarism are not known, but GH excess (in the context of acromegaly) is associated with diminished bone density in the axial skeleton [36]. Although this is in large part due to reduced sex steroids [36], the potential adverse consequences on bone metabolism of excess GH dosing should not be underestimated. This, and the long time course over which changes in bone metabolism occur, make the use of these markers impractical for the monitoring of GH dosing in hypopituitarism.

Cardiac dimensions and cardiovascular performance

Various indices of cardiac structure and function (exercise capacity, left ventricular wall thickness and fractional shortening) are adversely affected in GHD [37]. This has been most extensively documented in childhood-onset GHD, which is associated with echocardiographic and radionuclide evidence of reduced cardiac output and impaired diastolic function [38,39]. The evidence for similar abnormalities of cardiac function in adult-onset GHD is conflicting; certainly, the effects are less marked, although this may simply reflect the duration of GHD at the time of study. There is good evidence that GH replacement in hypopituitary adults improves exercise capacity, and that this improvement is associated with an increase in left ventricular wall thickness, stroke volume, fractional shortening and diastolic function [40]. Individual response to a uniform (weight-based) GH treatment regimen is, however, extremely variable. Again, the argument about the avoidance of overtreatment is important, particularly with respect to the induction of left ventricular hypertrophy (LVH). GH hypersecretion (in the context of acromegaly) leads to a specific cardiomyopathy in which, after an initial phase of cardiac hyperkinesis, myocardial hypertrophy and diffuse interstitial fibrosis gradually lead to diastolic dysfunction and, ultimately, congestive cardiac failure [41]. The detection of subtle signs of LVH and impaired diastolic relaxation may be difficult using standard transthoracic echocardiography, particularly outside the setting of clinical trials, where interobserver variation in measurements may be considerable. The main concern with the use of measurements of cardiac function as indices of GH efficacy in hypopituitarism is therefore that overtreatment may be associated with adverse cardiac consequences, and we consider this strategy to be inappropriate for dose selection and dose titration.

In summary, the various measurable clinical abnormalities that can be monitored during treatment with GH all have too narrow a therapeutic window for clinical use as a means of monitoring dosage, and a major criticism of all is that they do not avoid the possibility of inadvertent overtreatment. A further, more practical argument, perhaps, is that all such measurements are sufficiently time-consuming (and therefore expensive) and that translating the intensive monitoring that accompanies carefully controlled scientific studies into clinical practice may not be straightforward.

Safety issues

The concern about improvements in the clinical parameters that characterize the GHD state at the expense of elevation of serum markers of GH action raises the question of whether long-term elevation of serum IGF-I is acceptable in the context of GH replacement therapy. No long-term data exist on the effects of supraphysiological levels of serum IGF-I in hypopituitarism, although clinical experience in acromegaly would suggest this should be avoided. Long-term excess GH exposure has been shown to induce left ventricular hypertrophy [40], exacerbate insulin resistance [42] and predispose to colonic neoplasia [43]. The argument for caution in this regard is further strengthened by the reported decreased incidence of mortality from malignant disease in male hypopituitary patients [6,7].

Potential markers of GH efficacy: biochemical

More recently, attention has shifted towards the use of biochemical markers of GH status during the treatment of hypopituitarism. GH exerts many of its biological actions through IGF-I. IGF-I, in turn, circulates as a ternary complex of about 150 kDa in association with IGFBP-3 and ALS, both of which are known to be GH-dependent. All three might be regarded as potential candidates for the monitoring of GH replacement therapy.

De Boer *et al.* [44] randomized 46 male patients in a 12-month double-blind, placebo-controlled trial to one of four groups: placebo for six months followed by GH 2 IU/m^2/day; and three different doses of GH (1, 2 and 3 IU/m^2/day) for the full 12 months. Some dose adjustment was performed on the basis of side-effects suggestive of overdosing. A significant finding of this study was that the minimum mean GH dose required to normalize serum IGF-I was 0.66 IU/m^2/day—considerably lower than many of the weight-based and surface area-based regimens used in the early trials of GH replacement. A further important finding was that the vast majority of patients receiving 3 IU/m^2/day had serum IGF-I levels above the age-related reference range. Serum IGF-I rose outside the age-related reference range in significantly more patients than did IGFBP-3 and ALS (Fig. 10.1), suggest-

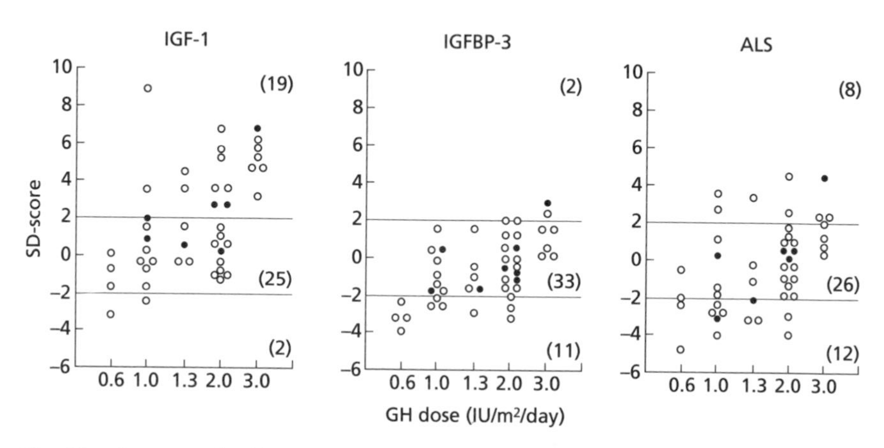

Fig. 10.1 Serum insulin-like growth factor (IGF-I), insulin-like growth factor-binding protein-3 (IGFBP-3) and acid-labile subunit (ALS) levels (standard deviation scores) in growth hormone-deficient patients after six months of growth hormone (GH) treatment, ranked in ascending order according to the dose of GH used. Open circles: no symptoms of GH excess. Closed circles: patients with symptoms of GH excess. Reproduced with permission from de Boer *et al.* [44].

ing that the latter parameters are less sensitive markers than serum IGF-I of GH action in hypopituitarism. It is also clear from Figure 10.1 that the absence of symptoms is a poor guide to GH overexposure, as judged by serum IGF-I levels.

A limitation of the study by De Boer *et al.* is that one of the inclusion criteria was a subnormal serum IGF-I level, although it is known that a considerable proportion of patients with established GHD have serum IGF-I levels within the age-related reference range [45]. Basing the GH dose solely on normalization of serum IGF-I levels may leave some patients relatively undertreated, although it is unlikely that significant excess GH exposure will result from this strategy. The data from de Boer *et al.* suggest that measurement of IGFBP-3 and ALS levels adds little information for the purposes of dose selection and dose monitoring.

Janssen and colleagues [46] randomized 51 GH-deficient adults to receive one of three dosing schedules of GH (0.6, 1.2 and 1.8 IU/day, based on estimates of GH production rates) for 12 weeks, and monitored serum IGF-I and IGFBP-3 levels over this time. They concluded that a mean GH dose of 1.2 IU daily was required to normalize serum IGF-I overall (Fig. 10.2), although the individual response to treatment was highly variable, most notably on the basis of sex. At a dose of 0.6 IU/day, almost all men had a normal serum IGF-I level, whereas 40% of women still had serum IGF-I levels below the age-related reference range. The higher doses of 1.2 and 1.8 IU daily were sufficient to normalize serum IGF-I levels in most women,

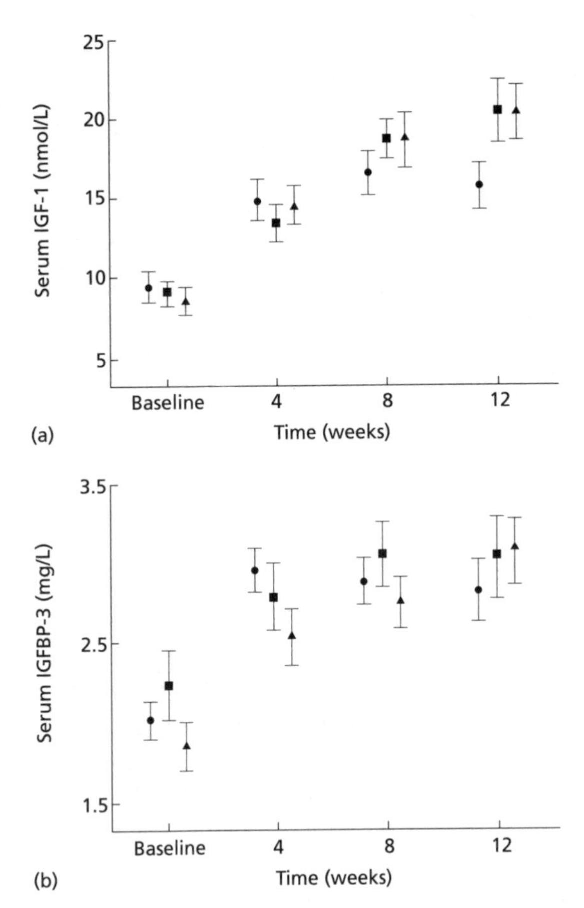

Fig. 10.2 Serum insulin-like growth factor (IGF-I) (a) and insulin-like growth factor-binding protein-3 (IGFBP-3) (b) levels during treatment with three different dosing regimens of recombinant human growth hormone (rhGH). Circles: 0.6 IU/day for 12 weeks. Rectangles: 0.6 IU/day for four weeks, followed by 1.2 IU/day. Triangles: 0.6 IU/day for four weeks, followed by 1.2 IU/day for four weeks and 1.8 IU/day thereafter. Reproduced with permission from Janssen *et al.* [46].

although nearly half the men randomized to these doses had serum IGF-I levels above the age-related reference range.

Measurements of serum IGFBP-3 were highly variable in this study (Fig. 10.2). Both subnormal and supranormal levels were recorded in all groups at baseline and, at the conclusion of the study, no statistically significant difference could be found in IGFBP-3 levels between the groups. As with the data of de Boer [44], this suggests that serum IGFBP-3 levels are a poor index of GH efficacy in the treatment of hypopituitarism.

Dose selection

Many of the early placebo-controlled trials on the effects of rhGH replacement used weight-based and surface area-based dosing, largely on the basis of extensive paediatric experience with such regimens, where linear growth provides a physiological marker of GH action. In retrospect, the use of such dosing schedules was simplistic and took no account of individual susceptibility to GH replacement. With time, it has become clear that this approach is not satisfactory, partly because GH production rates in health vary considerably according to age, sex and other less well-defined factors; but also because individual susceptibility to GH replacement therapy is highly variable. In addition, the side-effect profile associated with this method of GH dosing (mainly ankle oedema and arthralgia due to sodium and water retention) was considerable [47]. More recently, it has become apparent that the dose of GH needs to be individually tailored in order both to maximize the potential benefits of GH replacement whilst minimizing the side-effects and avoiding potential excess GH exposure.

Physiological considerations

It is difficult to conceive of a method of replacement of GH in which the normal physiological profile is restored (Fig. 10.3). GH secretion in health is pulsatile, with long periods of very low-level secretion punctuated by surges of GH secretion, mainly at night. An additional complicating factor is the known sex difference in the pattern and quantity of GH secretion, due largely to an amplitude-specific divergence in the pulsatile mode of GH secretion, such that women of reproductive age produce approximately three times as much GH as men of identical age and weight, yet serum IGF-I levels are identical [48]. Furthermore, there are marked changes in the level of GH secretion through life, with increased secretion around puberty and

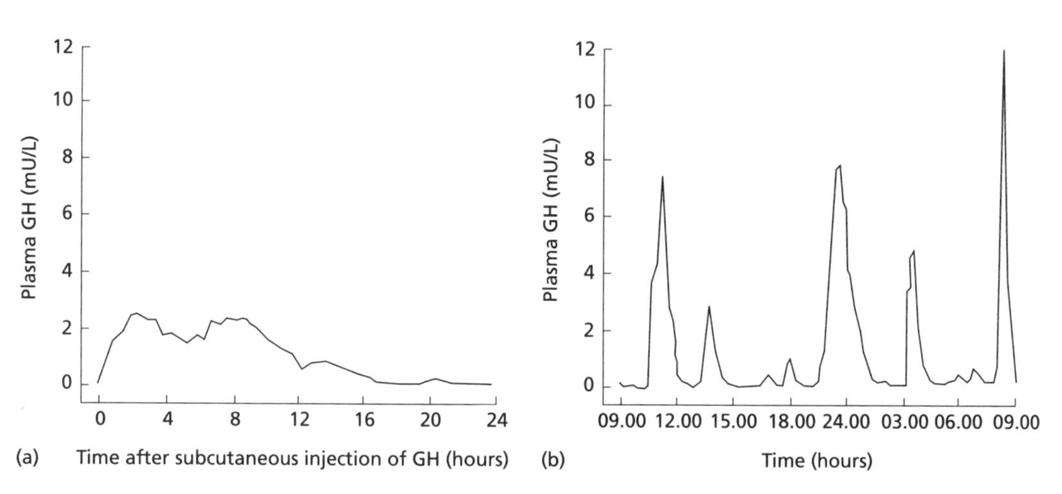

Fig. 10.3 Plasma-time curve for 1.2 IU growth hormone (GH), given by injection to an adult with GH (a), differs markedly from the natural physiological pattern of GH secretion (b). Reproduced with permission from Janssen *et al.* [46].

decreased secretion after the age of 40 [49]. Immediately it becomes apparent, then, that GH replacement therapy should be tailored to the individual patient, based somehow on the available markers of GH action.

The sex differences in GH production rates perhaps form the most compelling argument against simplistic weight-based or surface area-based dosing and in favour of individual dose selection. Burman *et al.* [50] found that such regimens cause significantly lower increments in IGF-I in women (Fig. 10.4), with greater benefit in men, judged by measurements of total body fat, markers of bone remodelling and serum levels of Apolipoprotein B (Fig. 10.5). Conversely, many more of the men had serum IGF-I levels above the age-related reference range. In other words, there exists, with the use of weight-based or surface area-based dosing, a potential for significant undertreatment in women and overtreatment in men.

Within the sexes there is considerable individual variation in GH susceptibility. This is most marked in women, for reasons that are not clear, although oestrogen modulation is an obvious candidate. Oestrogen therapy is associated with a reduction in the severity of the symptoms and signs of acromegaly [51], while oestrogen administration to men alters the pattern of GH secretion from male to female [52]. Furthermore, oral oestrogen administration results in a fall in serum IGF-I and a rise in GH secretion in postmenopausal women [53,54]. Although a correlation between oestrogen status and GH requirement in hypopituitarism has not been proven, it seems probable that gonadal steroids play some modulatory role in determining GH susceptibility.

A further, important, factor to consider in dose selection is the effect of visceral obesity on GH bioavailability and susceptibility. Growth hormone-

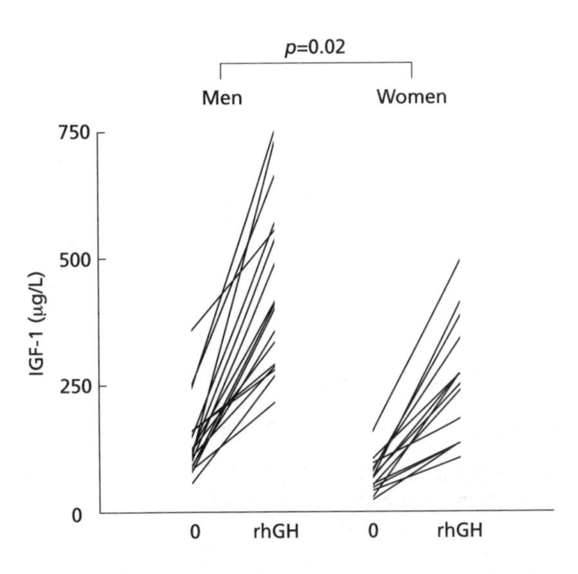

Fig. 10.4 Serum insulin-like growth factor (IGF-I) in 21 men and 15 women with growth hormone deficiency before and after nine months of treatment with recombinant human growth hormone (rhGH). The p value refers to the difference in response to treatment between men and women. Reproduced with permission from Burman *et al.* [50].

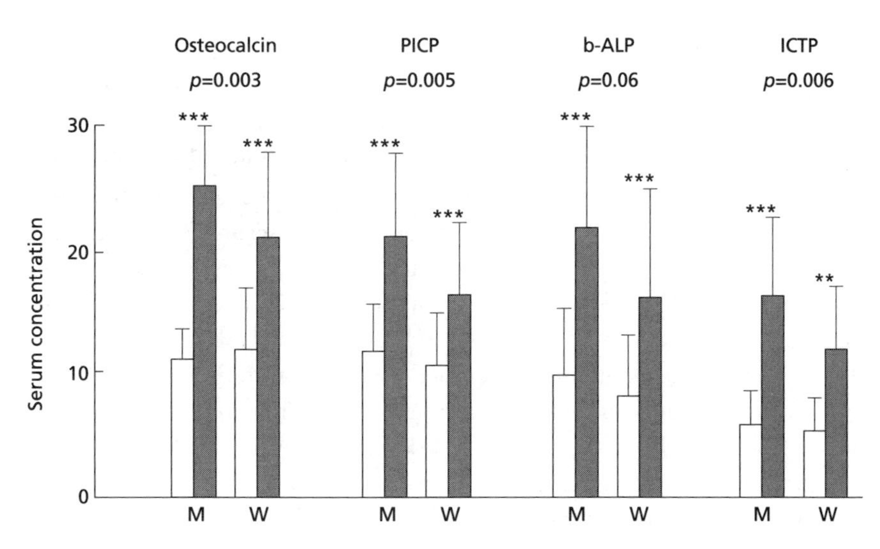

Fig. 10.5 Concentrations of osteocalcin (µg/L), carboxyterminal cross-linked telopeptide of type I collagen (ICTP; µg/L), carboxyterminal propeptide of type I procollagen (PICP; µg/L × 10^{-1}) and bone-specific alkaline phosphatase (bALP; µka/L × 10^{-1}) in 21 men and 15 women with growth hormone deficiency before and after nine months of treatment with recombinant human growth hormone (rhGH). Results are shown as the mean ± standard deviation. ** $p < 0.01$; * $p < 0.001$. p values in letters refer to the differences in the response to treatment between men and women. Reproduced with permission from Burman *et al.* [50].

binding protein (GHBP) is derived from the extracellular region of the GH receptor by proteolytic cleavage and, once bound, the GH–GHBP complex is less susceptible to degradation. Intuitively, one might expect GHD to be associated with low levels of GHBP, but this has been shown not to be the case. It appears that a more important determinant of GHBP levels is abdominal fat [55]. Hence, in patients with GHD (with, characteristically, increased visceral adiposity), a greater proportion of the administered GH is 'protected' from degradation and is therefore able to exert its diverse physiological effects. Given that obese patients, in weight-based regimens, receive larger doses of GH than patients of identical height with lower BMI, this may explain the higher incidence of early side-effects (mainly peripheral oedema) in obese patients, and provides a further argument for individual tailoring of GH dosing in hypopituitarism.

Individual dose titration

A combination of the concern surrounding the potential adverse effects of excess GH exposure in hypopituitarism and an appreciation of the marked variation between patients in their susceptibility to GH replacement has led to the clinical practice of individual dose titration. Most authorities now accept that the incidence of side-effects can be significantly reduced by starting with low doses of GH and using subsequent dose increments to find the 'optimum' dose for a given patient based on various markers of GH action. In a recent series of 50 consecutive hypopituitary adults [26], individual dose titration was performed, using increments from a starting dose of 0.8 IU/day, based on the defined objective of achieving a serum IGF-I level between the median and the upper end of the age-related reference range. This was empirically defined as the 'optimum' GH dose, to take account of the fact that many patients with established adult-onset GHD have a serum IGF-I within their age-related reference range. This study confirmed the known sex difference in GH susceptibility. The maintenance dose requirement to achieve the 'target' IGF-I was higher, the time taken to reach a maintenance dose was significantly longer and the increment in serum IGF-I was significantly less in women. However, the clinical effects of GH replacement (judged by AGHDA scores and waist : hip ratio) were identical in males and females. In other words, this type of titration regimen results in a lower maintenance dose in men without compromising efficacy.

How should we treat GHD?

It is clear that there is no ideal clinical or biochemical marker for monitoring treatment with GH in hypopituitarism, and this is likely to remain the case until accurate data in healthy volunteers become available that correlate GH production rates throughout life with measurements of GH-dependent serum markers, body composition and nutritional state. In the absence of such data, it will not be possible to assess, physiologically or biochemically, when optimum GH status has been restored in hypopituitary patients. In the meantime, any dosage schedule used must have as one of its main criteria the avoidance of excess GH exposure and, in this regard, regimens based on the measurement of serum IGF-I are likely to prove the safest. Such a regimen can take account of the fact that serum IGF-I is normal in many GHD patients, and it can also accommodate the marked variability in GH susceptibility that requires careful individual tailoring of GH dose.

Acknowledgements

The authors are grateful for research support from Pharmacia and Upjohn (Peptide Hormones).

References

** 1 Jorgensen JOL, Pedersen SA, Thuesen L *et al.* Beneficial effects of growth hormone treatment in GH-deficient adults. *Lancet* 1989; **i**: 1221–5.

** 2 Salomon F, Cuneo RC, Hesp R, Sönksen PH. The effects of treatment with recombinant human growth hormone on body composition and metabolism in adults with growth hormone deficiency. *N Engl J Med* 1989; **321**: 1979–83.

* 3 Binnerts A, Swart GR, Wilson JH *et al.* The effect of growth hormone administration in growth hormone deficient adults on bone, protein, carbohydrate and lipid homeostasis, as well as on body composition. *Clin Endocrinol* 1992; **37**: 79–87.

* 4 Whitehead HM, Boreham C, McIlwrath EM *et al.* Growth hormone treatment of adults with growth hormone deficiency: results of a 13 month placebo-controlled cross-over study. *Clin Endocrinol* 1992; **36**: 45–52.

* 5 Bengtsson BÅ, Eden S, Lonn L *et al.* Treatment of adults with growth hormone deficiency with recombinant human growth hormone. *J Clin Endocrinol Metab* 1993; **76**: 309–17.

** 6 Rosén T, Bengtsson BÅ. Premature mortality due to cardiovascular disease in hypopituitarism. *Lancet* 1990; **336**: 285–8.

** 7 Bates AS, Van't Hoff W, Jones PJ, Clayton RN. The effect of hypopituitarism on life expectancy. *J Clin Endocrinol Metab* 1996; **81**: 1169–72.

** 8 Bülow B, Hagmar L, Mikoczy Z, Nordstrom CH, Erfurth EM. Increased cerebrovascular mortality in patients with hypopituitarism. *Clin Endocrinol* 1997; **46**: 75–81.

9 Bates AS, Van't Hoff W, Jones JM, Clayton RN. An audit of outcome of treatment in acromegaly. *Q J Med* 1993; **86**: 293–9.

10 Plotz CM, Knowlton AI, Ragan C. The natural history of Cushing's syndrome. *Am J Med* 1952; **41**: 597–613.

11 De Boer H, Blok GJ, Voerman HJ, Phillips M, Schouten JA. Serum lipid levels in growth hormone deficient men. *Metabolism* 1994; **43**: 199–203.

* 12 Weaver JU, Monson JP, Noonan K *et al.* The effect of low dose recombinant human growth hormone replacement on regional fat distribution, insulin sensitivity and cardiovascular risk factors in hypopituitary adults. *J Clin Endocrinol Metab* 1995; **80**: 153–9.

13 Johansson JO, Landin K, Tenborn L, Rosén T, Bengtsson BÅ. High fibrinogen and plasminogen activator inhibitor activity in growth hormone-deficient adults. *Arterioscler Thromb* 1994; **14**: 434–7.

* 14 Markussis V, Beshyah SA, Fisher C *et al.* Detection of premature atherosclerosis by high-resolution ultrasonography in symptom-free hypopituitary adults. *Lancet* 1992; **340**: 1188–92.

15 Pfeifer M, Verhovec R, Zizek B, Prezelj J, Poredos P. GH replacement reverses intima-media thickening of carotid arteries in GH deficient adults [abstract]. *J Endocrinol* 1998; **156** (Suppl): OC28.

** 16 Johannson G, Rosén T, Bengtsson BÅ. Individualised dose titration of growth hormone replacement. *Clin Endocrinol* 1997; **47**: 571–81.

* 17 Thorén M, Hilding A, Baxter RC *et al.* Serum insulin-like growth factor I (IGF-I), IGF-binding protein-1 and -3 and the acid-labile subunit as serum markers of body composition during growth hormone (GH) therapy in adults with GH deficiency. *J Clin Endocrinol Metab* 1997; **82**: 223–8.

** 18 Burman P, Broman JE, Hetta J *et al.* Quality of life in adults with growth (GH) deficiency: response to treatment with recombinant human GH in a placebo-controlled 21-month trial. *J Clin Endocrinol Metab* 1995; **80**: 3585–90.

** 19 Rosén T, Wirén L, Wilhelmsen L, Wiklund I, Bengtsson BA. Decreased psychological well-being in adult patients with growth hormone deficiency. *Clin Endocrinol* 1994; **40**: 111–16.

20 Hunt SM, McKenna SP, McEwen J, Wilson J, Papp E. The Nottingham Health Profile: subjective health status and medication consultations. *Soc Sci Med* 1981; **15A**: 221–9.

21 Dupuy HJ. The psychological general well-being (PGWB) index. In: Wenger

NK, ed. *Association of Quality of Life in Clinical Trials of Cardiovascular Therapies*. New York: Le Jacq Publications, 1984: 170–83.

22 Verhelst J, Abs R, Vandeweghe M *et al.* Two years of replacement therapy in adults with growth hormone deficiency. *Clin Endocrinol* 1997; **47**: 485–94.

*23 Carroll PV, Littlewood R, Weissberger AJ *et al.* The effects of two doses of replacement growth hormone on the biochemical, body composition and psychological profiles of growth hormone-deficient adults. *Eur J Endocrinol* 1997; **137**: 146–53.

*24 Doward LC. The development of the AGHDA score: a measure to assess quality of life of adults with growth hormone deficiency. *Qual Life Res* 1995; **4**: 420–1.

25 Wirén L, Willhemsen L, McKenna S, Hember-Stahl EA. A comparison of the quality of life of GH-deficient adults and a random sample population in Sweden. Paper presented at the 22nd International Symposium on Growth Hormone and Growth Factors, Vienna, 1996.

26 Drake WM, Kaltsas G, Korbonits M *et al.* Optimising growth hormone replacement therapy by dose titration in hypopituitary adults. *J Endocrinol Metab* 1998; **83: 3913–19.

27 Holmes SJ, Shalet SM. Characteristics of adults who wish to enter a trial of growth hormone replacement. *Clin Endocrinol* 1995; **42**: 613–18.

28 Holmes SJ, Shalet SM. Factors influencing the desire for long-term growth hormone replacement in adults. *Clin Endocrinol* 1995; **43**: 151–7.

29 Holmes SJ, Economou G, Whitehouse RW, Adams JE, Shalet SM. Reduced bone mineral density in patients with adult onset growth hormone deficiency. *J Clin Endocrinol Metab* 1994; **87**: 669–74.

*30 Beshyah SA, Thomas E, Skinner E, Sharp P, Johnston DG. Total body and spine bone mineral density in women with hypopituitarism. In: Ring EFJ, ed. *Current Research in Osteoporosis and Bone Mineral Measurement, II*. London: The British Institute of Radiology, 1992: 17–18.

31 Johannsson G, Rosén T, Bosaeus I, Sjöstrom L, Bengtsson BÅ. Two years of growth hormone (GH) treatment increases bone mineral content and density in hypopituitary patients with adult-onset GH deficiency. *J Clin Endocrinol Metab* 1996; **81: 2865–73.

*32 Weaver JU, Monson JP, Noonan K *et al.* The effects of low dose recombinant human growth hormone replacement on indices of bone remodelling and bone mineral density in hypopituitary growth hormone deficient adults. *Endocrinol Metab* 1996; **3**: 55–61.

33 Finkstedt G, Gasser RW, Hofle G, Watfah C, Fridrick L. Effects of growth hormone (GH) replacement on bone metabolism and mineral density in adult onset of GH deficiency: results of a double-blind placebo-controlled study with open follow-up. *Eur J Endocrinol* 1997; **136**: 282–9.

*34 Beshyah SA, Thomas E, Kyd P *et al.* The effect of growth hormone replacement therapy in hypopituitary adults on calcium and bone metabolism. *Clin Endocrinol* 1994; **40**: 383–91.

35 Vandeweghe M, Taelman P, Kaufman JM. Short and long term effects of growth hormone treatment on bone turnover and bone mineral content in adult growth hormone deficiency males. *Clin Endocrinol* 1993; **39**: 409–15.

36 Diamond T, Nery L, Rosen S. Spinal and peripheral bone densities in acromegaly: the effects of excess growth hormone and hypogonadism. *Ann Intern Med* 1989; **111**: 567–73.

*37 Shahi M, Beshyah SA, Hackett D *et al.* Myocardial dysfunction in treated adult hypopituitarism: a possible explanation for increased cardiovascular mortality. *Br Heart J* 1992; **67**: 92–6.

*38 Amato G, Carella C, Fazio S *et al.* Body composition, bone metabolism, heart structure and function in growth hormone deficient adults before and after growth hormone replacement therapy at low doses. *J Clin Endocrinol Metab* 1993; **77**: 1671–6.

39 Cittadini A, Cuocolo A, Merola B *et al.* Impaired cardiac performance in growth hormone deficient adults and its improvement after growth hormone replacement. *Am J Physiol* 1994; **267**: E219–E225.

*40 Fort S, Weaver JU, Monson JP, Mills PG. The effects of low-dose recombinant human growth hormone on cardiovascular structure and function in hypopituitary growth hormone-deficient adults. *Endocrinol Metab* 1995; **2**: 119–26.

*41 Sacca L, Cittadini A, Fazio S. Growth hormone and the heart. *Endocr Rev* 1994; **15**: 555–73.

*42 Fowelin J, Attrall S, Lager I, Bengtsson BÅ. Effects of treatment with recombinant human growth hormone on insulin sensitivity and glucose metabolism in adults with growth hormone deficiency. *Metabolism* 1993; **42**: 1443–7.

43 Jenkins PJ, Fairclough P, Lowe DG *et al.* Acromegaly, colonic polyps and neoplasia. *Clin Endocrinol* 1997; **47**: 17–22.

44 De Boer H, Blök GJ, Popp-Smjders C *et al.* Monitoring of growth hormone replacement therapy in adults based on measurement of serum markers. *J Clin Endocrinol Metab* 1996; **81: 1371–7.

45 Hoffman DM, O'Sullivan AJ, Baxter RC, Ho KY. Diagnosis of growth hormone deficiency in adults. *Lancet* 1994; **343: 1064–8.

46 Janssen YJ, Frölich M, Roelfsema F. A low starting dose of genotropin in growth hormone deficient adults. *J Clin Endocrinol Metab* 1997; **82: 129–35.

47 Mardh G, Laudin K, Borg G, Jonsson B, Lindeberg A. Growth hormone replacement therapy in adult hypopituitary patients with growth hormone deficiency: combined data from 12 European placebo-controlled clinical trials. *Endocrinol Metab* 1994; **1** (Suppl A): 43–9.

48 Van den Berg G, Veldhuis D, Fröhlich M, Roelfsema F. An amplitude-specific divergence in the pulsatile mode of GH secretion underlies the gender difference in mean GH concentrations in men and premenopausal women. *J Clin Endocrinol Metab* 1996; **81**: 2460–7.

49 Corpas E, Harman SM, Blackman MR. Human growth hormone and human ageing. *Endocr Rev* 1993; **14**: 20–39.

50 Burman P, Johansson AG, Siegbahn A, Vessby B, Karlsson FA. Growth hormone (GH)-deficient men are more responsive to GH replacement therapy than women. *J Clin Endocrinol Metab* 1997; **82: 550–5.

51 Clemmons DR, Underwood LE, Ridgay EC *et al.* Estradiol treatment of acromegaly: reduction of immunoreactive somatomedin C and improvement of metabolic status. *Am J Med* 1980; **69**: 571–5.

52 Frantz AG, Rabkin MT. Effects of estrogen and sex difference on secretion of human growth hormone. *J Clin Endocrinol Metab* 1965; **25**: 1470–80.

53 Ho KK, O'Sullivan AJ, Weissberger AJ, Kelly JJ. Sex steroid regulation of growth hormone secretion and action. *Horm Res* 1996; **45**: 67–73.

54 Goodman-Gruen D, Barrett-Connor E. Effect of replacement estrogen on insulin-like growth factor-I in postmenopausal women: the Rancho Bernardo Study. *J Clin Endocrinol Metab* 1996; **81**: 4268–71.

*55 Fisker S, Vahl N, Jorgensen JOL, Christiansen JS, Orskov H. Abdominal fat determines growth hormone-binding protein levels in healthy nonobese adults. *J Clin Endocrinol Metab* 1997; **81**: 123–8.

11: What is the impact of GH deficiency and GH replacement on quality of life in childhood-onset and adult-onset GH deficiency?

Stephen P. McKenna and Lynda C. Doward

Introduction

'We should set the highest value, not on living, but on living well.'
Socrates (quoted by Plato) [1]

The question of how life gains its quality has been an issue of concern to philosophers throughout history. As early as the fourth century BC, Socrates declared that there were some things he feared more than death, and that it is not life itself but the quality of that life that counts most. In the health-care sector, interest in the concept of 'quality of life' (QoL) has increased steadily since the late 1940s and particularly over the last 20 years. During that time, QoL has developed from an imprecise concept into a topic of scientific enquiry. The first clinical publications incorporating the term QoL appeared in the 1960s, and QoL was introduced as a heading on MEDLINE in 1975. The number of MEDLINE publications using QoL as a keyword has grown exponentially since then, with 8 136 such articles published between 1991 and 1996 [2].

This vast interest in QoL belies the fact that the construct is still beset by theoretical disagreement. Two main approaches to the measurement of QoL can be found: the function-based health-related quality of life (HRQL) approach and the needs-based approach.

Health-related quality of life

The recent adoption of the term HRQL has grown out of the functionalist approach to QoL. Patrick [3] defines HRQL as the capacity to perform the usual daily activities for a person's age and major social role. This role might be paid employment, school, housework, or simply self-care. Deviation from normality results in a reduced QoL. Within HRQL, there is general agreement that the concept is multidimensional, but there is less consensus about the full range of domains that should be measured. Despite this, most HRQL instruments include assessments of physical, social and emotional functioning.

Recently, clear distinctions have been made between the HRQL dimensions and QoL [4–6]. HRQL instruments tend to focus on symptoms and disability (in functional terms) and, as such, provide information about the level of impairment or disability experienced by the patient. However, it is argued that they do not provide information about the impact of the condition on the patient's QoL. For example, the salience of limited mobility is dependent on environmental, employment and social factors. HRQL can be seen as equating to health status, and the terms are often used interchangeably [7,8]. Indeed, almost any measure that assesses some aspect of health status may be referred to as an HRQL instrument.

Needs-based QoL

One of the earliest advocates of the importance of needs in relation to QoL was Thomas More [9]. In 1516, More argued that the quality of human life is dependent on the satisfaction of certain basic needs—for example, those for health, mobility, adequate nutrition and shelter. Major advances in needs theory in the 1940s and 1950s resulted from investigations into human motivation. Researchers in this field proposed that individuals are driven or motivated by their needs [10,11]. The relation between needs and QoL continued to be explored within the social indicators movement [12].

The needs-based approach to QoL measurement draws on these theories. Hunt and McKenna's model [13], which is the most widely implemented, arose during the development of a QoL instrument specific to depression. Analysing the transcripts of patient interviews, they noticed that the respondents described their experiences in terms of needs that were (or were not) being met, rather than in functional terms. Patients who had recovered from their illness referred to needs that they had become able to satisfy as their health improved.

This led to the adoption of a model [13,14] that postulated that life gains its quality from the ability and capacity of individuals to satisfy their needs (either inborn or learned during socialization processes). Functions such as employment, hobbies and socializing are important only insofar as they provide the means by which these needs can be fulfilled. It is taken as axiomatic that QoL is high when most human needs are fulfilled, and low when few needs are being satisfied.

Impact of growth hormone deficiency on quality of life

Growth hormone deficiency and quality of life in childhood

Enhanced growth velocity is an established benefit of replacement GH (rGH) therapy in GH-deficient children. Indeed, the administration of rGH in such circumstances is the classic example of the use of QoL to justify pharmaceutical therapy. However, there is an assumption that the benefits of the therapy are large enough to justify the cost to health services. Several recent studies have suggested that short stature in itself is not a major determinant of QoL [15–18], and consequently would not provide sufficient justification for replacement therapy. However, such a conclusion would assume that GH-deficiency does not have additional influences on QoL.

Unfortunately, the measures available to assess the QoL of GH-deficient children are rather primitive, lacking the quality of disease-specific measures developed for use with adults. Despite this, there is some limited evidence of the adverse effects of GH-deficiency on the health status of children. Erling *et al.* [19] showed that short children did not differ from a control group of children of average height in terms of their physical abilities, well-being, and relationships with family members and others. In contrast, there were indications that GH-deficient children had lower scores than average children for alertness, mood and stability. Unfortunately, the study suffered from low sample sizes and the use of visual analogue scales, which lack reliability and sensitivity.

It is clear that should questions be raised in the future about the value of rGH therapy for the treatment of GH-deficient children, it will be necessary to develop measurement instruments capable of assessing the real impact of the deficiency on their QoL. However, assessing QoL in children presents problems related to their ability to complete the type of questionnaires employed with adults and to their rapid intellectual development. For example, the needs of children below the age of eight years are likely to be very different from those of individuals who have reached sexual maturity.

Adult growth hormone deficiency and quality of life

Individuals who acquired growth hormone deficiency (GHD) as children have been reported to have normal educational attainment, but it was found that they are more likely to be unemployed and less likely to be married than members of the general population. Relatively few hold a driving licence and low involvement in leisure activities has been reported [20].

Several small-scale and larger international studies of the effects of GHD on health status in adulthood have been conducted. A majority of the patients studied had structural pituitary disease and varying degrees of hypopituitarism on conventional replacement therapy. Therefore, the aetiological role of GHD in this situation is inferred rather than proven. These studies have generally employed either the Nottingham Health Profile (NHP) or the Psychological General Well-Being Schedule (PGWB), or both. The NHP assesses physical, social and emotional distress and is widely used throughout Europe and North America [21]. It provides information about problems experienced in six areas: physical mobility, pain, energy level, sleep, emotional reactions and social isolation. As its name suggests, the PGWB is a measure of well-being [22]. It was developed in the United States but has been adapted for use in a number of European countries [23]. Unfortunately, the adapted versions of the PGWB differ from country to country.

Comparisons of GH-deficient patients with control subjects

Björk *et al.* [24] compared 23 childhood-onset GH-deficient adults with a control group consisting of 47 healthy individuals, using the NHP, the PGWB and other indicators. They found that the GH-deficient patients reported more problems with sleep, social isolation and physical mobility.

Rosén *et al.* [25] studied 86 patients with GHD, all but three of whom developed the deficiency in adulthood. This sample was matched by gender, marital status and socio-economic group with a random sample from the local population, for whom NHP data were available. They found that the GH-deficient patients had more problems in the energy level and social isolation sections of the profile.

Changes in health status following treatment with replacement growth hormone

McGauley [26] assessed the perceived health of 24 adults aged 18–55 years before and after treatment with rGH, again using the NHP and PGWB. Prior to treatment, patients had greater distress than members of a control group matched for age, gender and other sociodemographic variables, particularly in the energy level, social isolation and emotional reactions sections of the NHP. The GH-deficient patients had lower well-being, measured by the PGWB. Unfortunately, nine of the patients were GH-deficient secondary to treatment of Cushing's disease, which is likely to have influenced the results.

The GH-deficient patients were randomly allocated to treatment or placebo groups. No difference in health status was found between the two groups following one month's treatment. After six months, the energy level and emotional reaction scores of treated patients showed significant improvements compared with placebo patients, although there were no significant differences on the PGWB.

In a study involving 14 patients, Whitehead *et al.* [27] failed to find a change in the NHP or PGWB scores after six months treatment with GH.

Bengtsson and his colleagues [28] used the Comprehensive Psychological Rating Scale (CPRS) in a placebo-controlled six-month crossover trial involving 10 GH-deficient patients. Prior to treatment, the patients exhibited problems of lack of energy, lack of initiative, poor concentration and memory difficulties. After 26 weeks' treatment, there was a significant improvement in scores on the CPRS.

The findings from these studies suggest that GHD does have an impact on the health status and well-being of patients. Furthermore, treatment appears to improve well-being. However, these studies have a number of shortcomings that may explain the lack of consistency in the findings. In particular, the studies used small groups of patients and unspecified or inappropriate control groups.

More recently, better-designed studies have been carried out. Sample sizes have been increased by recruiting patients from a number of European countries. Mårdh and colleagues [29] combined data from seven European trials in which 124 patients were included. All the trials consisted of a six-month double-blind, placebo-controlled treatment period, followed by open-labelled treatment for six or 12 months. The patient-completed measures were the NHP and the PGWB. After six months, significant improvements were found in the energy level and sleep sections of the NHP for the treated group (controlling for changes in the control group). A relative improvement in well-being was also found.

These larger studies still fail to overcome another problem with research in this field. The health status measures used (including the NHP and PGWB) are generic measures of distress and well-being. Consequently, they lack the responsiveness necessary to show the depth and range of the impact that GHD might have on the lives of patients.

The role of adult short stature on health status

Busschbach *et al.* [18] compared the NHP scores of short adults relative to GHD, childhood-onset renal failure, Turner syndrome and idiopathic short stature. Short adults who had not been presented to a paediatrician as children were also included in the study. The researchers were unable to show differences between the groups in scores on the NHP sections. Furthermore, the scores of the patient groups did not differ from those of the control population. The authors concluded that short stature does not influence QoL.

However, there are a number of problems with the study. The use of the NHP, together with the small sample sizes employed (ranging from 17 to 44 patients), means that the probability of finding real differences between the groups was extremely small. Furthermore, the application of new scaling techniques makes it clear that it is not valid to make direct comparisons between different diagnostic groups using a generic measure [30]. Finally, as has been seen above, short stature is only one of the influences on the QoL of patients.

Health status as an indicator of the need for replacement growth hormone

Holmes and Shalet [31] invited adults with GHD attending the endocrinology out-patient department at Christie Hospital National Health Service Trust in Manchester to participate in a 12-month study of GH replacement. Approximately two-thirds of those invited chose to enter the trial. There was no indication that the patients who entered the study had a greater degree of GH deficiency than those who did not. However, participants were shown to have significantly higher distress in energy level and emotional reactions, as measured by the NHP. PGWB scores did not differ between the groups. The study suggests that it is not the extent of the deficiency itself but the associated distress that determines whether or not patients seek treatment.

In a follow-up study by the same authors [32], participants were asked whether they wished to continue treatment at the end of the 12-month trial. Of those who completed the trial, approximately half expressed a desire to continue on treatment. There were no clear differences between the groups in terms of their clinical or perceived health status, although there was a tendency for those who chose to continue on GH replacement to have a greater severity of GHD and to have had more marked energy-related distress at entry to the study.

The inadequacy of the standard instruments was confirmed by the study by Carroll *et al.* [33]. They concluded that the NHP and PGWB were not useful for predicting who would choose rGH therapy in the long term. In contrast, Wallymahmed *et al.* [34] found that patients who chose to continue on rGH therapy reported statistically significant improvements in energy level, as assessed by the NHP.

Measures specific to adult growth hormone deficiency

The lack of sensitivity of the health status instruments commonly used to assess the impact of GHD on adult patients has led researchers to develop more accurate instruments. The major problem with the older instruments is that they were designed to be used with a wide range of health problems. Their generic nature means that they miss areas that are important to patients with GHD. Similarly, they include items that are of little relevance to this population. Consequently, the measures are far less likely to identify changes in the QoL of patients associated with effective replacement therapy. This lack of sensitivity is compounded by the instruments' relatively poor psychometric properties. For example, they have inadequate reliability, and the resultant high levels of measurement error again limit their ability to detect changes in QoL. The full impact of the deficiency on QoL and the benefits of replacement therapy cannot be determined without using measures specifically designed to assess the concerns of these individuals.

Two groups have attempted to develop measures that are specific to this patient population. Wallymahmed *et al.* [35] report on the development of a QoL 'model' for adult GHD. Rather than developing a new measure, the bulk of their instrument battery consists of pre-existing measures, including the NHP. They also include a generic measure of life fulfilment, which appears to be the main aspect of the model. Unfortunately, the instruments included in the battery have inadequate reliability, limiting their ability to measure change in health status.

The lack of responsiveness of the battery was confirmed by its use in a three-year follow-up of GHD adults receiving replacement therapy [34]. Although a large number of measures were included in the study, no consistent changes were found for the treatment group. Despite the inadequacy of the measuring instruments employed, the authors suggest that the beneficial effects of rGH therapy on QoL decline after two years. Unfortunately, such a conclusion is unjustified given the data presented.

The second approach involved a major programme of research that led to the development of the Quality of Life-Assessment of GH Deficiency in Adults (QoL-AGHDA). This measure, which adopted the needs-based model of QoL, was required to be patient-based and to have very good psychometric properties, making it suitable for use in clinical trials and for monitoring individual patients in routine clinical practice.

The remainder of this chapter reports on the development and testing of the QoL-AGHDA, and considers its value in research into adult GHD.

Development of the QoL-AGHDA

Thirty-six GHD adults attending the Christie Hospital in Manchester were invited to be interviewed in their own homes [36]. Fifteen of the respondents were men and 21 women, and their ages ranged from 21 to 52 years. Of the sample, 21 were in full-time employment. Fifteen of the interviewees had acquired their deficiency in childhood.

An in-depth qualitative approach was used, whereby the interviewees were encouraged to talk at length about their daily life and any problems they perceived themselves to have. The interviews, which were tape-recorded, took the form of a focused conversation with minimum input from the interviewer.

The interviews confirmed that the questionnaires used previously had not addressed the range and nature of concerns of the patients. Most respondents were in fact troubled by lack of energy, but their social problems were more complex than had been suggested by previous research. They had problems with memory and attention, over-stimulation, irritability, lack of stamina and the consequences of their body shape, making it difficult to find clothes that would fit them. Consequently, they were reluctant to join or accept invitations to social gatherings, leading to social isolation. Other problems identified included low motivation and sexual drive, poor self-confidence and self-esteem, depressed mood and lack of strength and stamina.

There were clear differences between those who had acquired GH deficiency as children and those who had become deficient as adults. The former group appeared to be better adapted to their limitations. Most were unmarried, and had found ways of coping with the way they felt. In contrast, patients who had developed the deficiency as adults were very aware of the ways in which they had changed. They were mostly married with children and in employment, but found themselves no longer able to participate in family, social and occupational life in the way to which they had been accustomed.

The tapes of the interviews were transcribed, and item analyses conducted to produce a pool of items from which the draft measure was constructed. As far as possible, the actual words used by the interviewees were maintained. As it was intended that the new measure should be available in several languages, a meeting was held to consider the relevance of the items derived to the different cultures and whether they could be translated into the target languages. The countries represented were the United Kingdom, Sweden, Germany, Italy and Spain. This meeting led to a refinement of the item pool, which went forward for translation.

The major issues for translation related to conveying conceptual equivalence of the English items in the other languages. Two panels were employed in each country. The first consisted of local people who were bilingual or who had very good English. They agreed conceptually equivalent translations, which were passed to the second panel. This group consisted of people of average educational level who worked in the target language only. This second panel was used to ensure that the level of language in the new measure would be suitable for potential respondents.

Testing the QoL-AGHDA

Face and content validity

The face and content validity of the new instrument was tested through interviews with GHD patients in each country. Access to between 13 and 20 relevant patients was arranged through local clinicians. Patients were asked to complete the questionnaire and then to comment on the ease of completion, relevance of the content and whether any important aspects of their experience had been omitted.

Overall, patients found the QoL-AGHDA easy to understand and complete. It was felt by respondents that the QoL-AGHDA was much more relevant than others they had completed in the past (primarily the NHP and the PGWB). A majority said that the response system was easy to understand and very clear. However, it was necessary to remove three items that were reported to be of questionable relevance, and to amend another. Consequently, a second draft of the questionnaire was prepared and tested with 10 more people in each country.

The final version of the QoL-AGHDA has 25 items which are answered 'yes' or 'no'. Each positive response is scored 1, with a high score indicating poor QoL. Examples of the items in the measure are shown in Table 11.1 [37].

Reliability and construct validity

In the UK, 52 GHD patients from three hospitals in the Manchester area were recruited to a postal survey designed to assess the new instrument's reliability and construct validity. None of those recruited were receiving rGH. Patients from one of the hospitals had turned down the opportunity

Table 11.1 Example items from the QoL-AGHDA.

It is difficult for me to make friends
I often forget what people have said to me
I find it difficult to plan ahead
I have difficulty controlling my emotions
I often feel too tired to do the things I ought to do

to enter a GH replacement trial. It was hypothesized that these patients would have better QoL than those from the other two hospitals.

All patients completed the QoL-AGHDA on two occasions, approximately three weeks apart. They also answered the General Well-Being Index (GWBI, the British version of the PGWB [23]) on the first occasion. Background demographic information was available from medical records.

All but one of the patients contacted returned completed questionnaires on both occasions, and one failed to return the second questionnaire. This represented an overall response rate of 98.2%, which is remarkably high for this type of research and indicates the acceptability of the measure to the respondents. The rate of missing responses was 0.7%, again a very good figure for a postal survey.

Similar methodologies were followed in the other countries, with minor variations. The main difference between the national studies was in the choice of the comparator instrument. The PGWB was used in Sweden and Germany and the NHP in Spain and Italy, where validated versions of the PGWB were not available. Details of the samples are shown in Table 11.2.

Reliability of the QoL-AGHDA

Any measure of health must have good reliability (that is, low random measurement error). The appropriate method of determining the reliability of an instrument intended for use in a clinical trial is to apply it twice to the same population and correlate the scores, as stability over time (or reproducibility) is the crucial variable. Where reliability is low (that is, where the correlation coefficient is below 0.85 [38]), an instrument will lack sensitivity.

The reliability of the five language versions of the QoL-AGHDA ranged from 0.86 to 0.92, indicating that all versions of the measure produce low levels of random error. Many authors of measures report internal consistency (alpha values) rather than test–retest reliability. This statistic indicates the interrelatedness of items, and does not include error associated with time between administrations, which is crucial where an instrument is

Table 11.2 Patient samples included in the Quality of Life-Assessment of GH Deficiency in Adults (QoL-AGHDA) validation studies.

Country	*n*	Male (%)	Female (%)	Mean age (years)	Paediatric onset (%)	Married (%)
UK	55	46	54	46.9	18	59
Sweden	82	55	45	44.4	23	59
Germany	34	82	18	31.1	69	28
Italy	44	66	34	39.5	30	55
Spain	38	66	34	33.9	*	34

*Information not available.

intended for use in longitudinal studies. However, the QoL-AGHDA also has excellent internal consistency, with the values ranging from 0.88 to 0.93 for the five versions of the measure.

Construct validity of the QoL-AGHDA

The validity of any questionnaire is established over time as it is used. However, a number of assessments of the validity of the different versions of the QoL-AGHDA have been undertaken.

In the absence of a gold standard for QoL, scores on the QoL-AGHDA were related to those on the comparator measures completed by respondents. Previous studies have suggested that QoL instruments have a correlation of between 0.5 and 0.8 with the PGWB/GWBI. Psychological well-being is not the same construct as QoL, but would be expected to have a relatively strong association with it. The correlation between the QoL measure and the PGWB/GWBI was 0.74 in the UK, 0.78 in Sweden and 0.75 in Germany. These consistent findings indicate that the measures have between 55 and 60% variance in common.

There was less correspondence with the NHP, which was used in Italy and Spain. Despite this, there was a moderately strong relation between QoL and the energy level, emotional reactions and physical mobility sections. As noted above, these sections of the NHP have been shown to be important in previous research.

There were clear trends between the scores of treated and untreated patients in the different countries, with the latter group having worse QoL. These differences could only be tested formally in Sweden, where there was a larger sample size in the validation study and a more even split between the number of treated and untreated patients.

GHD patients in the UK who had declined the opportunity to receive replacement therapy had a mean score of 6.0 on the measure, compared with 11.1 for those waiting to start treatment. This difference was statistically significant ($p < 0.02$), and suggests that patients' willingness to receive treatment is related to their QoL, confirming the view of Shalet [39].

A final piece of evidence of the validity of the new measure came from Spain. Here, patients were asked to rate their general health on each occasion that they completed the questionnaires. It was expected that differences in reported health status would be associated with QoL. There was a strong relation between self-rated general health and QoL, with those rating their health fair or bad having significantly worse QoL on both occasions.

Responsiveness of the QoL-AGHDA

Any instrument intended for use in clinical trials or for monitoring individual patient care must be responsive—that is, able to detect changes in the measured construct associated with a change in the condition of the patient. Measures with poor reliability have poor responsiveness [40]. This is true for generic health status measures such as the NHP and the Short Form Health Survey (SF-36) [41], which both have relatively poor reliability [42]. Dixon *et al.* [43] suggest that up to 20 000 cases would be required to show differences in effectiveness between two treatments using the SF-36.

In contrast, the QoL-AGHDA has very good reliability and would, consequently, be expected to have better responsiveness than the generic health status measures. Responsiveness can only truly be determined by comparing the performance of different instruments employed in the same study. Various formulae are used to determine responsiveness, but the most widely used is effect size, which is the change in the mean score of the group resulting from an intervention, divided by the baseline standard deviation. Evidence of the QoL-AGHDA's responsiveness comes from studies of GH replacement in the Netherlands and Spain [44,45], where the NHP was also employed. The effect sizes for the QoL-AGHDA were 0.47 and 0.68, respectively. The comparative effect sizes for the NHPD (the index of distress embedded in the NHP) were 0.27 and 0.44. Thus, it is clear that the new measure is more likely to detect improvements in QoL resulting from GH replacement therapy.

New language versions of the QoL-AGHDA

Work is continuing on the development and validation of new language versions of the QoL-AGHDA. Validation has recently been completed on the US, French, Dutch and Danish versions and measures for Belgium, Norway and Iceland are due to be completed in the near future. Table 11.3 shows the test–retest reliability and alpha values for the completed versions. The consistently high reliability and internal consistency are indicative of the suitability of the measure for adaptation into other languages.

The QoL-AGHDA is currently included in KIMS, an international outcomes research database, monitoring the long-term efficacy and safety of GH replacement therapy [46].

Comparison of QoL-AGHDA scores between GHD and normal adults

A study was conducted in Sweden to investigate the scaling properties of

Country	Test–retest reliability	Internal consistency
United Kingdom	0.93	0.93
Sweden	0.93	0.92
Germany	0.89	0.90
Italy	0.85	0.89
Spain	0.91	0.88
USA	0.88	0.89
The Netherlands	0.94	0.88
Denmark	0.93	0.93
France	0.88	0.95

Table 11.3 Test–retest correlation and internal consistency of the different language versions of the QoL-AGHDA.

the QoL-AGHDA [47]. The study also allowed the comparison of scores for GHD adults with those of matched age and gender cohorts in the Swedish population.

The QoL-AGHDA was administered to a sample of 1618 people from the general population in Sweden. The mean age of this sample was 45.8 years (range 25–65 years), and 52% were women. QoL-AGHDA scores were also available for a sample of 111 untreated GHD adults, with a mean age of 49.9 years (range 20–75 years), 35% of whom were women.

The data from both samples were fitted to the dichotomous Rasch model [48] to investigate the scaling properties of the measure and to see whether it would be valid to make comparisons between the two groups.

The results of these analyses confirmed the unidimensionality of the QoL-AGHDA, providing additional evidence of its construct validity. The data also suggested that it would be valid to make direct comparisons between scores obtained by the two groups, thus allowing a comparison to be made between the QoL of GH-deficient adults and those without the deficiency.

The Rasch analyses indicated that the Swedish QoL-AGHDA provided data at the ordinal level. Consequently, the use of median scores and nonparametric statistics are recommended, unless raw scores are transformed through Rasch analysis.

Table 11.4 shows the age-specific and sex-specific scores for the two groups. There were no significant differences in QoL-AGHDA scores by age for the Swedish population. However, women scored significantly higher than men (Mann–Whitney U test, $p < 0.01$). There were no significant differences in score associated with age or sex in the patient sample.

The data from the untreated GHD population show significant differences from those of the Swedish population, as determined by non-overlapping confidence intervals. These are the first reliable data indicating that the QoL of GH-deficient adults is inferior to that of an average population.

Table 11.4 QoL-AGHDA scores for growth hormone-deficient adults and a random population sample in Sweden.

Group	*n*	Mean	Median	95% confidence interval
Male GHD	72	8.5	7.5	6.9–10.2
Male population sample	710	3.0	2.0	2.7–3.3
Female GHD	39	9.5	10.0	7.8–11.3
Female population sample	738	4.3	3.0	4.0–4.6
Total GHD	111	8.9	9.0	7.7–10.1
Total population sample	1448	3.7	2.0	3.4–3.9

GHD: growth hormone deficiency.

It remains necessary to test whether Swedish patients receiving GH replacement therapy obtain scores on the QoL-AGHDA that approach or are equal to those of individuals without the deficiency. Preliminary data from KIMS indicate that this is the case. Similarly, data from a UK study indicate the sensitivity of QoL-AGHDA in documenting improvement in hypopituitary patients treated with GH using a careful dose titration regimen [49].

Conclusions

It has long been accepted that GHD children should receive replacement GH therapy. The justification for such treatment is clearly based on the impact of short stature on QoL. However, the case for the provision of replacement therapy is not so clear-cut in adults.

Recent evidence has shown that adult GHD is associated with a number of adverse changes in body composition, physical performance and health status. Holmes and Shalet [31,32] argue that it will be some time before clinical benefits such as reductions in mortality from cardiovascular disease or reductions in the incidence of bone fractures will be able to justify replacement therapy. Consequently, the most compelling reason for prescribing such therapy at present is to improve the QoL of patients. Given the relatively high cost of the therapy, it is essential to be able to show accurately these improvements in QoL. Until recently, researchers have been limited to the use of rather dated generic measures of health status. The development of the QoL-AGHDA has made it possible to explore the impact of replacement therapy in adults with much greater precision. Evidence now suggests that there are clear differences between the QoL of adult GHD patients and normal populations.

It remains necessary to show, using the QoL-AGHDA, that QoL improves significantly with replacement therapy. Such evidence may well be required to justify the costs of the therapy to severely stretched health service budgets. The measure may also be useful in predicting those GHD adults who will benefit most from receiving replacement therapy.

References

1 Plato. *Euthyphro, Apology, Crito,* trans. Church FJ. New York: Bobbs-Merrill, 1956: 57.

** 2 Doward LC, McKenna SP. Evolution of quality of life assessment. In: Rajagopalan R, Sheretz EF, Anderson RT, eds. *Care Management of Skin Diseases: Life Quality and Economic Impact.* New York: Dekker, 1997: 9–33.

3 Patrick DL, Bush JW, Chen MM. Methods for measuring levels of well-being for a health status index. *Health Serv Res* 1973; 8: 228–45.

* 4 Tennant A, McKenna SP. Conceptualising and defining outcome. *Br J Rheumatol* 1995; **34**: 899–900.

* 5 Heinemann AW, Whiteneck GG. Relationships among impairment, disability, handicap and life satisfaction in persons with traumatic brain injury. *J Head Trauma Rehabil* 1995; **10**: 54–63.

6 Whiteneck GG. Measuring what matters: key rehabilitation outcomes. *Arch Phys Med Rehabil* 1994; **75**: 1073–6.

7 Patrick DL, Bergner M. Measurement of health status in the 1990s. *Ann Rev Public Health* 1990; **11**: 165–83.

8 Guyatt GH, Feeny DH, Patrick DL. Measuring health-related quality of life. *Ann Intern Med* 1993; **118**: 622–9.

9 More T. *Utopia* [1516]. London: Chatto and Windus, 1908.

10 Maslow AH. *Motivation and Personality.* New York: Harper and Row, 1954.

11 McCelland DC. *Personality.* New York: Sloane, 1951.

12 McCall S. Quality of life. *Soc Indic Res* 1975; **2**: 229–48.

** 13 Hunt SM, McKenna SP. The QLDS: a scale for the measurement of quality of life in depression. *Health Policy* 1992; **22**: 307–19.

14 Hörnquist JO. The concept of quality of life. *Scand J Soc Med* 1982; **10**: 57–61.

15 Sandberg DE, MacGillivray MH, Clopper RR *et al.* Quality of life among formerly treated childhood-onset growth hormone-deficient adults: a comparison with unaffected siblings. *J Clin Endocrinol Metab* 1996; **83**: 1134–42.

16 Pilpel D, Leiberman E, Zadik Z, Carel CA. Effect of growth hormone treatment on quality of life of short-stature children. *Horm Res* 1995; **44**: 1–5.

17 Downie AB, Mulligan J, McCaughey ES *et al.* Psychological response to growth hormone treatment in short normal children. *Arch Dis Child* 1996; **75**: 32–5.

18 Busschbach JJV, Rikken B, Grobbee DE, De Charro FT, Wit JM. Quality of life in short adults. *Horm Res* 1998; **49**: 32–8.

19 Erling A, Wiklund I, Albertsson-Wikland K. The well-being of short statured children. *Qual Life Res* 1995; **4**: 422–3.

20 Blizzard R, Joyce S, Mitchell T *et al.* Psychosocial impact of long-term growth hormone therapy. In: Raita S, Tolman R, eds. *Human Growth Hormone.* New York: Plenum Medical, 1986: 93–106.

21 Hunt SM, McEwen J, McKenna SP. *Measuring Health Status.* London: Croom Helm, 1986.

22 Monk M. Blood pressure awareness and psychological well-being in the Health and Nutrition Examination Survey. *Clin Invest Med* 1981; **4**: 183–9.

23 McKenna SP, Hunt SM. The General Well-Being Index: adapting and retesting an American measure for use in the United Kingdom. *Br J Med Econ* 1992; **4**: 41–50.

24 Björk S, Jonsson B, Westphal O, Levin JE. Quality of life of adults with growth hormone deficiency: a controlled study. *Acta Paediatr Scand* 1989; **356** (Suppl): 55–9.

25 Rosén T, Wirén L, Wilhelmsen L, Wiklund I, Bengtsson BÅ. Decreased psychological well-being in adult patients with growth hormone defi-

ciency. *Clin Endocrinol* 1994; **40**: 111–16.

26 McGauley G. Quality of life assessment before and after growth hormone treatment in adults with growth hormone deficiency. *Acta Paediatr Scand* 1989; **356** (Suppl): 70–2.

27 Whitehead HM, Boreham C, McIlrath EM *et al.* Growth hormone treatment of adults with growth hormone deficiency: results of a 13-month placebo controlled cross-over study. *Clin Endocrinol* 1992; **36**: 45–52.

28 Bengtsson BÅ, Edén S, Lönn L *et al.* Treatment of adults with growth hormone deficiency with recombinant human growth hormone. *J Clin Endocrinol Metab* 1993; **76**: 309–17.

*29 Mårdh G, Lundin K, Borg G, Jonsson B, Lindeberg A. Growth hormone replacement therapy in adult hypopituitary patients with growth hormone deficiency: combined data from 12 European placebo-controlled clinical trials. *Endocrinol Metab* 1994; **1** (Suppl): 43–9.

30 Whalley D. Determining cross-cultural validity of quality of life measures. *Qual Life Res* 1997; **6**: 742.

31 Holmes SJ, Shalet SM. Characteristics of adults who wish to enter a trial of growth hormone replacement. *Clin Endocrinol* 1995; **42**: 613–18.

32 Holmes SJ, Shalet SM. Factors influencing the desire for long-term growth hormone replacement in adults. *Clin Endocrinol* 1995; **43**: 151–7.

33 Carroll PV, Littlewood R, Weissberger AJ *et al.* The effects of two doses of replacement growth hormone on the biochemical, body composition and psychological profiles of growth hormone-deficient adults. *Eur J Endocrinol* 1997; **137**: 146–53.

34 Wallymahmed ME, Foy P, Shaw D *et al.* Quality of life, body composition and muscle strength in adult growth hormone deficiency: the influence of growth hormone replacement therapy for up to 3 years. *Clin Endocrinol* 1997; **47**: 439–46.

35 Wallymahmed ME, Baker GA, Humphrist G, Dewey M, MacFarlane IA. The development, reliability and validity of a disease-specific quality of life model for adults with growth hormone deficiency. *Clin Endocrinol* 1996; **44**: 403–11.

*36 Hunt SM, McKenna SP, Doward LC. Preliminary report on the development of a disease-specific instrument for assessing quality of life of adults with growth hormone deficiency. *Acta Endocrinol* 1993; **128** (Suppl 2): 37–40.

37 Holmes SJ, McKenna SP, Doward LC, Shalet SM. Development of a questionnaire to assess the quality of life of adults with growth hormone deficiency. *Endocrinol Metab* 1995; **2: 63–9.

38 Weiner E, Stewart B. *Assessing Individuals*. Boston: Little, Brown, 1984.

39 Shalet SM. Growth hormone deficiency and replacement in adults [editorial]. *Br Med J* 1996; **313**: 314.

40 Streiner D, Norman G. *Health Measurement Scales*. Oxford: Oxford University Press, 1989.

41 Ware JE, Sherbourne CD. The MOS 36-Item Short-Form Health Survey (SF-36). *Med Care* 1992; **30**: 474–83.

42 McKenna SP. Quality of life assessment in the conduct of economic evaluations of medicines. *Br J Med Econ* 1995; **8**: 33–8.

43 Dixon P, Heaton J, Long A, Warburton A. Reviewing and applying the SF-36. *Outcomes Brief* 1994; 4 (Aug): 3–25.

44 McKenna SP, Koppeschaar HPF, Zelissen PMJ, *et al.* The impact of replacement GH on the quality of life of patients: results of a clinical trial in the Netherlands. *Endocrinol Metab* 1997; **4** (Suppl B): 167.

45 McKenna SP, Vazquez JA, Pico A *et al.* The quality of life of adult GHD patients in Spain and the effects of replacement therapy with Genotropin. *Endocrinol Metab* 1997; **4** (Suppl B): 168.

46 Pharmacia & Upjohn. An introduction to KIMS: an international database for follow-up on the long-term efficacy and safety of Genotropin therapy. Oxford: Oxford Clinical Communications, 1994.

*47 Wirén L, Wilhelmsen L, McKenna SP, Hernberg-Ståhl E. The quality of life of adult growth hormone deficient patients in Sweden compared to a random population sample assessed by QoL-AGHDA. *Endocrinol Metab* 1997; **4** (Suppl B): 167.

48 Wright BD, Masters GN. *Rating Scale Analysis*. Chicago: MESA Press, 1982.

49 Drake WM, Coyte D, Camacho-Hübner C *et al.* Optimising growth hormone replacement therapy by dose titration in hypopituitary adults. *J Clin Endocrinol Metab* 1998; **83**: 3913–19.

12: Is there an upper age limit for GH replacement in hypopituitary adults?

Andrew A. Toogood and Stephen M. Shalet

Introduction

Growth hormone (GH) deficiency is a frequent finding in patients with organic disease of the hypothalamic–pituitary axis, and is associated with many adverse pathophysiological changes [1]. Body composition is abnormal in GH-deficient adults, in whom the fat mass is increased and lean body mass is reduced. Bone mineral density is reduced, particularly in younger adults who developed GH deficiency during childhood, and markers of bone formation and bone resorption are abnormal, suggesting that bone turnover is also affected. One retrospective study has suggested that GH-deficient adults are at greater risk of fracture compared with the general population [2]. Quality of life is impaired in adults with GH deficiency. Patients typically describe lack of energy, poor concentration and memory, reduced drive and stamina and difficulty coping with stress. GH deficiency causes abnormal cardiac function, an adverse serum lipid profile and reduced insulin sensitivity compared with healthy, age-matched controls. These findings may play a role in the increased cardiovascular mortality that has been observed in patients with hypopituitarism and attributed to GH deficiency [3].

GH replacement therapy has been shown to have considerable benefits in adults with GH deficiency, correcting many of the abnormalities, particularly altered body composition, osteopaenia and reduced quality of life. Although many of the adverse cardiovascular risk factors are favourably influenced by GH replacement, long-term monitoring of patients receiving GH replacement therapy is required to determine its impact on mortality [1].

Many of the changes that occur in adults with GH deficiency are also found in the ageing population [4,5]. As we grow older, there is a gain in fat mass and loss of lean body mass, and the skeleton loses minerals and is more prone to fracture. Cardiovascular risk factors evolve, and the risk of dying from cardiovascular disease increases. These changes occur in the face of declining GH secretion, leading some authors to suggest that the normal elderly are functionally GH-deficient and that GH treatment may reverse some of the age-related changes [6]. This raises several questions related to the GH status of the elderly who have organic disease of the hypothalamic–pituitary axis, e.g. a pituitary adenoma or craniopharyngioma. Do they have GH deficiency distinct from the hyposomatotropism of ageing? If the answer is yes, does it result in changes similar to those seen in younger adults with GH deficiency? Do elderly patients with organic GH deficiency benefit from GH replacement therapy, or should GH replacement therapy be restricted to patients below a certain age, e.g. those under 60 years of age? These are the questions that will be considered in this chapter.

GH status in the elderly with hypothalamic–pituitary disease

Early studies, using radioimmunoassays with low sensitivity, suggested that GH secretion actually ceased during the seventh decade of life [7]. The introduction of ultrasensitive GH assays demonstrated clearly that GH secretion continues throughout life, but there is a significant decline with time. It has been estimated that, following the increased secretion of GH associated with growth and puberty, GH secretion falls by approximately 14% per decade of adult life [8].

GH status has been studied in the elderly with hypothalamic–pituitary disease and compared with healthy controls of similar age, sex and body mass index (BMI) (Table 12.1) [9]. Twenty-four-hour GH profiles, sampling blood every 20 minutes, were recorded in 24 patients and 24 control

Table 12.1 Growth hormone status in 24 patients (16 male) with hypothalamic–pituitary disease and 24 healthy controls (17 male). AUC_{GH} = area under the 24-hour profile. Values are median (range). Data from Toogood *et al.* [9].

	Patients	Controls	*p*
Age	66.0 (61.0–85.7)	70.6 (60.8–87.5)	0.04
BMI	28.2 (22.6–37.3)	26.1 (20.1–37.0)	N.S.
AUC_{GH} (μg/L/min)	119.25 (7.27–843.60)	968.54 (227.20–4625.00)	<0.00001
Peak GH response to arginine (μg/L)	0.41 (0.03–6.89)	8.87 (1.77–42.3)	<0.00001
IGF-I* (μg/L)	102 (14–162)	142 (59–298)	<0.0001
IGFBP-3* (mg/L)	2.29 (0.81–3.75)	2.59 (1.00–3.52)	0.009

*Normative data for serum IGF-I and IGFBP-3 were obtained from 125 adults (77 male) aged between 60 and 87 years. BMI: body mass index; GH: growth hormone; IGF-I: insulin-like growth factor-I; IGFBP-3: insulin-like growth factor binding protein-3.

individuals. Initially, the profiles were assayed in a standard immunoradiometric assay, the sensitivity of which was 0.4 µg/L. Using this assay, 16 of the 24 patients were found to have no measurable GH throughout the 24-hour period. In contrast, however, each control exhibited a detectable GH level at some point during the same period. Because the assay sensitivity was relatively poor, it was not possible to quantify GH secretion accurately in either the patients or controls.

In order to obtain quantitative data from the 24-hour GH profiles, it was necessary to use a chemiluminescence GH assay that was modified to increase the sensitivity to 0.002 µg/L. Using this assay, GH levels were detectable in all the samples from all the profiles in both patients and controls. It was therefore possible to determine that organic disease in the elderly with hypothalamic–pituitary disease resulted in a 90% reduction in GH secretion (Fig. 12.1) [10].

This considerable reduction in GH secretion, similar in severity to that seen in younger adults with GH deficiency [11], demonstrates that

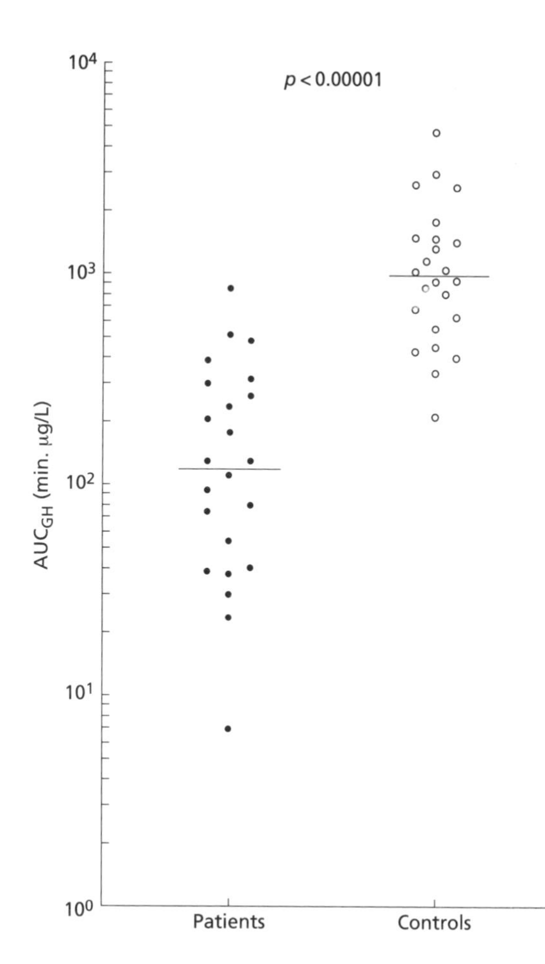

Fig. 12.1 Area under the curve of the 24-hour growth hormone (GH) profile in 24 patients with hypothalamic–pituitary disease and 24 controls, all aged over 60 years. Blood was drawn every 20 minutes and assayed in an ultrasensitive chemiluminescence GH assay, sensitivity 0.002 µg/L. Reproduced with permission from Toogood *et al.* [10].

hypothalamic–pituitary disease in an elderly patient does result in GH deficiency distinct from the hyposomatotropism of increasing age. Such a deficit warrants consideration of GH replacement therapy. However, as the pathophysiological changes associated with GH deficiency in younger adults are similar to those observed in normal ageing, it is important to determine that this degree of GH deficiency does cause significant deficits in the elderly patient.

Changes associated with organic GH deficiency in the elderly

Serum IGF-I and IGFBP-3

The serum concentrations of insulin-like growth factor-I (IGF-I) and insulin-like growth factor binding protein-3 (IGFBP-3) gradually decrease with increasing age. It is likely that this decline is, in part, related to the decline in GH secretion [12]. It is therefore important to consider IGF-I and IGFBP-3 concentrations from GH-deficient patients in the context of an age-specific normal range. Thus, it is common practice to express the serum IGF-I level as a standard deviation (SD) score, the normal range being from –2 to +2.

Serum IGF-I levels are reduced in adults with GH deficiency. IGF-I is used by some practitioners as a diagnostic marker of severe GH deficiency, although it is not suitable for all groups of patients. It is a particularly useful marker in young adults who have childhood-onset severe GH deficiency, the majority of whom have an IGF-I SD score below –2. As the age of the GH-deficient patient increases, the proportion of patients with an abnormally low IGF-I SD score diminishes [13]. In the elderly with GH deficiency, serum IGF-I levels are significantly lower than in healthy, age-matched controls; with respect to group data, however, only 12.5% of patients have an SD score below –2 (Fig. 12.2).

IGFBP-3 is the major binding protein of IGF-I, and is also regulated by GH [14]. The serum levels of IGFBP-3 also decline with increasing age, but to a lesser degree than IGF-I [15,16]. In the elderly with organic GH deficiency, the serum IGFBP-3 is significantly lower than in healthy subjects, but the overlap between the two groups is 96% (Fig. 12.2) [17].

Body composition

In GH-deficient adults, not only is body composition abnormal, with an increase in fat mass and a reduction in lean body mass [18,19], but the distribution of body fat is also abnormal, with an increase in central fat

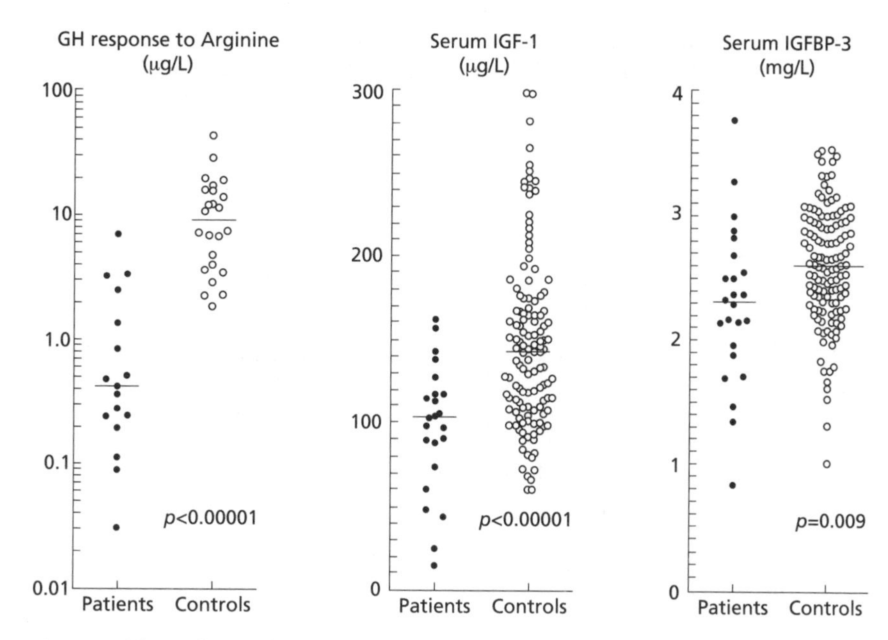

Fig. 12.2 The peak growth hormone (GH) responses to arginine infusion in 24 GH-deficient adults and 24 healthy controls aged over 60 years. Serum IGF-I and IGFBP-3 levels in the 24 GH-deficient patients compared with normative samples obtained from 125 subjects aged between 60 and 87 years. Reproduced with permission from Toogood *et al.* [17].

deposition [20]. Ageing is associated with similar changes in body composition; lean body mass falls and fat mass rises [21,22]. Computed tomography has demonstrated a change in the distribution of fat mass associated with increasing age. The amount of subcutaneous fat in the upper thigh and the abdomen decreases, and the quantity of intra-abdominal fat increases [23].

In the elderly with organic GH deficiency, body composition is abnormal compared with that of healthy controls matched for age and BMI [24]. Using dual-energy X-ray absorptiometry (DEXA) to determine body composition in elderly patients with organic GH deficiency, we were able to demonstrate that fat mass was significantly increased. Lean body mass, however, did not differ significantly between patients and controls (Table 12.2). The DEXA scans were divided into regions of the body, but this did not reveal any significant differences between the patients and controls. Despite this, the waist : hip ratio, a crude marker of central obesity, was greater in the patients than in the controls. The determination of body composition by DEXA is limited by its inability to distinguish subcutaneous fat from visceral fat. It is possible, therefore, that increased visceral fat in the GH-deficient patients was not detected in our study because of this technical limitation.

Table 12.2 Body composition in patients with growth hormone (GH) deficiency compared with healthy controls aged over 60 years. Values are expressed as median (range). Data from Toogood *et al.* [24].

	Patients ($n = 21$, 15 male)	Controls ($n = 24$, 17 male)	*P*
BMI (kg/m²)	27.8 (22.7–37.3)	26.1 (20.1–37.4)	0.15
Waist : hip ratio	0.93 (0.7–1.17)	0.89 (0.69–1.02)	< 0.05
Total fat mass (kg)	27.76 (19.245–50.243)	21.233 (8.812–49.151)	< 0.005
Total lean body mass (kg)	51.19 (26.96–69.18)	51.55 (32.35–60.525)	0.99

GH secretion is reduced by obesity *per se* [25] and, as expected, a significant negative correlation between GH secretion and fat mass was observed in healthy elderly subjects. Such a relationship was not present, however, amongst the patients with pituitary disease, indicating that the reduction in GH secretion in this group was not caused by the presence of increased fat mass.

The lean body mass determined by DEXA comprises cell mass and extracellular fluid, and is therefore influenced by hydration. Cell mass can be determined from estimates of total body potassium, and extracellular fluid can be measured by various dilution techniques. Several studies have shown a reduction in lean body tissue in younger GH-deficient adults, but in GH-deficient adults over 50 years of age, the reduction in body cell mass was no longer apparent. Thus, it appears that the effects of GH deficiency on body composition are attenuated in the elderly, the major impact being on fat mass.

The effects of GH deficiency on bone

Mineralization of the human skeleton occurs predominantly during adolescence, and continues at a diminished rate into the third decade of adult life [26,27]. Once peak bone mass has been attained, there is a gradual decline in bone mineral with increasing age [28,29]. GH-deficient adults have been shown to have reduced bone mineral density (BMD) and to be at increased risk of fracture compared with healthy controls [2]. At the present time, osteopenia is one of the major indications for GH replacement therapy in adult life. The severity of osteopenia in GH-deficient adults appears to be related to the age at which GH deficiency develops. Young adults who developed GH deficiency during childhood appear to be more osteopenic than patients who develop GH deficiency during adult life [30,31]. In our group of patients with organic GH deficiency, aged over 60, BMD was not significantly reduced at the hip or in the spine compared with age-matched controls [32]. A similar finding has been reported by a group from Japan who studied a slightly younger group with a mean age of 52 years [33].

Although BMD in elderly patients with GH deficiency was not reduced, markers of bone formation and resorption, serum osteocalcin and urine deoxypyridinoline cross-links, respectively, were reduced, suggesting that bone turnover is reduced. There is evidence to suggest that reduced bone turnover actually diminishes the risk of fracture in an elderly population [34]. It is therefore possible that the risk of fracture is decreased in the elderly with GH deficiency, rather than increased as is seen in younger GH-deficient adults. Further studies are required to determine whether the reduction in bone turnover alters the risk of fracture in the elderly with GH deficiency, and whether GH replacement therapy has a clinically significant impact on bone.

Distinguishing GH deficiency from the hyposomatotropism of age

Hypothalamic–pituitary disease in the elderly results in a 90% reduction in GH secretion, which in turn impacts significantly on serum IGF-I, body composition and bone turnover. These findings suggest that the elderly with GH deficiency may benefit from GH replacement therapy. Before we consider the effects of GH replacement therapy in this group of patients, it is important to determine which individuals are likely to derive the most benefit from GH therapy. In young adults with pituitary disease, the patients who perceive the greatest benefit from GH therapy are those who have the greatest degree of GH deficiency [35]. In the elderly, in whom the effects of GH deficiency are attenuated, it is likely that only those with very severe GH deficiency will derive benefit from GH therapy. It is important that these patients are identified effectively and with minimal risk.

It would be ideal to use a single measurement of a serum marker to determine GH status in all patients with GH deficiency and particularly in the elderly. We have already seen, however, that the degree of overlap between GH-deficient patients and controls in terms of serum IGF-I and IGFBP-3 concentrations precludes their use as diagnostic markers for GH deficiency. Measurement of spontaneous GH secretion was not able to completely separate patients from controls (Fig. 12.1); furthermore, the 24-hour GH profile is not a practical method of determining GH status in clinical practice. The most effective method of determining GH status at the present time is a dynamic test. The test that is frequently quoted in the literature as the 'gold standard' is the insulin tolerance test (ITT), although this may be superseded by alternative tests that are currently being evaluated. The ITT is, however, contraindicated in patients who have cardiovascular disease or a history of epileptiform seizures. It is therefore potentially hazardous in the elderly, who may have significant but occult ischaemic heart disease.

The arginine stimulation test (AST) provides a safe alternative to the ITT, although arginine is not as potent a secretagogue as insulin-induced hypoglycaemia. The GH response to arginine is not influenced by age. The side-effects associated with the infusion of arginine are limited to mild dizziness, which settles rapidly when the infusion rate is reduced, and the test is well tolerated even by those with overt ischaemic heart disease. For these reasons, we chose to use the arginine stimulation test to study the elderly with GH deficiency [17].

There was an excellent correlation between spontaneous GH secretion and the GH response to arginine in both the patients with hypothalamic–pituitary disease and controls. However, the arginine stimulation test was not able to distinguish completely between patients and controls (Fig. 12.2). The lowest GH peak in the controls was 1.8 μg/L. If a peak GH value of less than 1.5 μg/L is used as the diagnostic threshold for GH deficiency, then the sensitivity of the AST is 77%, with a specificity of 100%. If the three patients with isolated GH deficiency are excluded, the sensitivity of the AST rises to 86%. We have therefore suggested that a GH peak of 1.5 μg/L to this particular stimulus should be used to indicate impaired GH secretion in patients over the age of 60 years.

The severity of GH deficiency in patients with hypothalamic–pituitary disease increases as the number of anterior pituitary hormone deficiencies increases [36]. Ninety per cent of patients with two or three additional anterior pituitary hormone deficits had a GH peak during an ITT of less than 2 μg/L. Using similar methods in the elderly with hypothalamic–pituitary disease, 89% of patients with two or three additional hormone deficiencies had a peak GH response to arginine of less than 0.8 μg/L. A GH response to arginine of less than 0.8 μg/L in the elderly with hypothalamic–pituitary disease is indicative of severe GH deficiency [17].

Growth hormone replacement in the elderly with organic GH deficiency

The first studies of GH replacement therapy in adults with GH deficiency were published in 1989 [37,38]. Since then, there have been many studies in the literature reporting the effects of GH therapy in patients with organic GH deficiency, the majority involving adults less than 60 years of age. We have clearly demonstrated that hypothalamic–pituitary disease in adults over the age of 60 years may be associated with a 90% reduction in GH secretion that is distinct from the hyposomatotropism of ageing. This degree of GH deficiency results in significant changes in serum IGF-I, body composition and bone turnover, although the severity of the changes appears to be attenuated compared with that seen in younger GH-deficient adults. None the less, it seems likely that GH-deficient adults over 60 years of age may derive some benefit from GH replacement therapy.

Determining the dose of GH replacement therapy in the elderly

Many of the early studies of GH replacement therapy in adults used a dosage schedule defined by the patient's weight, a strategy which failed to take into account a number of important issues. GH secretion in healthy subjects falls with increasing obesity and with increasing age. One also has to take into consideration sex and other differences in susceptibility, and the interindividual variation in absorption from the GH injection site. Using a dosage schedule based on weight resulted in many patients being overtreated and suffering from unwanted side-effects [39]. It is now recognized that GH therapy should be commenced at a low dose and increased with knowledge of the serum IGF-I and clinical response of the individual patient [40]. The aim of therapy is to maintain the serum IGF-I within the age-specific normal range. This may be an important consideration in the light of two recent studies that have demonstrated an increased risk of breast and prostate cancer in healthy subjects who have serum IGF-I concentrations in the upper part of the normal range [41,42]. The adoption of this strategy has led to a considerable reduction in the dosages of GH used to treat GH-deficient adults [43].

In the knowledge that GH secretion falls with age and that the effects of organic GH deficiency are attenuated in the elderly, it therefore seems likely that the dose of GH required by this patient group will be lower than that in younger adults. We have recently completed a dose-finding study in 12 GH-deficient adults over the age of 60 years, using three doses of GH: 0.17, 0.33 and 0.5 mg/day. Each dose was taken for three months, in ascending order. We monitored serum IGF-I, as this has been shown to be the most sensitive marker for detecting GH over-replacement [40].

The serum IGF-I level rose in a dose-dependent manner over the course of the study in all 12 patients. A GH dose of 0.17 mg/day was sufficient to increase the IGF-I concentration within the normal range in all subjects. However, when the dose was increased to 0.33 mg/day and then further to 0.5 mg/day, the IGF-I level became abnormally high (> + 2 SD) in two subjects and in six subjects, respectively (Table 12.3).

Beneficial effects associated with GH replacement therapy

In order to justify the use of GH replacement in the elderly with GH deficiency, it is important to demonstrate that a number of biological end points other than the serum IGF-I are affected in a beneficial manner. In our study, we used DEXA to determine body composition changes and a disease-specific questionnaire to study quality of life in the patients receiving GH replacement therapy (Table 12.3).

Body composition

Over the course of the nine-month study, there were significant changes in body composition. There was a reduction in fat mass and a rise in lean body mass. Fat mass did not fall significantly until the sixth month of the study, but the reduction was maintained until the end of the study. Somewhat to our surprise, lean mass, which was not found to be reduced in the elderly with GH deficiency, rose significantly after three months' therapy and again at six months, and then remained stable until the end of the study (Table 12.3). It is likely that the rise in lean mass to some extent reflects the increase in extracellular fluid seen during GH therapy, which cannot be distinguished from cell mass by DEXA.

Quality of life

One of the major indications for GH replacement therapy in adult life is diminished quality of life. In order to determine the effect of GH therapy in

Table 12.3 The effects of three doses of growth hormone (GH) therapy on serum insulin-like growth factor-I (IGF-I), insulin-like growth factor binding protein-3 (IGFBP-3), body composition and quality of life, determined using the Adult Growth Hormone Deficiency Assessment (AGHDA) score, in patients with GH deficiency aged over 60 years. Data from Toogood *et al.* [45].

	Baseline	12 weeks	24 weeks	36 weeks	*p*
GH dose (mg/day)	–	0.17	0.33	0.5	
IGF-I (μg/L)	101 (49–148)	149 (49–227)	200 (70–453)	239 (122–502)	< 0.0001
IGF-I standard deviation score	–1.09 (–1.97 to –0.04)	–0.17 (–1.96 to +1.27)	0.84 (–1.52 to +5.52)	1.85 (–0.56 to +6.46)	< 0.0001
IGFBP-3 (mg/L)	2.04 (0.67–2.98)	2.53 (0.97–4.18)	2.94 (1.70–4.28)	2.84 (1.24–4.57)	< 0.0001
Total fat mass (kg)	23.256 (15.508–46.394)	22.598 (14.476–45.087)	21.539 (14.276–45.777)	20.701 (15.868–44.509)	0.0003
Total lean body mass (kg)	59.803 (35.618–69.986)	60.023 (36.178–70.518)	62.226 (37.186–72.925)	60.990 (35.756–71.319)	0.0001
AGHDA Score	4.5 (0–18)	3.5 (0–15)	3 (0–17)	2.5 (0–16)	0.002

the elderly, we used a questionnaire developed specifically for use in adults with GH deficiency, the Adult Growth Hormone Deficiency Assessment (AGHDA) [44]. The patients answer 25 questions and score one mark for each question answered with 'yes'. In this assessment of quality of life, a high degree of distress (or low quality of life) is indicated by a high score. In our study, there was a significant, beneficial change in quality of life after three months of GH therapy, a change that was still present at nine months (Table 12.3).

There are two problems with these data. Firstly, the baseline AGHDA scores achieved by the patients were low, indicating that they did not perceive a great degree of distress despite being severely GH-deficient. Secondly, the degree of change, as determined by the questionnaire, was small (Table 12.3). The reason for these observations may be related to the age of the patients in the study. The questionnaire was developed in a group of patients with a mean age of 35 years, whereas the average age of the elderly GH-deficient patients was 67 years. Ageing itself is associated with a diminished quality of life, so that any effects that GH deficiency in the elderly has on quality of life may be less readily detectable with the AGHDA. Normative data for the AGHDA are not available in the elderly.

Despite these limitations, the changes in quality of life detected by the questionnaire appeared to have real significance for the patients. Perhaps the most telling observation was that 75% of the patients who took part in the study asked to restart GH therapy when they were reassessed three months after stopping GH. Thus, although the changes in quality of life detected by the AGHDA were subtle, the patients perceived sufficient benefit to wish to continue self-administering a daily injection.

Adverse effects

During the course of our study, three patients developed side-effects that were attributable to GH therapy. Two patients developed headaches; one required a permanent reduction of the GH dose from 0.5 to 0.33 mg/day, the second reduced the dose to 0.33 mg/day for seven days before increasing it back to 0.5 mg/day. A third patient developed arthralgia that settled when the GH dose was cut to 0.33 and 0.5 mg/day on alternate days.

Is there an upper age limit for GH replacement in hypopituitary adults?

Adults over the age of 60 years who have organic disease of the hypothalamic–pituitary axis exhibit the features of GH deficiency observed in younger GH-deficient adults. The effects of GH deficiency in this age group, however, do appear to be attenuated. GH replacement therapy in this age group was well tolerated, with a low incidence of short-term side-effects. The majority of patients who received GH therapy perceived sufficient improvement in their quality of life to desire continuing therapy on a long-term basis. The oldest participant in the study was 87 years old [45].

It is important that we pay particular attention to monitoring GH therapy in this age group. The primary goal of monitoring is to maintain the serum IGF-I within the age-specific normal range for each patient. The elderly are a particularly difficult group to study, as they are more prone to develop diseases such as cancer. Although there has been no direct link established between GH replacement therapy in adults and the development of malignancy, it is important that we should remain vigilant. All GH-deficient patients commencing GH replacement therapy should be enrolled in a long-term follow-up study. These studies will determine the long-term safety of GH therapy, allowing us to compare the frequency of various disease states in patients with that of the general population.

At present, there is no evidence to suggest that GH therapy should be restricted to patients below a certain age. If an elderly patient with pituitary disease and severe GH deficiency exhibits typical symptoms of GH deficiency, he or she should be offered a trial of therapy. It is our practice to offer GH therapy for six months, with the patient receiving the appropriate titrated dose for this period of time. At the end of the trial period, the patient is reassessed and a decision is taken, in conjunction with the patient, on whether or not to continue GH replacement therapy. Only large multicentre studies, with long-term follow-up periods, in elderly patients with GH deficiency receiving GH treatment can determine the true benefits of GH in this group.

References

** 1 Carroll P, Christ E, Bengtsson BÅ *et al.* Growth hormone deficiency in adulthood and the effects of growth hormone replacement: a review. *J Clin Endocrinol Metab* 1998; **83**: 382–95.

* 2 Rosen T, Wilhelmsen L, Landin-Wilhelmsen K, Lappas G, Bengtsson B. Increased fracture frequency in adult patients with hypopituitarism and GH deficiency. *Eur J Endocrinol* 1997; **137**: 240–5.

3 Rosen T, Bengtsson B. Premature mortality due to cardiovascular disease in hypopituitarism. *Lancet* 1990; **336**: 285–8.

** 4 Corpas E, Harman SM, Blackman MR. Human growth hormone and human aging. *Endocr Rev* 1993; **14**: 20–39.

* 5 Ho KKY, Hoffman DM. Aging and growth hormone. *Horm Res* 1993; **40**: 80–6.

** 6 Rudman D, Kutner MH, Rogers CM *et al.* Impaired growth hormone secretion in the adult population: relation to age and adiposity. *J Clin Invest* 1981; **67**: 1361–9.

* 7 Finklestein JW, Roffwarg HP, Boyar RM, Kream J, Hellman L. Age-related change in the twenty-four-hour spontaneous secretion of growth hormone. *J Clin Endocrinol Metab* 1972; **35**: 665–70.

* 8 Iranmanesh A, Lizarralde G, Veldhuis JD. Age and relative obesity are specific negative determinants of the frequency and amplitude of growth hormone (GH) secretory bursts and the half-life of endogenous GH in healthy men. *J Clin Endocrinol Metab* 1991; **73**: 1081–8.

** 9 Toogood AA, O'Neill PA, Shalet SM. Beyond the somatopause: growth hormone deficiency in adults over the age of 60 years. *J Clin Endocrinol Metab* 1996; **81**: 460–5.

* 10 Toogood AA, Nass RM, Pezzoli SS *et al.* Preservation of growth hormone pulsatility despite pituitary pathology, surgery and irradiation. *J Clin Endocrinol Metab* 1997; **82**: 2215–21.

* 11 Hoffman DM, O'Sullivan AJ, Baxter RC, Ho KKY. Diagnosis of growth hormone deficiency in adults. *Lancet* 1994; **343**: 1064–8.

* 12 Landin-Wilhelmsen K, Wilhelmsen L, Lappas G *et al.* Serum insulin-like growth factor I in a random population sample of men and women: relation to age, sex, smoking habits, coffee consumption and physical activity, blood pressure and concentrations of plasma lipids, fibrinogen, parathyroid hormone and osteocalcin. *Clin Endocrinol* 1994; **41**: 351–7.

* 13 Shalet S, Toogood A, Rahim A, Brennan B. The diagnosis of growth hormone deficiency in children and adults. *Endocr Rev* 1998; **19**: 203–23.

14 Blum WF, Ranke MB, Kietzmann K *et al.* A specific radioimmunoassay for the growth hormone-dependent somatomedin-binding protein: its use for diagnosis of growth hormone deficiency. *J Clin Endocrinol Metab* 1990; **70**: 1292–8.

15 Donahue LR, Hunter SJ, Sherblom AP, Rosen C. Age-related changes in serum insulin-like growth factor-binding proteins in women. *J Clin Endocrinol Metab* 1990; **71**: 575–9.

16 Corpas E, Harman SM, Blackman MR. Serum IGF-binding protein-3 is related to IGF-I, but not to spontaneous GH release, in healthy old men. *Horm Metab Res* 1992; **24**: 543–5.

** 17 Toogood A, Jones J, O'Neill P, Thorner M, Shalet S. The diagnosis of severe growth hormone deficiency in elderly patients with hypothalamic–pituitary disease. *Clin Endocrinol* 1998; **48**: 569–76.

18 Rosen T, Bosaeus I, Tolli J, Lindstedt G, Bengtsson BY. Increased body fat mass and decreased extracellular fluid volume in adults with growth hormone deficiency. *Clin Endocrinol* 1993; **38**: 63–71.

19 Beshyah SA, Freemantle C, Thomas E *et al.* Abnormal body composition and reduced bone mass in growth hormone-deficient hypopituitary adults. *Clin Endocrinol* 1995; **42**: 179–89.

* 20 Weaver JU, Monson JP, Noonan K *et al.* The effect of low dose recombinant human growth hormone replacement on regional fat distribution, insulin sensitivity and cardiovascular risk factors in hypopituitary adults. *J Clin Endocrinol Metab* 1995; **80**: 153–9.

21 Forbes GB, Reina JC. Adult lean body mass declines with age: some longitudinal observations. *Metabolism* 1970; **19**: 653–63.

* 22 Novak LP. Aging, total body potassium, fat-free mass and cell mass in males and females between ages 18 and 85. *J Gerontol* 1972; **27**: 439–43.

23 Borkan GA, Hults DE, Gerzof SG, Robbins AH, Silbert CK. Age-related changes in body composition revealed by computed tomography. *J Gerontol* 1983; **38**: 673–7.

** 24 Toogood AA, Adams JE, O'Neill PA, Shalet SM. Body composition in growth hormone deficient adults over the age of 60 years. *Clin Endocrinol* 1996; **45**: 399–405.

25 Veldhuis JD, Iranmanesh A, Ho KKY *et al.* Dual defects in pulsatile growth hormone secretion and clearance subserve the hyposomatotropism of obesity in man. *J Clin Endocrinol Metab* 1991; **72**: 51–9.

26 Bonjour JP, Theintz G, Buchs B, Slosman D, Rizzoli R. Critical years and stages of puberty for spinal and femoral bone mass accumulation during adolescence. *J Clin Endocrinol Metab* 1991; **73**: 555–63.

27 Theintz G, Buchs B, Rizzoli R *et al.* Longitudinal monitoring of bone mass accumulation in healthy adolescents: evidence for a marked reduction after 16 years of age at the levels of lumbar spine and femoral neck in female subjects. *J*

Clin Endocrinol Metab 1992; 75: 1060–5.

28 Raisz LG. Local and systemic factors in the pathogenesis of osteoporosis. *N Engl J Med* 1988; **318**: 818–28.

29 Smith RW Jr. Dietary and hormonal factors in bone loss. *Fed Proc* 1967; **26**: 1737–46.

30 O'Halloran DJ, Tsatsoulis A, Whitehouse RW *et al.* Increased bone density after recombinant human growth hormone therapy in adults with isolated GH deficiency. *J Clin Endocrinol Metab* 1993; **76**: 1344–8.

31 Holmes SJ, Economou G, Whitehouse RW, Adams JE, Shalet SM. Reduced bone mineral density in patients with adult onset growth hormone deficiency. *J Clin Endocrinol Metab* 1994; **78**: 669–74.

** 32 Toogood AA, Adams JE, O'Neill PA, Shalet SM. Elderly patients with adult-onset growth hormone deficiency are not osteopenic. *J Clin Endocrinol Metab* 1997; **82**: 1462–6.

** 33 Kaji H, Abe H, Fukase M, Chihara K. Normal bone mineral density in patients with adult onset GH deficiency. *Endocrinol Metab* 1997; **4**: 163–6.

34 Garnero P, Sornay-Rendu E, Chapuy MC, Delmas PD. Increased bone turnover in late postmenopausal women is a major determinant of osteoporosis. *J Bone Miner Res* 1996; **11**: 337–49.

35 Holmes SJ, Shalet SM. Factors influencing the desire for long-term growth hormone replacement in adults. *Clin Endocrinol* 1995; **43**: 151–7.

36 Toogood AA, Beardwell CG, Shalet SM. The severity of growth hormone deficiency in adults with pituitary disease is related to the degree of hypopituitarism. *Clin Endocrinol* 1994; **41**: 511–16.

* 37 Jørgensen JOL, Pedersen SA, Thuesen L *et al.* Beneficial effects of growth hormone treatment in GH-deficient adults. *Lancet* 1989; **i**: 1221–5.

* 38 Salomon F, Cuneo RC, Hesp R, Sönksen PH. The effects of treatment with recombinant human growth hormone on body composition and metabolism in adults with growth hormone deficiency. *N Engl J Med* 1989; **321**: 1797–803.

39 Holmes SJ, Shalet SM. Which adults develop side-effects of growth hormone replacement? *Clin Endocrinol* 1995; **43**: 143–9.

** 40 Growth Hormone Research Society. Consensus guidelines for the diagnosis and treatment of adults with growth hormone deficiency: summary statement of the Growth Hormone Research Society workshop on adult growth hormone deficiency. *J Clin Endocrinol Metab* 1998; **83**: 379–81.

** 41 Hankinson S, Willett W, Colditz G *et al.* Circulating concentrations of insulin-like growth factor-I and risk of breast cancer. *Lancet* 1998; **351**: 1393–6.

** 42 Chan J, Stampfer M, Giovannucci E *et al.* Plasma insulin-like growth factor-I and prostate cancer risk: a prospective study. *Science* 1998; **279**: 563–6.

43 Janssen YJH, Frolich M, Roelfsema F. A low starting dose of genotropin in growth hormone-deficient adults. *J Clin Endocrinol Metab* 1997; **82**: 129–35.

44 Holmes SJ, McKenna SP, Doward LC, Hunt SM, Shalet SM. Development of a questionnaire to assess the quality of life of adults with growth hormone deficiency. *Endocrinol Metab* 1995; **2**: 63–9.

** 45 Toogood A, Shalet S. Growth hormone replacement therapy in the elderly with hypothalamic–pituitary disease: a dose-finding study. *J Clin Endocrinol Metab* 1999; **84**: 131–6.

13: Is open-ended GH replacement a safe therapy?

Roger Abs and Johan Verhelst

Introduction

Safety during medical therapy has always been of major concern to physicians, especially when the treatment of a condition is not yet well characterized or optimized. All occurrences of adverse events (AEs) should be actively sought in order to recognize potentially health-threatening changes. Growth hormone (GH) replacement therapy in GH-deficient (GHD) hypopituitary adult patients has been used increasingly during the past 10 years. This has followed the widespread recognition of a GHD syndrome in adults, with characteristic signs and symptoms, and increasing evidence that GH replacement therapy is effective in reversing many aspects of this syndrome [1–3]. Unfortunately, from the original reports of this treatment in 1989 it was evident that GH administration induced significant adverse events in a large number of patients [4,5]. All studies reporting favourable effects of GH replacement also expressed concerns about safety. The high frequency of adverse events related to fluid retention and the subsequent high drop-out rate caused further research to focus on defining the inclusion criteria and the exact dosage of GH replacement. It rapidly became clear that adverse events were related to fluid retention occurring during the first months of treatment, and that these complaints subsided with time or when the GH dose was reduced. However, there are few data concerning adverse events in patients followed up for more than 18 months, and in these studies the numbers of patients have been relatively small [6–8]. One larger multicentre study addressed not only the question of efficacy but also safety, the drop-out rate, and the reason for discontinuation after two years of therapy [9]. The Pharmacia and Upjohn International Metabolic Study (KIMS) is a world-wide outcomes research database that was launched in 1994 and created with the principal aim of monitoring the long-term safety and efficacy of GH replacement therapy in a very large cohort of GHD hypopituitary adult patients. In this chapter, we will review the data on adverse events reported in the literature and present the information retrieved from the KIMS database concerning safety.

Methods

A broad definition of the term 'adverse event' has been used to describe all unfavourable changes in structure (signs), function (symptoms) or chemistry (laboratory tests or radiographs), regardless of whether the adverse event is considered to be drug-related. Consequently, the majority of adverse events relate to concomitant disorders or involve coincidental symptoms unrelated to GH replacement therapy. However, analysis of all adverse events is necessary to ensure that any unexpected adverse drug reaction following GH replacement therapy is detected.

Two sources of information are available for the analysis of safety during GH replacement therapy in GHD hypopituitary patients. All publications in the literature relating to GH therapy were scrutinized for reports of adverse events. The KIMS database was used to identify this safety issue. Adverse events in KIMS are coded according to the list of 'preferred terms' for adverse drug reactions developed by the World Health Organization (WHO) Collaborating Centre for International Drug Monitoring. Each preferred term is automatically classified according to the appropriate organ system. Since many adverse events are related to concomitant conditions, they cannot be coded from the list of adverse drug reactions. Such events are coded according to the manual of the International Statistical Classification of Diseases, Injuries and Causes of Death (ICD-9), and are entered into the appropriate organ system class.

Some adverse events that may be indicative of an excess of GH have received particular scrutiny. For example, acromegaly has been associated with glucose intolerance and an increased incidence of certain cancers. The KIMS database has therefore been carefully audited for information regarding changes in glucose tolerance and for new cases of malignancies. Furthermore, as most cases of GHD arise from hypothalamic or pituitary tumours, the recurrence of such tumours has also been closely examined.

Type and frequency of adverse events

Since recombinant human GH has the same sequence homology as native human GH, adverse events are expected to result from excess replacement alone, either by overdosing GH or by exposing the body to the full replacement dose too rapidly. In all placebo-controlled studies, significantly more fluid retention-related adverse events were noted in the GH-treated group, whereas no significant differences were seen for other adverse events. Results from the Belgian study of 148 GHD adult patients treated with 0.25 IU/kg/week GH (1.5 IU/m²/day or 0.08 mg/kg/week) are shown in Table 13.1 [9]. In this study, which can serve as an example of several others, most patients (23.6%) had their first fluid retention-related adverse events within the first three months of therapy. Between three months and 24 months, another 10.2% noticed fluid retention-related adverse events, making a total of 33.8% of patients experiencing adverse symptoms. The most prominent adverse events were peripheral or generalized oedema (25%), arthralgia (12.1%), myalgia (7.4%), stiffness in extremities (4.1%), carpal tunnel syndrome (4.1%) and paraesthesia (1.4%). Carpal tunnel syndrome occurred usually after more than six months of therapy (mean 9.5 months). In a similar study covering 52 adult-onset GHD patients [10], corresponding percentages of complaints were noted during the six-month placebo-controlled period: peripheral or generalized oedema (28.8%), arthralgia (9.6%), myalgia (7.7%), joint disorders (5.8%), paraesthesia (3.9%) and headaches (5.8%). In a combined analysis of several European trials including a total of 333 GHD patients [11], a similar incidence of side-effects was reported: peripheral or generalized swelling and oedema (25.8%), arthralgia (11.4%), myalgia (5.1%), pain and stiffness in the extremities (11.1%), paraesthesia

Table 13.1 Percentage of adverse events (AEs) occurring within the six months of the placebo-controlled period.

AEs	% in GH treated group ($n = 71$)	% in placebo group ($n = 77$)
Related to fluid retention		
Arthralgia	15.4	2.4
Peripheral oedema	12.6	1.2
Generalized oedema	5.6	0.0
Myalgia	4.2	0.0
Paraesthesia	2.8	0.0
Stiffness in extremities	2.8	1.2
Carpal tunnel syndrome	2.8	0.0
Not related to fluid retention		
Depression	2.8	1.3
Dyspepsia	2.8	0.0
Nervousness	2.8	1.3
Hyperuricemia	1.4	1.3
'Flu'	1.4	1.3
Hypertension	1.4	1.4
Headache	1.4	1.4
Tendinitis	1.4	1.2
Tiredness	0.0	1.3
Insomnia	0.0	2.6
Rash	0.0	2.8

From: Verhelst *et al.* [9].

(6.3%) and carpal tunnel syndrome (1.8%). Over half the patients with fluid retention had more than one fluid-related adverse event. In the Belgian study, no action was taken in 40% of patients with fluid retention-related adverse events, the GH dose was reduced in 40%, GH treatment was interrupted in 10%, and GH treatment was stopped in only 10% of patients [9]. Most of these adverse events had disappeared by the time of the patient's next visit.

In a first study of the KIMS database in relation to the demographic and clinical characteristics of GHD patients, a dosage and safety analysis was also performed [12]. During the 818 patient-years covered by the analysis, 883 adverse events were reported in 29.6% of the 1034 patients. The incidence of adverse events was higher in women than in men (1.19 vs. 0.97 AEs/year; relative risk, 1.05–1.45, 99% CI) and higher in patients with childhood-onset GHD than in patients with adult-onset GHD (1.00 vs. 0.85 AEs/year; relative risk, 0.93–1.49). The incidence of adverse events reported in KIMS was also compared with the incidence reported during the six-month double-blind period in 1145 GHD patients from 43 clinical trials of GH replacement. The total incidence of adverse events reported in KIMS patients was significantly lower than that reported in patients on GH from the clinical trials (1.07 vs. 6.3 AEs/year; relative risk, 0.15–0.19). Furthermore, signs and symptoms of fluid retention were reported less frequently in newer patients enrolled in KIMS (0.45/year) than in trial patients receiving either GH (2.26/year; relative risk, 0.13–0.30) or placebo (0.60/year; relative risk, 0.50–1.13), due to a lower dose of GH, the technique of dose titration and slower dose increment.

The most recent analysis of the KIMS database was performed in March 1998, and included 2083 patients and ≈2000 patient-years of GH therapy. The numbers of reported adverse events categorized according to the WHO International Drug Monitoring List are shown in Table 13.2. The relatively high percentage of patients reporting adverse events (44.2%), and the relatively high rate of adverse events per patient (1.38) suggest that most adverse events are being reported. The most frequently reported WHO category of adverse events was 'body as a whole/general disorders', which accounted for 29.4% of the reports. In this category, oedema was the most common adverse event, and was reported in 193 patients (9.3%). Also in the same category, pain and fatigue were commonly reported. The WHO category of respiratory system disorders accounted for 10.4% of all adverse events reported. The most frequent adverse event in this category was upper respiratory tract infection, with 211 patients (10.1%), accounting for 9.7% of the reported adverse events. The WHO category of musculoskeletal system disorders accounted for 10.3% of the reported adverse events. The most common adverse event in this category was arthralgia,

Table 13.2 Adverse events (AEs) reported in the Pharmacia and Upjohn International metabolic study (KIMS), categorized according to the World Health Organization (WHO) International Drug Monitoring List. The WHO codes for the adverse event category are given in parentheses.

AE category	Number of AEs	Number of patients reporting the AE
Body as a whole/general disorders (1810)	842	476
Respiratory system disorders (1100)	299	173
Musculoskeletal system disorders (0200)	296	230
Metabolic and nutritional disorders (0800)	200	134
Central and peripheral nervous system disorders (0410)	191	142
Gastrointestinal system disorders (0600)	169	121
Skin and appendages disorders (0100)	99	82
Psychiatric disorders (0500)	96	86
Body system unknown (0000)	93	67
Endocrine disorders (0900)	92	83
Resistance mechanism disorders (1830)	89	65
Autonomic nervous system disorders (0420)	64	48
Liver and biliary system disorders (0700)	39	29
Reproductive disorders, female (1420)	36	32
Urinary system disorders (1300)	35	30
Red blood cell disorders (1210)	34	28
Vision disorders (0431)	31	30
Cardiovascular disorders, general (1010)	27	24
Vascular (extracardiac) disorders (1040)	21	19
White cell and reticuloendothelial disorders (1220)	21	18
Secondary terms (2000)	19	17
Neoplasms (1700)	15	10
Myo-, endo-, pericardial and valve disorders (1020)	13	12
Hearing and vestibular disorders (0432)	10	9
Heart rate and rhythm disorders (1030)	9	9
Reproductive disorders, male (1410)	9	8
Platelet, bleeding and clotting disorders (1230)	9	5
Application site disorders (1820)	6	5
Special senses other, disorders (0433)	2	2
Fetal disorders (1500)	1	1
Total	2867	1995

with 108 reports in 86 patients. The most common adverse event in KIMS in the WHO category of central and peripheral nervous system disorders was headache/migraine, with 107 reports in 91 patients.

When listed according to ICD-9 categories (Table 13.3), upper respiratory tract infection (10.3%) was the most common adverse event reported in KIMS. The high incidence of respiratory system disorders reported is not unexpected. This is a common observation in pharmaco-vigilance studies,

Table 13.3 The 10 most common adverse events (AEs) reported in KIMS, according to the ICD-9 category.

		Number of patients reporting the AE	
AE	Number of AEs	Females	Males
Upper respiratory tract infection	278	111	100
Oedema	216	119	74
Arthralgia	108	42	44
Headache/migraine	107	54	37
Myalgia	65	30	31
Gastroenteritis	63	26	23
Hypercholesterolaemia	63	14	27
Back pain	60	24	32
Pain	55	20	29
Fatigue	44	23	19

as physicians are obliged to report all viral and bacterial infections unrelated to therapy. The second most commonly reported adverse event was oedema (9.3%), and this probably reflects the true incidence of adverse events related to GH therapy. Following in decreasing frequency are: headache/migraine (4.4%), arthralgia (4.1%) and myalgia (3.1%).

Factors influencing the occurrence of adverse events

Dose of GH replacement

In a dose-ranging study, 18% of patients showed fluid retention with a GH dose of 1 IU/m^2/day, 35% with 2 IU/m^2/day, and 67% with 3 IU/m^2/day [13]. These data fit well with results from most other centres as well as the Belgian study, in which 26.8% of patients had GH-related adverse events during the first six months of therapy with a dose of 1.5 IU/m^2/day [9]. Older studies using a GH dose of 3 IU/m^2/day reported an average of 61% of their patients complaining of fluid retention and 40% of them needing dose reduction [2]. In addition to a high GH dose, most studies exposed patients to the full replacement dose within a short time span, which is now known to increase further the incidence of adverse events. In addition, adult patients usually received GH in a weight-dependent dose, following the tradition of GH therapy in GHD children, without the considerable interindividual difference in GH sensitivity being taken into account. The search for the optimal dosage of GH has therefore been replaced by guidelines advising a gradual increase in GH dosage, depending on insulin-like growth factor-I (IGF-I) levels and clinical effects [12,14,15]. Patients treated in this way report less than half the incidence of GH-related side-effects compared to patients in trials with fixed doses.

Serum IGF-I levels

The pretreatment serum IGF-I level has not been found to be a potential marker for patients at risk of developing GH-related adverse events [9,10]. If patients are split up into two groups, depending on the occurrence of fluid retention or not, a significantly greater increase in circulating IGF-I is found after three or six months of therapy in patients developing adverse events [9,10,16]. However, because of a substantial overlap, the increase in serum IGF-I level is not a useful indicator for predicting adverse events in the individual patient. Measuring serum IGF-I levels during therapy, in addition to the clinical evaluation, has tended to replace standard doses depending on weight or body surface, and has become an important tool in optimizing the dose of GH.

Body weight

Heavier patients are known to have more frequent adverse events [10,16]. However, taking into account the fact that heavier patients received significantly higher doses of GH, weight has probably no independent influence on the number of adverse events if corrected for the dose of GH.

Childhood-onset vs. adult-onset GHD

Adverse events were reported more frequently in adult-onset compared to childhood-onset patients during GH replacement [10,17]. These data suggested that adult-onset patients, as compared with childhood-onset ones, are more sensitive to GH. Another study did not support this finding [12]. The fact that in this analysis no distinction was made between GH-related and other adverse events might account for this discrepancy.

Pretreatment stimulated peak serum GH concentration

Although a higher initial peak serum GH level during stimulation tests has been reported to be predictive of subsequent adverse events in some studies [16,18], others could not confirm these data [9,10]. In the first two studies, the number of patients with higher GH peaks was probably too small for useful conclusions to be drawn.

Age

In contrast to one study [16], others [9,10] have failed to observe fluid retention more frequently in older patients. One explanation for this might be the low incidence (7%) of childhood-onset GHD in the Belgian study and the fact that childhood-onset GHD patients were considered to be a different group in the other study. This suggests that the influence of age can be explained by the younger age of childhood-onset GHD patients alone. A higher sensitivity for adverse events in patients above 60 years cannot be excluded from these data, as patients older than 60 years were usually not included in the studies.

Sex

A correlation between adverse events and sex was not found in the Belgian study [9]. Similarly, the incidence of adverse events in KIMS was generally similar in men and women, with the exception of oedema, which was reported in 12.2% of women but only 6.7% of men (Table 13.3). This may suggest that women are more sensitive to GH-induced fluid-related adverse events compared to men, whereas the opposite is seen for IGF-I generation and changes in body composition. Further data are required to establish whether the higher incidence of oedema in women is a consistent and significant observation.

Specific adverse events

Metabolic disorders

In the Belgian study, a small but significant increase in fasting blood glucose levels was observed, which was maintained over the whole study period [9]. Similar findings were demonstrated by other groups [4,19–21], although not all [22]. The increase in fasting blood glucose resulted in a limited increase in glycated haemoglobin, which disappeared after 18 months of treatment. In contrast with our own results, glycated haemoglobin did not change significantly in most of the earlier and smaller studies [4,23]. These findings are in line with current insights that glucose intolerance is only temporarily worsened by GH therapy, and returns to baseline as soon as the decrease in visceral fat stores counterbalances the antagonistic effect of GH on insulin sensitivity [24].

A total of 14 cases (eight men and six women) with abnormal glucose tolerance have been reported in KIMS following 1.0–113.5 months of GH replacement therapy (Table 13.4). Although it appears from these data that GH replacement therapy does not induce diabetes mellitus, glucose levels should be monitored closely in patients with hyperglycaemia at the start of therapy, because of the possibility of temporary worsening due to increased insulin resistance [25].

Lipoprotein (a) is an independent risk factor for the development of cardiovascular disorders, and in most studies, serum levels increase during GH replacement therapy [26]. However, GH is also known to influence beneficially other cardiovascular risk factors such as low-density lipoprotein cholesterol and blood pressure. Effects on insulin sensitivity are probably neutral in the medium term, and the net consequences on overall cardiovascular risk appear favourable, although long-term observations are required to determine whether this is translated into a beneficial effect on morbidity and mortality.

Cardiovascular disorders

A total of 12 cases (eight men and four women) of cardiovascular disease have been reported in KIMS following 1–62 months of GH replacement therapy (Table 13.5). Coronary heart disease was the most frequently recorded adverse event, while stroke and claudication were mentioned once and twice, respectively.

Neurological disorders

In KIMS, episodes of convulsions have been reported in 17 cases (10 men and seven women) following 54.9 months (range 1.0–226.8 months) of GH replacement therapy. Mean age was 34.3 years (range 18.7–55.8 years). Mean GH dose was 2.0 IU/day (range 0.4–3.5 IU/day). A diagnosis of epilepsy had been made in 13 patients before GH replacement therapy, while no information was available for the remaining four patients.

Recurrence of cranial tumours

The recurrence rate of cranial tumours is difficult to estimate, because the probability of recurrence is likely to vary depending on the surgical approach used to remove the original tumour and whether postoperative irradiation was given. In one study involving 73 patients with nonsecreting pituitary adenomas who had undergone transsphenoidal surgery alone, the five-year recurrence-free survival rate was 90%, which suggests that the annual recurrence rate is ≈2% [27].

Table 13.4 Incidence of abnormal glucose tolerance occurring in patients included in KIMS. Values are given as means, with the range in parentheses.

Diagnosis/symptoms	Number of patients		Age (years)	Dose of GH (IU/day)	Duration of GH treatment (months)		Number of patients receiving glucocorticoids	Number of patients discontinuing GH treatment
	Females	Males			In KIMS	Total		
Diabetes mellitus	3	2	41.8 (21.5–53.4)	1.5 (0.5–2.0)	9.5 (6.0–14.6)	35.2 (7.4–113.5)	3	0
Abnormal glucose tolerance	2	4	56.2 (49.3–63.6)	2.2 (1.5–3.0)	11.5 (6.1–16.1)	36.0 (10.6–48.3)	5	1
Hyperglycaemia	1	2	47.8 (30.4–58.6)	2.0 (1.0–3.0)	3.4 (1.0–8.1)	14.0 (1.0–39.7)	2	2

GH: growth hormone.

Table 13.5 Incidence of cardiovascular diseases occurring in patients included in KIMS. Values are given as means, with the range in parentheses.

Diagnosis/symptoms	Number of patients		Age (years)	Dose of GH (IU/day)	Duration of GH treatment (months)		Number of patients discontinuing GH treatment
	Females	Males			In KIMS	Total	
Myocardial infarction	0	2	43.0, 44.7	3.0, 3.0	4, 5	16, 36	0
Angina pectoris	2	5	59.1 (42.4–78.9)	1.3 (0.4–3.5)	5.3 (1.0–23.0)	16.1 (1.0–62.0)	1
Claudication	1	1	50.0, 53.7	0.4, 1.2	16, 18	24, 43	0
Stroke	1	0	51.3	1.6	6	16	0

GH: growth hormone.

Recurrence of cranial tumours in KIMS occurred in 22 patients (14 men and eight women) following a total of 2–43 months of GH replacement therapy (Table 13.6). Recurrences of pituitary tumours were reported in 19 of 1752 patients (1.1%), whereas recurrences of craniopharyngioma were reported in two of 419 patients (0.5%). Although there is no clear evidence of an enhanced risk of recurrence of such tumours during GH replacement, the limited number of patient-years accumulated so far precludes any firm conclusion, and further analyses are needed.

Neoplasia

The GH–IGF-I axis plays an important role in regulating fetal and childhood somatic growth, and there is substantial evidence that it could be critical in maintaining neoplastic growth. It may also be a determinant of cancer incidence, as suggested by the strong association between circulating IGF-I concentrations and the relative risk of breast cancer in premenopausal women [28]. A similar association has been reported for prostate cancer [29]. The fact that circulating IGF-I, measured long before presentation of the tumour, is so strongly associated with the risk of breast and prostate cancers raises important questions about the cause of these diseases and strategies for risk assessment and reduction. On the other hand, elevated GH concentrations, as in acromegaly, only seem to be associated with a limited increase in the incidence of colorectal and breast cancers [30,31]. The reports linking IGF-I with breast and prostate cancers indicate that the association between cancer risk and IGF-I concentrations is strengthened if these concentrations are adjusted for the amount of its main carrier protein, insulin-like growth factor binding protein-3 (IGFBP-3). Patients with high IGF-I and low IGFBP-3 concentrations incur the greatest risk. By the nature of the disease in acromegaly, and through overtreatment with GH in GHD adults, circulating IGF-I concentrations are supranormal, but both conditions also show increased IGFBP-3 concentrations. IGFBP-3 may have a protective effect and account for the absence of a clearly increased incidence of breast and prostate cancers in patients with acromegaly. This line of thinking may imply that the risk of these cancers from GH replacement therapy is limited. Extensive world-wide experience of GH treatment in GHD children indicates no increased incidence in cancers [32].

In KIMS, *de novo* neoplasms were reported in 10 patients (four men and six women) who had received GH replacement therapy for a total of 17–270 weeks (Table 13.7). Three of these patients developed benign neoplasms. A further three cases were of basal-cell carcinoma, all occurring in patients in Australia, a country known to have a high incidence of skin cancers attributable to environmental factors. No evidence was found for

Table 13.6 Recurrence of cranial tumours in patients included in KIMS. Ranges are given in parentheses.

Neoplasm	Number of patients: Females	Number of patients: Males	Age (years)	Symptoms	Time to recurrence (years)	Dose of GH (IU/day)	Duration of GH treatment at recurrence (months)	Number of patients discontinuing GH treatment
Pituitary adenoma, clinically non-secreting	5	9	Mean, 53 (39–72)	11 asymptomatic, 1 with headache, 1 with oculomotor paresis	Mean, 7.4 (1–19)	Mean, 1.5 (0.8–3.5)	Mean, 24 (2–43)	6
Pituitary adenoma, prolactin-secreting	2	0	55, 59	Visual field defect	6, 11	2.0, 2.5	8, 24	0
Pituitary adenoma, gonadotrophin-secreting	0	1	51	Bilateral hemianopia	6.5	3.5	24	1
Pituitary adenoma, ACTH-secreting	0	1	44	Asymptomatic	4	1.6	6	1
Pituitary adenoma, unknown type	0	1	54	Asymptomatic	4	1.0	35	0
Craniopharyngioma	0	2	23, 45	1 with headache	14, 22	1.5, 2.4	7, 19	2
Dysgerminoma	1	0	39	Loss of memory	23	2.3	38	1

ACTH: adrenocorticotrophic hormone; GH: growth hormone.

Table 13.7 *De novo* neoplasms reported in patients included in KIMS.

Neoplasm	Age (years)	Sex	Aetiology of GHD	Dose of GH (IU/day)	Duration of GH treatment at occurrence of AE (weeks)		GH treatment
					In KIMS	Total	
Abnormal cervical smear	34	Female	Lymphocytic hypophysitis	3.0	83	104	Continued
Basal-cell carcinoma	56	Male	Pituitary adenoma, ACTH-secreting	1.5	21	21	Continued
Basal-cell carcinoma	64	Male	Irradiation (other than pituitary area)	1.5	107	156	Continued
Basal-cell carcinoma	49	Female	Pituitary adenoma, unknown type	2.5	26	79	Continued
Rectal polyp	45	Female	Pituitary adenoma, prolactin-secreting	2.0	103	134	Discontinued
Lipoma, breast	41	Female	Pituitary adenoma, clinically non-secreting	2.0	52	109	Continued
Lipoma, breast	60	Female	Pituitary adenoma, clinically non-secreting	0.5	17	17	Discontinued
Meningioma	50	Female	Pituitary adenoma, clinically non-secreting	1.5	37	270	Continued
Pituitary neoplasm (not otherwise specified)	49	Male	Irradiation (other than pituitary area)	3.5	108	165	Discontinued
Testis teratoma	26	Male	Pituitary adenoma, prolactin-secreting	1.0	0	94	Discontinued

ACTH: adrenocorticotrophic hormone; AE: adverse event; GH: growth hormone; GHD: growth hormone deficiency.

an increased incidence of prostate or breast cancer. At this stage there are, however, unanswered questions, and more work is required to establish the extent of any risk through large, long-term surveillance studies in GHD patients receiving GH replacement.

Withdrawals

Cumulative drop-out rates during the Belgian study were 10.1% after six months, 29% after 12 months, and 38.5% after 24 months [9]. Twenty patients left the study because of adverse events (five fluid-related and 15 not fluid-related) and 35 because of insufficient perceived improvement or non-compliance. No significant difference for any baseline parameter could be found between patients completing the study and patients withdrawing from the study [9,33]. These results are at variance with the data from a single-centre trial in which patients who preferred to continue GH therapy showed a tendency to greater severity of GHD, lower baseline levels of energy and vitality, and a better improvement in energy during therapy [34]. Since the first two studies were multicentre trials, the opinion and wishes of the clinician approaching the patient could have influenced the patient's decision and weakened possible correlations.

The overall discontinuation rate in KIMS (6%) is low, and does not appear to correlate with the percentage of patients with adverse events or the number of adverse events per patient. The number of deaths in KIMS is 17. Three causes of death were reported twice: relapse of central nervous system tumour, cerebral bleeding, and myocardial infarction. Single causes of death were: cerebral aneurysm, melanoma, new central nervous system tumour, ischaemic colitis, aortic aneurysm, adrenal insufficiency, and suicide. No information could be obtained concerning the remaining four deaths.

Conclusions

Ten years after the introduction of GH replacement therapy in GHD hypopituitary adult patients, there are no solid arguments that would imply that open-ended GH replacement therapy is unsafe. A detailed review of the literature and an analysis of the KIMS database do not provide compelling evidence for serious adverse events during GH treatment.

The majority of adverse events are not the result of GH replacement, but are secondary to concomitant conditions such as upper respiratory tract infections. The most commonly described adverse events that are recognized to be associated with GH treatment, such as oedema, arthralgia and myalgia, are the consequences of fluid retention. Moreover, oedema is known to be a dose-related effect that is generally transient and can usually be avoided by using low starting doses and careful dose titration.

The KIMS database is an excellent instrument for the scrutiny of serious or rare adverse events in large cohorts of GHD patients treated with GH for longer periods of time. In some 2000 treatment years, only 14 cases of abnormal glucose tolerance, 22 recurrences of pituitary tumours, and 10 *de novo* neoplasms have been reported, indicating that concerns that GH replacement therapy might increase the incidence of diabetes mellitus or neoplasms are not supported by the current data. The incidences of these adverse events appear to be within population reference ranges, although further data are required before definitive statements can be made.

Continuing accumulation and analysis of safety data may lead to the identification of factors that predispose individuals to particular adverse events. The identification of individuals most likely to suffer adverse events, and individuals with greater or lesser responsiveness to GH, will allow the risks associated with GH replacement therapy to be minimized.

Acknowledgement

The authors are indebted to the members of the KIMS International Board and all participating physicians for the opportunity to present KIMS data, and in particular to Björn Westberg for data management.

References

* 1 Cuneo RC, Salomon F, McGauley GA, Sönksen PH. The growth hormone deficiency syndrome in adults. *Clin Endocrinol* 1992; **37**: 387–97.
* 2 De Boer H, Blok GJ, van der Veen EA. Clinical aspects of growth hormone deficiency in adults. *Endocr Rev* 1995; **16**: 63–86.
* 3 Carroll PV, Christ ER, Growth Hormone Research Society Scientific Committee. Growth hormone deficiency in adulthood and the effects of growth hormone replacement: a review. *J Clin Endocrinol Metab* 1998; **83**: 382–95.
* 4 Salomon F, Cuneo RC, Hesp R, Sönksen PH. The effects of treatment with recombinant human growth hormone on body composition and metabolism in adults with growth hormone deficiency. *N Engl J Med* 1989; **321**: 1797–803.
* 5 Jørgensen JOL, Pedersen SA, Thuesen L *et al.* Beneficial effects of growth hormone treatment in GH-deficient patients. *Lancet* 1989; **i**: 1221–5.
* 6 Jørgensen JOL, Thuesen L, Müller J *et al.* Three years of growth hormone treatment in growth hormone-deficient adults: near normalization of body composition and physical performance. *Eur J Endocrinol* 1994; **130**: 224–8.
* 7 Beshyah SA, Freemantle C, Shahi M *et al.* Replacement treatment with biosynthetic human growth hormone in growth hormone-deficient hypopituitary patients. *Clin Endocrinol* 1995; **42**: 73–84.
* 8 Johannsson G, Rosén T, Lindstedt G, Bosaeus I, Bengtsson BÅ. Effect of 2 years of growth hormone treatment on body composition and cardiovascular risk factors in adults with growth hormone deficiency. *Endocrinol Metab* 1996; **3** (Suppl A): 3–12.
** 9 Verhelst J, Abs R, Vandeweghe M *et al.* Two years of replacement therapy in 148 adults with growth hormone

deficiency. *Clin Endocrinol* 1997; **47**: 485–94.

** 10 Chipman JJ, Attanasio AF, Birkett MA *et al.* The safety profile of GH replacement therapy in adults. *Clin Endocrinol* 1997; **46**: 473–81.

** 11 Mårdh G, Lindeberg A. Growth hormone replacement therapy in adult hypopituitary patients with growth hormone deficiency: combined clinical safety data from clinical trials in 665 patients. *Endocrinol Metab* 1995; **2** (Suppl B): 11–16.

** 12 Abs R, Bengtsson BÅ, Hernberg-Ståhl E *et al.* GH replacement in 1034 growth hormone deficient hypopituitary adults: demographic and clinical characteristics, dosing and safety. *Clin Endocrinol* [in press].

** 13 De Boer H, Bolk GJ, Voerman B, Derriks P, van der Veen EA. Changes in subcutaneous and visceral fat mass during growth hormone replacement therapy in adult men. *Int J Obes* 1996; **20**: 580–7.

** 14 Janssen YJH, Fröhlich M, Roelfsema F. A low starting dose of genotropin in growth hormone-deficient adults. *J Clin Endocrinol Metab* 1997; **82**: 129–35.

* 15 [Anonymous]. Consensus guidelines for the diagnosis and treatment of adults with growth hormone deficiency: summary statement of the Growth Hormone Research Society workshop on adult growth hormone deficiency. *J Clin Endocrinol Metab* 1998; **83**: 379–81.

** 16 Holmes SJ, Shalet SM. Which adults develop side-effects of growth hormone replacement? *Clin Endocrinol* 1995; **43**: 143–9.

* 17 Attanasio AF, Lamberts SWJ, Matranga AMC *et al.* Adult growth hormone (GH)-deficient patients demonstrate heterogeneity between childhood onset and adult onset before and during human GH treatment. *J Clin Endocrinol Metab* 1997; **82**: 82–8.

* 18 Johansson G, Bjarnason R, Bramnert M *et al.* The individual responsiveness to growth hormone (GH) treatment in GH-deficient adults is dependent on the level of GH-binding protein, body-mass index, age, and gender. *J Clin Endocrinol Metab* 1996; **81**: 1575–81.

* 19 Binnerts A, Swart GR, Wilson JHP *et al.* The effect of growth hormone administration in growth hormone deficient adults on bone, protein, carbohydrate and lipid homeostasis, as well as of body composition. *Clin Endocrinol* 1992; **37**: 79–87.

* 20 Degerblad M, Elgindy N, Hall K, Sjöberg HE, Thorén M. Potent effect of recombinant growth hormone on bone mineral density and body composition in adults with panhypopituitarism. *Acta Endocrinol* 1992; **126**: 387–93.

* 21 Rosén T, Johansson G, Hallgren P *et al.* Beneficial effects of 12 months' replacement therapy with recombinant human growth hormone to growth hormone-deficient adults. *Endocrinol Metab* 1994; **1** (Suppl A): 55–66.

* 22 Whitehead HM, Boreham C, McIlrath EM *et al.* Growth hormone treatment of adults with growth hormone deficiency: results of a 13-month placebo-controlled cross-over study. *Clin Endocrinol* 1992; **36**: 5–52.

* 23 Bengtsson BÅ, Edén S, Lonn L *et al.* Treatment of adults with growth hormone (GH) deficiency with recombinant GH. *J Clin Endocrinol Metab* 1993; **76**: 309–17.

* 24 Fowelin J, Attvall S, Lager I, Bengtsson BÅ. Effects of treatment with recombinant human growth hormone on insulin sensitivity and glucose metabolism in adults with growth hormone deficiency. *Metabolism* 1993; **42**: 1443–7.

* 25 Møller J, Jørgensen JOL, Lauersen T *et al.* Growth hormone dose regimens in adult GH deficiency: effect on biochemical and metabolic parameters. *Clin Endocrinol* 1993; **39**: 403–8.

* 26 Weaver JU, Monson JP, Noonan K *et al.* The effect of low dose recombinant human growth hormone replacement on regional fat distribution, insulin sensitivity and cardiovascular risk factors in hypopituitary adults. *J Clin Endocrinol Metab* 1995; **80**: 153–9.

* 27 Bradly KM, Adams CB, Potter CP *et al.* An audit of selected patients with non-functioning pituitary adenoma treated by transsphenoidal surgery without irradiation. *Clin Endocrinol* 1994; **41**: 655–9.

* 28 Hankinson SE, Willett WC, Colditz GA *et al.* Circulating concentrations of insulin-like growth factor-I and risk of breast cancer. *Lancet* 1998; **351**: 1393–6.

* 29 Chan JM, Stampfer MJ, Giovanucci E *et al.* Plasma insulin-like growth factor-I and prostate cancer risk: a prospective study. *Science* 1998; **279**: 563–6.

30 Jenkins P, Fairclough P, Lowe D, *et al.* Acromegaly, colonic polyps and neoplasia. *Clin Endocrinol* 1997; **47**: 17–22.

* 31 Orme SM, McNally R, Staines A, Cartwright RA, Belchetz PE. Mortality and cancer incidence in acromegaly: a retrospective cohort study. *J Clin Endocrinol Metab* 1998; **83**: 2730–4.

* 32 Shalet SM, Brennan BM, Reddingius RE. Growth hormone therapy and malignancy. *Horm Res* 1997; **48** (Suppl 4): 29–32.

* 33 Cuneo RC, Judd S, Wallace JD *et al.* The Australian multicenter trial of growth hormone (GH) treatment in GH-deficient adults. *J Clin Endocrinol Metab* 1998; **83**: 107–16.

* 34 Holmes SJ, Shalet SM. Factors influencing the desire for long-term growth hormone replacement in adults. *Clin Endocrinol* 1995; **43**: 151–7.

Part 3: Adolescent–adult Transition

14: How should we evaluate GH-treated patients at completion of linear growth?

Maïthé Tauber and Pierre Moulin

Introduction

Growth hormone deficiency (GHD) is now a recognized clinical syndrome in adults, with childhood or adult onset, and it is associated with significantly increased cardiovascular morbidity and mortality [1]. Adult patients should be considered for treatment with GH replacement therapy, as many of the abnormal parameters of this syndrome are reversible [2–6]. Based on these data, the question has been raised that it might be advisable not to stop growth hormone (GH) treatment in GHD adolescents who have reached their final height, in order to prevent the development of the adult GHD syndrome. However, in most countries at the present time, there is no established indication for the continuation of GH replacement therapy in adolescents aged under 20 years after completion of growth, thus rendering them GH-deficient for a variable number of years.

During the last five years, most of the literature on this topic has focused on the re-evaluation of GH secretion in this population after completion of GH treatment [7–16], as a prior diagnosis of GHD in childhood is insufficient evidence to conclude that a patient will be GHD for life. Consequently, how should we re-evaluate GH-treated patients at the completion of linear growth?

There is still no consensus among paediatric endocrinologists on the best way of evaluating GH secretion in children [17–20], and furthermore, our knowledge concerning GH secretion in adolescents is at present limited. Moreover, we not only need to re-evaluate GH secretion, but also to evaluate the clinical and biological parameters of the so-called adult GHD syndrome. This type of re-evaluation has important objectives: firstly, to decide which patients should continue GH treatment for life, and secondly, to increase our knowledge concerning the GH status in this population of patients during this particular period.

Analysis of data on the re-evaluation of GH secretion after completion of GH treatment

All paediatric endocrinologists now agree that all GH-treated children should be systematically re-evaluated with regard to GH secretion after the completion of treatment, as GHD can be transient, particularly in isolated GHD (IGHD). Table 14.1 shows that the GH test normalized in 56% of patients (GH peak above 10 µg/L) when they were re-tested at the end of treatment, or as young adults. The biochemical criteria for the diagnosis of GHD in adults differ from those used to make the diagnosis in children. Most of the studies carried out by paediatric endocrinologists reported a cut-off level of 10 µg/L used to define GHD in childhood, and these patients were subsequently classified into severe GHD, complete GHD, or partial GHD depending on GH peak values below 3.5 µg/L, below 5 µg/L, and between 5 and 10 µg/L, respectively.

Re-evaluation of GH secretion in GHD-treated patients has shown conflicting results, and between 12% and 100% of these patients have low GH secretion. Table 14.2 shows the results and discrepancies in the major studies that can be analysed that were conducted on this topic between 1987 and 1997.

Table 14.1 Normalization of growth hormone (GH) responsiveness after GH treatment in patients treated for isolated growth hormone deficiency.

First author (ref.)	Patients with normalization of GH secretion (%)	Number of patients (*n*)	Time of GH retesting	GH retesting: Type of test	GH retesting: Number of tests	Cut-off
Clayton [7]	26	19	End of growth	ITT/AG	1 or 2	> 7.5 µg/L
Carel [11]	81	142	End of growth	–	1	> 10 µg/L
BGSPE*	36	33	End of growth	–	1	> 10 µg/L
Tauber [14]	67	121	End of growth	Clonidine + betaxolol	1	> 10 µg/L
Wacharasindhu [16]	88	8	End of growth	ITT	1	> 7.5 µg/L
Longobardi [10]	53	23	Adult	ITT/GHRH + PD	1 or 2	> 10 µg/L
Juul [13]	45	62	Adult	Clonidine	1	> 7.5 µg/L
Rutherford	50	6	Adult			
Total	56	414				

*These data were reported at the Seminaire d'Endocrinologie Pédiatrique, Paris 20–21 January 1997, by Thomas *et al.* for the collaborative Belgian Study Group for Paediatric Endocrinology (BGSPE). ARG: arginine; ITT: insulin tolerance test. GHRH + PD: growth hormone-releasing hormone plus pyridostigmine.

Table 14.2 Percentage of patients remaining growth hormone (GH) deficient with a cut-off of 5 μg/L or 15 mU/L, after completion of linear growth.

	IGHD		MPHD (*n*)		Radiation-induced GHD		Cranio-pharyngioma, other tumors	
	%	*n*	%	*n*	%	*n*	%	*n*
Clayton [7]	63	19	100	(3)	100	(15)	100	(3)
Carel [11]			100	(33)				
BSGPE	24	(33)						
Tauber [14]	23	(121)		–	100	(2)	100	(5)
Wacharasindhu [16]	12	(8)	–			–		–
Nicolson* [9]	53	(32)			56	(43)	70	(10)
Total	35	213		–	85	(60)	90	(18)

Retesting was performed in young adults within two years of completing GH therapy in 84% of the patients. BSGPE: Belgian Study Group for Paediatric Endocrinology; GHD: growth hormone deficiency; IGHD: isolated growth hormone deficiency; MPHD: multiple pituitary hormone deficiencies.

Factors explaining the differences observed

Aetiology of GHD. The results are very similar in the group of patients with organic or radiotherapy-induced GHD and patients with multiple pituitary hormone deficiencies (MPHD), ranging from 56% to 100% of GHD, most of these patients having persistent severe GHD. As expected, the major discrepancies were observed in the IGHD patients, with wide variation in the frequency of GHD after the completion of treatment, ranging from 33% to 80% of patients re-evaluated if the cut-off for GH peak was 10 μg/L and 12–63% if the cut-off limit was reduced to 5 μg/L.

Severity of GHD at initial diagnosis. Not all published studies mentioned the GH peak at the initial evaluation. In our study [14], there was 49% complete GHD after completion of treatment in the complete GHD group, while in the partial GHD group, 29% had partial and 14% had complete GHD. It is particularly interesting to note that the peak GH value at the initial diagnosis cannot predict the result of re-evaluation. Even in complete GHD patients, approximately half can have transient hormonal deficiency. Interestingly, combining the aetiology and severity of GHD at the initial diagnosis, it has been shown that in patients who received radiotherapy, GH secretion tends to deteriorate, while partial idiopathic GHD tends to improve [9]. Patients with MPHD, even if idiopathic, had lower GH secretion at re-evaluation, as shown in Figure 14.1. Another point to emphasize is that pituitary deficiency has a tendency to worsen, with additional hormonal deficiencies at re-evaluation [13].

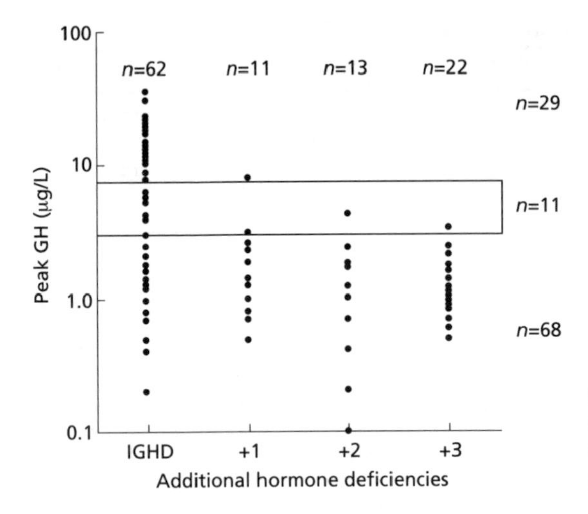

Fig. 14.1 Peak growth hormone (GH) levels during GH provocative testing relative to the degree of hypopituitarism. Reproduced with permission from Juul *et al.* [13].

Duration of the disease. In the study by Juul *et al.* [13], multiple regression analysis revealed that peak GH levels were significantly dependent on the duration of the disease, as well as on the severity of hypopituitarism.

Type of testing. Another point that might explain the discrepancies observed between the studies is the type of test used, and particularly the fact that the test used at the end of treatment was often not the same as that used for the initial diagnosis. The lack of reproducibility of pharmacological tests in children is well known [21–25], and the same is probably the case for adolescents. There are also other confounding factors in these studies: firstly, most patients are re-evaluated as adults using the insulin tolerance test (ITT) or GH-releasing hormone (GHRH) test alone, or combined with pyridostigmine (PD) or arginine (ARG) [26–28]; secondly, the amplitude of the GH peak could be different between adolescents and adults, even young adults, and the cut-off levels might be different at this age; thirdly, the GHRH test is not used frequently by paediatric endocrinologists [20], and the results at re-evaluation may therefore not be comparable with the first evaluation; and fourthly, the cut-off level depends on the test used.

In this particular population, our primary objective is to identify severe GHD patients who should receive GH replacement therapy to prevent the adult GHD syndrome, and the lack of reproducibility of the pharmacological tests is perhaps less important. Nevertheless, we do not actually know whether patients with a GH peak at re-testing between 5 and 10 μg/L are at higher risk for developing the GHD syndrome than strictly normal patients, and this population should be carefully followed up and regularly evaluated by adult endocrinologists.

The value of 24-hour GH secretion and/or urinary GH secretion tests in this situation has not yet been reported. This point should be evaluated, particularly in the specific population of GHD defined in childhood as neurosecretory dysfunction characterized by normal stimulated GH secretion on pharmacological tests and low spontaneous GH secretion. We do not know at present whether the abnormality of spontaneous GH secretion persists at the end of treatment, and if this is the case, what the consequences in adults might be.

How should we evaluate GH secretion after the completion of growth?

All patients should be re-evaluated at the end of growth. There is now a need to standardize the re-testing of adolescents with childhood-onset GHD with regard to the method used for re-testing and the peak GH level that is taken to represent continuing GH deficiency. As it has been postulated [29], it would be preferable for research purposes, as well as for clinical decision-making, if the number of provocative tests used at various centres were reduced to a minimum and if the biochemical methods for the diagnosis of GHD were thoroughly standardized world-wide.

Type of test

As the two tests most commonly used to assess GH secretion in adults are the ITT and GHRH combined with ARG or PD, it appears advisable to use the same test in adolescence [26–28]. The ITT is the standard test used for the diagnosis of GH deficiency in adults and has good sensitivity and reproducibility, even though a blunted GH response can be observed in normal individuals [30]. Clonidine is not a potent GH secretagogue in adults, and should be avoided. A consensus conference on the diagnosis of GHD in adults proposed a cut-off of 3 μg/L, below which adults are diagnosed as GH-deficient [31]. This value may be inappropriately low for adolescents, and until further data become available, a cut-off of 5 μg/L may be more appropriate as an indicator of the need for follow-up. GHRH combined with ARG or PD is also a reliable test in adults, with a higher cut-off value (Fig. 14.2). Whatever the tests performed, patients should have a washout period of around eight days without GH treatment before re-testing.

Number of tests

This issue has been extensively discussed in a consensus conference (in Pisa, 27–28 March 1998), and alternative strategies for isolated GHD and MPHD were proposed. Patients with isolated GHD should undergo two tests, given the degree of discordance at re-evaluation when two tests are used [9]. The

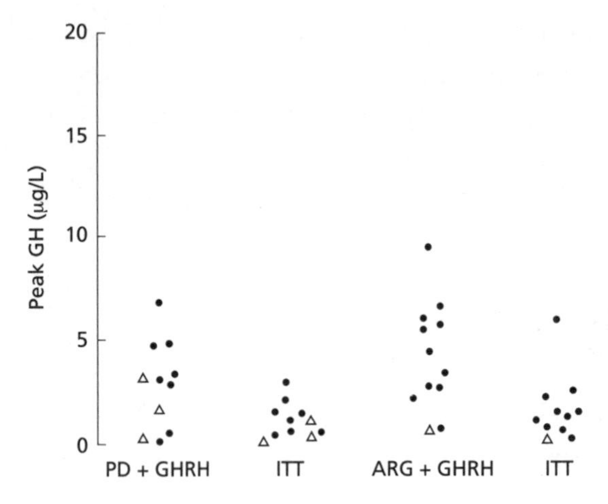

Fig. 14.2 Comparisons in the same hypopituitary individuals between peak growth hormone (GH) responses to the growth hormone-releasing hormone (GHRH) plus pyridostigmine (PD) test (age 20–40 years); to the GHRH plus arginine (ARG) test (age 20–80 years); and to the insulin tolerance test (ITT). Open triangle: GH levels in idiopathic growth hormone deficiency. Reproduced with permission from Ghigo *et al.* [28].

second test could be eliminated if insulin-like growth factor-I (IGF-I) and/or insulin-like growth factor binding protein-3 (IGFBP-3) are low and/or additional anterior pituitary hormone deficiency is found at re-evaluation. It seems justified to argue that patients with MPHD, whatever the origin, and patients with GHD due to hypothalamic pituitary disease or radiotherapy, require only one test of GH secretion.

Some paediatric endocrinologists have suggested that in patients with GHD due to hypophysectomy (often for craniopharyngioma), re-evaluation should not be performed at adolescence, allowing continuation of treatment for as long as possible and thereby enhancing peak bone mass, leaving the re-testing until later adult life. In France, two tests are required to obtain GH replacement therapy for GHD adults (above 20 years). This point should therefore be clarified, and it might be proposed that one test should be performed at the end of growth and another when the patient is an adult.

IGF-I and IGFBP-3 evaluation

In adults with childhood-onset GHD, the diagnostic value of serum IGF-I and IGFBP-3 measurements is comparable with that in GHD children if the results are compared to age-matched controls [13]. This has not always been done in reported studies, leading to controversial data. However, the same does not seem to apply in adult-onset GHD, for reasons that are unclear [26–28,32]. The combined use of these parameters improved the diagnostic value (Fig. 14.3): IGF-I and IGFBP-3 correlated significantly with peak GH levels, and were lower in patients with additional anterior pituitary hormonal deficiency [13].

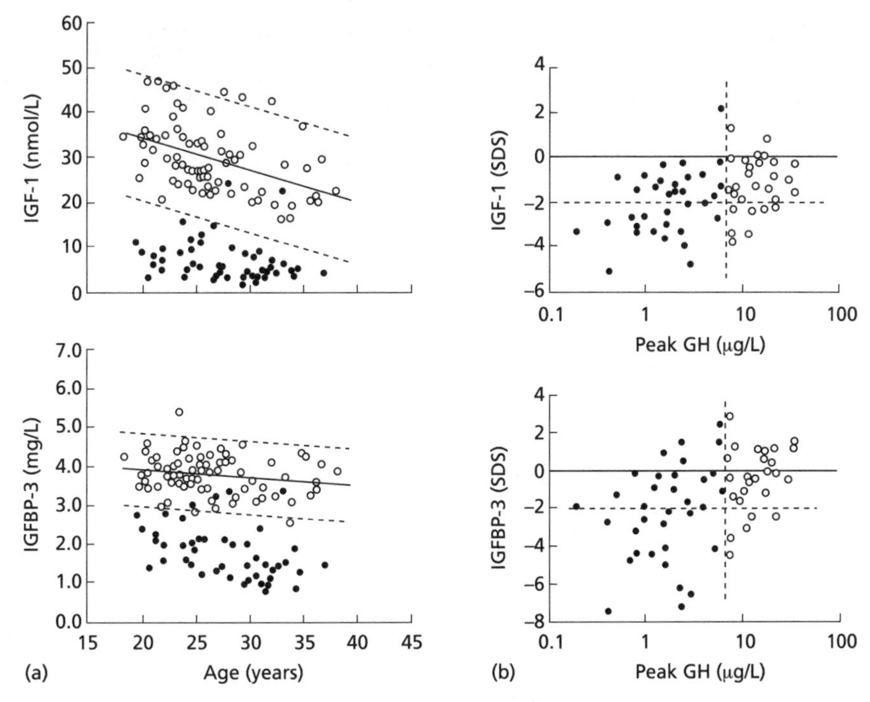

Fig. 14.3 Value of insulin-like growth factor-I (IGF-I) and insulin-like growth factor binding protein-3 (IGFBP-3) levels in the diagnosis of continuing growth hormone deficiency (GHD). (a) Serum IGF-I and IGFBP-3 in patients with established GHD (●) and healthy controls (○). Lines represent 2 standard deviation scores (SDS) from values in controls. Reproduced with permission from De Boer *et al.*, *Lancet* 1994; **343**: 1645–6 [27]. (b) Combined data for peak growth hormone (GH) and IGF-I (SDS) and IGFBP-3 (SDS). Peak GH level vs. IGF-I SD score (top panel) and IGFBP-3 score (bottom panel) in 62 patients previously treated with GH because of isolated GHD who were re-evaluated as young adults. The dotted lines represent the two cut-off values corresponding to 7.5 μg/L for peak GH and −2 SDS for IGF-I and IGFBP-3. Reproduced with permission from Juul *et al.* [13].

Other evaluations

A complete profile of pituitary function (prolactin, thyroid-stimulating hormone, luteinizing hormone and follicle-stimulating hormone, adrenocorticotrophic hormone) should be performed, as we know that secretory defects in other pituitary hormones can evolve, e.g. in patients with pituitary stalk section [33]. When re-testing confirms GHD, pituitary magnetic resonance imaging (MRI) should be performed if it has not been done previously, or if a previous MRI was reported as normal. This MRI should be analysed by trained neuroradiologists, who should measure the height of the pituitary gland [34] and the volume of the gland if possible.

In agreement with the conference consensus (Pisa, March 1998), we would make the following recommendations:

- All patients should be re-tested for GH secretion with two tests (isolated GHD) or one test (MPHD or organic GHD).
- The tests performed should be those used in adults to confirm GHD: ITT or GHRH + PD or ARG.
- IGF-I and IGFBP-3 should be evaluated and expressed as standard deviation scores (SDS) for age.
- Complete evaluation of pituitary hormones should be carried out.
- Pituitary MRI should be carried out if it has not been performed previously and if GHD is confirmed at re-evaluation.

What other factors should be evaluated in GHD patients after completion of linear growth?

Complete evaluation of GH status: tissue and metabolic targets of GH

In all GH-treated patients, it will be of interest to evaluate not only GH secretion but also functional GH status by analysing tissue and metabolic targets of GH such as body composition [35], bone mineral density (BMD) [36–39], physical performance, muscle strength, myocardial function, lipid levels [39] and quality of life [40], using adequate and validated questionnaires. These parameters should be studied firstly to provide baseline data on which to base precise follow-up of these patients and secondly to increase our knowledge of these populations. We do not yet know whether all of these patients reach normal status in terms of metabolic function on GH replacement therapy. In this regard, it has been reported that some GHD patients have low BMD at completion of growth and that GH should be continued until normal BMD values are reached [36].

Patients with GH peak of 5–10 μg/L (15–30 mU/L).

In addition to careful follow-up in those patients with confirmed GHD, it is also important to monitor regularly those patients with possible borderline reductions in GH secretion for a sufficiently long period, with special emphasis on body composition, serum lipid levels, physical performance and quality of life.

Follow-up

Finally, all patients should be followed up in order to obtain long-term outcome data that can be entered into a large multinational database in order to optimize our knowledge and ensure the dissemination of information, so that clinical practice can be modified if needed. This follow-up requires good and thorough collaboration between paediatric endocrinologists and adult endocrinologists, and patients must receive detailed explanations regarding the importance of long-term monitoring. At present, we are aware of the consequences of prolonged severe GH depletion, but we have little data regarding the effects of short-term GH depletion and no data on the short-term or long-term effects of borderline GHD (GH peak between 5 and 10 μg/L) after the completion of growth.

References

** 1 Rosén T, Bengtsson BÅ. Premature mortality due to cardiovascular disease in hypopituitarism. *Lancet* 1990; **336**: 285–8.

2 Bengtsson BÅ, Éden S, Lonn L *et al.* Treatment of adults with growth hormone (GH) deficiency with recombinant human GH. *J Clin Endocrinol Metab* 1993; **72**: 309–17.

3 Jørgenson JOL, Pederson SA, Thuesen L *et al.* Long-term growth hormone treatment in growth hormone-deficient adults. *Acta Endocrinol* 1991; **125**: 449–53.

4 Salomon F, Cueno RC, Hesp R, Sönksen PH. The effects of treatment with recombinant human growth hormone on body composition and metabolism in adults with growth hormone deficiency. *N Engl J Med* 1989; **321**: 1797–803.

5 Jørgenson JOL, Thuesen L, Muller J, Ovesen P, Christiansen JS. Three years of growth hormone treatment in growth hormone-deficient adults: near normalization of body composition and physical performance. *Eur J Endocrinol* 1994; **130**: 224–8.

6 Harrant I, Beauville M, Crampes F *et al.* Response of fat cells to growth hormone: effect of long-term treatment with recombinant human growth hormone in GH deficient adults. *J Clin Endocrinol Metab* 1994; **78**: 1392–5.

* 7 Clayton PE, Price DA, Shalet SM. Growth hormone state after completion of treatment with growth hormone. *Arch Dis Child* 1987; **62**: 222–6.

8 Cacciari E, Tassoni P, Parisi G *et al.* Pitfalls in diagnosing impaired growth hormone secretion: retesting after replacement therapy in 63 patients defined as GH deficient. *J Clin Endocrinol Metab* 1992; **74**: 1284–9.

** 9 Nicolson A, Toogood AA, Rahim A, Shalet SM. The prevalence of severe growth hormone deficiency in adults who received growth hormone therapy replacement in childhood. *Clin Endocrinol* 1996; **44**: 311–6.

** 10 Longobardi S, Merola B, Pivonello R *et al.* Reevaluation of growth hormone secretion in 69 adults diagnosed as GH-deficient during childhood. *J Clin Endocrinol Metab* 1996; **81**: 1244–7.

11 Carel JC, Gendrel C, Chatelain P, Rochiccioli P, Chaussain JL. Reevaluation of growth hormone secretion after treatment with human growth hormone. *Horm Res* 1996; **46 (12)**: 18.

12 Jocham A, Butenandt O, Sperlich M, Schwarz HP. Reevaluation of adults with childhood onset growth hormone deficiency. *Horm Res* 1996; **46 (12)**: 30.

** 13 Juul A, Kastrup KW, Pedersen SA, Skakkebaek NE. Growth hormone (GH) provocative retesting of 108 young adults with childhood-onset GH deficiency and the diagnosis value of insulin-like growth factor (IGF-I) and IGF-binding protein-3. *J Clin Endocrinol Metab* 1997; **82**: 1195–201.

** 14 Tauber M, Moulin P, Pienkowski C, Jouret B, Rochiccioli P. Growth hormone (GH) retesting and auxological data in 131 GH deficient patients after completion of treatment. *J Clin Endocrinol Metab* 1997; **82**: 352–6.

*15 Rutherford OM, Jones DA, Round JM, Buchanan CR, Preece MA. Changes in skeletal muscle and body composition after discontinuation of growth hormone treatment in GHD young adults. *Clin Endocrinol* 1991; **34**: 469–75.

*16 Wacharasindhu S, Cotterill AM, Camacho-Hübner C, Besser GM, Savage MO. Normal growth hormone secretion in growth hormone insufficient children retested after completion of linear growth. *Clin Endocrinol* 1996; **45**: 553–6.

17 Rosenfeld RG, Albertsson-Wikland K, Cassorla F *et al.* Diagnostic controversy: the diagnosis of childhood growth hormone deficiency revisited. *J Clin Endocrinol Metab* 1995; **80: 1532–40.

18 Rosenfeld RG. Is growth hormone deficiency a viable diagnosis? *J Clin Endocrinol Metab* 1997; **82: 349–51.

19 Tauber MT, Rochiccioli P. Exploration of the somatotropic axis. *Diabetes Metab* 1996; **22**: 240–4.

*20 Rochiccioli P, Pienkowski C, Tauber MT, Enjaume C. Combining pharmacological tests and 24-hour GH secretion ($n = 257$) for a new classification of GH deficiencies. *Horm Res* 1989; **31**: 27–35.

*21 Tassoni P, Cacciari E, Cau M *et al.* Variability of growth hormone response to pharmacological and sleep tests performed twice in short children. *J Clin Endocrinol Metab* 1990; **71**: 230–4.

*22 Rochiccioli P, Enjaume C, Tauber MT, Pienkowski C. Statistical study of 5743 results of nine pharmacological stimulation tests: a proposed weighting index. *Acta Pediatr* 1993; **82**: 245–8.

23 Rogers RS, Levin RJ, Uriatre M *et al.* The advantage of measuring stimulated as compared with spontaneous growth hormone levels in the diagnosis of growth hormone deficiency. *N Engl J Med* 1988; **319**: 201–7.

24 Donaldson DL, Howell JG, Pan F, Gifford RA, Moore WV. Growth hormone secretory profiles: variation on consecutive nights. *J Pediatr* 1989; **115**: 51–6.

25 Zadik Z, Chalew SA, Gilua Z, Zowarski AA. Reproducibility of growth hormone testing procedures: a comparison between 24-hour integrated concentration and pharmacological stimulation. *J Clin Endocrinol Metab* 1990; **71**: 1127–30.

26 Hoffman DM, O'Sullivan AJ, Baxter RC, Ho K. Diagnosis of growth hormone deficiency in adults. *Lancet* 1994; **343**: 1064–8.

27 De Boer H, Block GJ, Popp-Snijders C, Van Der Veen E. Diagnosis of growth hormone deficiency in adults. *Lancet* 1994; **343: 1645–6.

28 Ghigo E, Aimaretti G, Gianotti L *et al.* New approach to the diagnosis of growth hormone deficiency in adults. *Eur J Endocrinol* 1996; **134: 352–6.

29 De Boer H, Van Der Veen EA. Why retest young adults with childhood-onset growth hormone deficiency? [editorial]. *J Clin Endocrinol Metab* 1997; **82: 2032–5.

30 Hoeck HC, Vestergoard P, Jakobson PE, Laurberg P. Test of growth hormone secretion in adults: poor reproducibility of the insulin tolerance test. *Eur J Endocrinol* 1995; **133**: 305–12.

31 Thorner MO, Bengtsson BA, Ho KY *et al.* The diagnosis of growth hormone deficiency (GHD) in adults [letter]. *J Clin Endocrinol Metab* 1995; **80: 3097.

32 Attanasio AF, Lamberts SWJ, Matranga AMC *et al.* Adult growth hormone (GH)-deficient patients demonstrate heterogeneity between childhood onset and adult onset before and during human GH treatment. *J Clin Endocrinol Metab* 1997; **82**: 82–8.

33 Barbeau C, Jouret B, Gallegos D *et al.* Syndrome d'interruption de la tige pituitaire. *Arch Pediatr* 1998; **5**: 274–9.

34 Argiropolou M, Perignon F, Brauner R, Bruvelle F. Magnetic resonance imaging in the diagnosis of growth hormone deficiency. *J Pediatr* 1992; **120**: 886–91.

35 Colle M, Auzerie J. Discontinuation of GH therapy in GHD patients: assessment of body fat mass using bioelectric impedance. *Horm Res* 1993; **39**: 192–6.

36 Saggese G, Baroncelli GI, Bertelloni S, Barsanti S. The effect of long-term growth hormone (GH) treatment in bone mineral density in children with GH deficiency: role of GH in the attainment of peak bone mass. *J Clin Endocrinol Metab* 1996; **81: 3077–83.

37 Hyer SL, Rodin DA, Tobias JH, Leiper A, Nussey SS. Growth hormone deficiency during puberty reduces adult bone mineral density. *Arch Dis Child* 1992; **57**: 1472–4.

38 Kayman JM, Tallman P, Verneulen A,

Vandeweghe M. Bone mineral status in growth hormone deficient males with isolated and multiple pituitary deficiencies of childhood onset. *J Clin Endocrinol Metab* 1992; **74**: 118–23.
39 De Boer H, Blok GJ, Voerman HJ, Phillipos M, Schovten JA. Serum lipid levels in growth hormone-deficient men. *Metabolism* 1994; **43**: 199–203.
40 Deijen JB, De Boer H, Blok GJ, Van Der Veen EA. Cognitive impairments and mood disturbances in growth hormone deficient men. *Psychoneuroendocrinology* 1996; **21**: 313–22.

15: How should we transfer patients from paediatric to adult clinics?

John P. Monson and Pierre Chatelain

Introduction

The potential adverse effects of adult growth hormone deficiency (GHD) on body composition, cardiovascular risk factors, cardiac function, bone mineral density and psychological well-being have been extensively investigated and substantially confirmed in a large number of studies [1–6]. In consequence of this, the appropriateness of cessation of growth hormone (GH) therapy in childhood-onset GHD at the time of completion of linear growth has been called into question. Whilst there are insufficient data available at present to make definitive recommendations regarding continuation or discontinuation of therapy, it is none the less essential that robust mechanisms should be set in place to ensure follow-up of adolescents into adult life and, in addition, to accumulate information from controlled studies. The importance of maintaining the continuity of care from a wide spectrum of childhood-onset endocrine disorders into adult life is self-evident, but may be difficult to achieve unless formal arrangements for liaison and eventual transfer of clinical responsibility are established [7]. For the young person with GHD, as with insulin-dependent diabetes, a high level of commitment is required of the patient at a time when motivation and compliance with therapy may be suboptimal [8].

This chapter will address the rationale for ensuring adequate transition from paediatric to adult clinics in this patient group, and will outline the clinical scenarios involved, including the approach to re-testing, the definition of GHD in the young adult, the consequences of withdrawal of GH therapy on body composition, the effects of childhood GHD on achievement of peak bone mass, the approach to GH dosing in the young adult and practical aspects of achieving a seamless transition of care.

Which patients should be re-tested?

GHD associated with structural pituitary disease or genetic defects in GH expression can be assumed to be permanent. However, there is now accumulating evidence that isolated idiopathic GHD in childhood may not persist into adult life; various studies have demonstrated 'normal' GH responses on provocative testing in up to 80% of patients after cessation of GH therapy at the end of linear growth [9–11]. It is therefore mandatory to re-test the latter group, and in the establishment of future trials it is probably appropriate to re-test all patients, particularly to establish degrees of GHD in patients with structural disease for comparison with the experience in older adults. Because the duration of effect of exogenous GH (via insulin-like growth factor-I feedback) on spontaneous GH secretion is unknown, the earliest time for re-testing after cessation of GH replacement remains unclear. Unpublished observations (M.O. Savage and J.P. Monson) suggest that the GH response to insulin hypoglycaemia is unimpaired seven days after cessation. However, this may be too short an interval to use GH-dependent factors such as insulin-like growth factor-I (IGF-I), insulin-like growth factor binding protein-3 (IGFBP-3) and acid-labile subunit (ALS) for diagnosis, and the optimum timing for these latter measurements requires further investigation. These various issues are discussed in detail in Chapter 14.

How should GHD be defined in the transition phase?

This is a difficult question, not least because the current criterion for the diagnosis of GHD in adults (peak GH < 9 mU/L (3 ng/mL) after adequate hypoglycaemia) [12] is probably too strict for defining GHD in young adults. It is therefore essential that we should accumulate longitudinal information on the physical characteristics of young adults who have an apparently normal GH response on dynamic testing, in order to correlate clinical changes with GH secretory status. Important parameters to follow in this respect include indices of body fat and its distribution (e.g. measured by dual-energy X-ray absorptiometry and waist : hip measurements), bone mineral density, lipoprotein profiles, indices of insulin sensitivity and gonadal status and reproductive function. These observations on individuals with intermediate degrees of GH reserve should ideally be carried out in the context of controlled trials incorporating treated and untreated cohorts; additional information is likely to emerge from long-term follow up of treated and untreated individuals in the KIGS and KIMS (Pharmacia and Upjohn International Growth Study and Metabolic Surveillance, respectively) databases. Until these data are available, it is empirically appropriate to limit a definite diagnosis of persistent GHD to those individuals who demonstrate an unequivocally subnormal GH response on dynamic testing, but to continue follow-up of those young adults who demonstrate a peak GH of between 10 and 20 mU/L.

Consequences of withdrawal of GH on body composition and cardiac function

Although data are available documenting the consequences of GHD in adult patients with childhood-onset disease [13], there is a paucity of information on the more immediate consequences of withdrawal of GH therapy at the time of completion of linear growth. It is clear that adults with long-standing, persistent GHD of childhood onset share many of the features of the GH deficiency syndrome characterized in adult-onset GHD. However, there appear to be important differences, including significantly lower serum IGF-I concentrations, shorter stature, a lower body mass index, less obvious derangement in serum lipoprotein profiles and a more profound deficit in lean body mass [13]. Furthermore, quality-of-life assessments demonstrate less deviation from normal in adult GHD of childhood onset, although this may simply reflect long-term adaptation and the relative insensitivity of currently available questionnaires in this patient group.

The effects of the withdrawal of GH therapy in childhood-onset GHD on muscle function were systematically studied by Rutherford *et al.* [14]. They demonstrated significant reductions in quadriceps muscle strength and cross-sectional area (Fig. 15.1) after withdrawal of GH for 12 months. A reduction in forearm flexor cross-sectional area, but not strength, was also observed (Fig. 15.2). Rutherford *et al.* also demonstrated an increment in body fat, as determined by skinfold thickness, from a mean of 19.5% to 24.1%. In a subsequent study, Colle and Auzerie [15] documented a

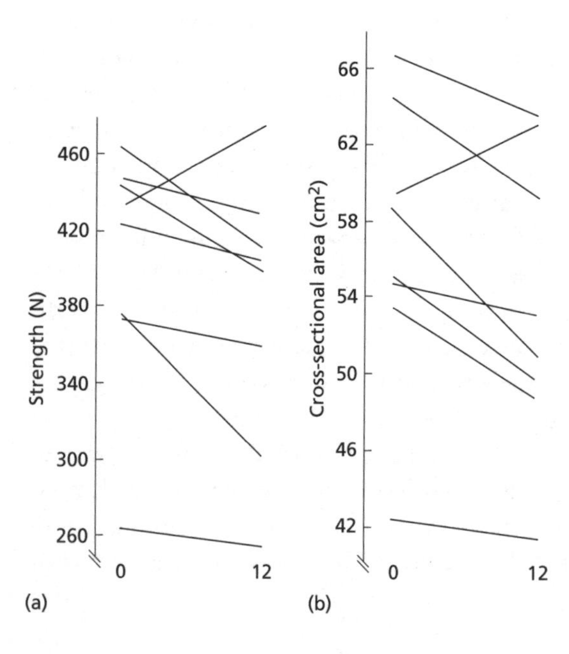

Fig. 15.1 Quadriceps strength (a) and (b) cross-sectional area at completion of growth hormone (GH) therapy and 12 months after discontinuation in eight patients with childhood-onset growth hormone deficiency (GHD). Reproduced with permission from Rutherford *et al.* [14].

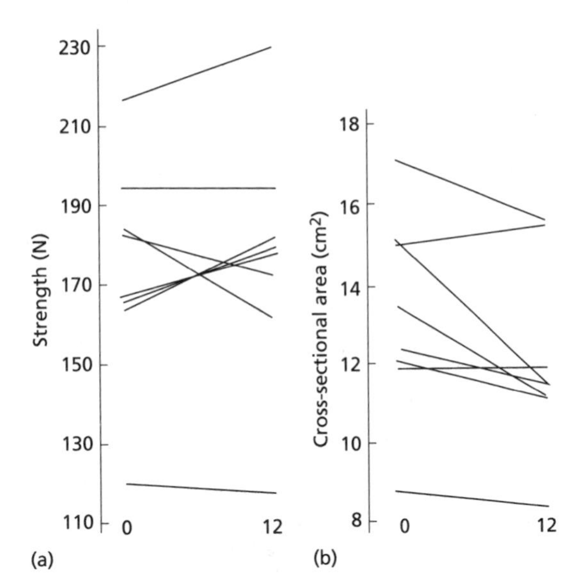

Fig. 15.2 Forearm flexor strength (a) and (b) cross-sectional area at completion of GH therapy and 12 months after discontinuation in eight patients with childhood-onset growth hormone deficiency (GHD). Reproduced with permission from Rutherford *et al.* [14].

significantly greater total fat mass, determined by bioelectrical impedance (BIA), at the time of discontinuation of GH therapy in males with continuing GHD compared with patients who had apparently normal GH reserve on re-testing, and a further rise in fat mass over the course of three months in the GHD group, which was not evident in the non-GHD subjects (Fig. 15.3). Although BIA may overestimate fat mass in GHD subjects, these studies provided initial evidence pointing towards the need to evaluate body composition after discontinuation of GH replacement. More recently, Johansson *et al.* [16] have presented data demonstrating substantial increments in total body fat, determined by dual-energy X-ray absorptiometry (DEXA), over a two-year period after withdrawal of GH therapy at completion of linear growth (Table 15.1). They also made the interesting observation that serum insulin concentrations decreased over this time scale, in contrast to the relative insulin resistance that characterizes adult-onset GHD.

Cardiac function has been examined by echocardiography in childhood-onset GHD some five years after cessation of GH replacement, and significant reductions were observed in all indices of cardiac function compared with matched controls [17] (Table 15.2). These phenomena may be at least partially reversible, and studies of cardiac function during GH replacement therapy in adult GHD would support this. In addition, Capaldo *et al.* [18] have demonstrated increased arterial intima–media thickness in young adults with childhood-onset GHD (Fig. 15.4), a phenomenon that has been demonstrated previously in GHD of adult onset [19] and which may reflect an increased predisposition to atherogenesis.

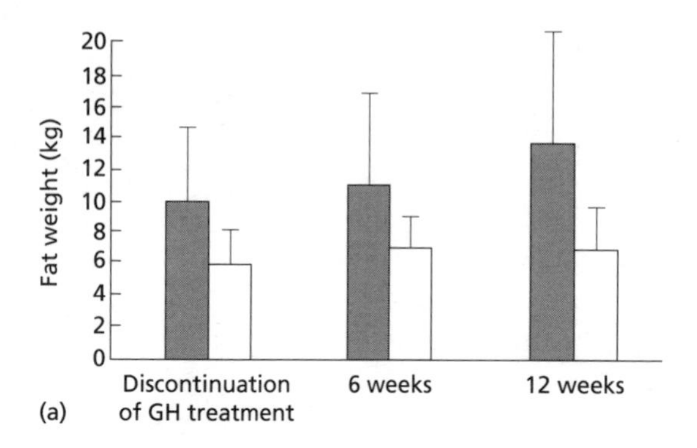

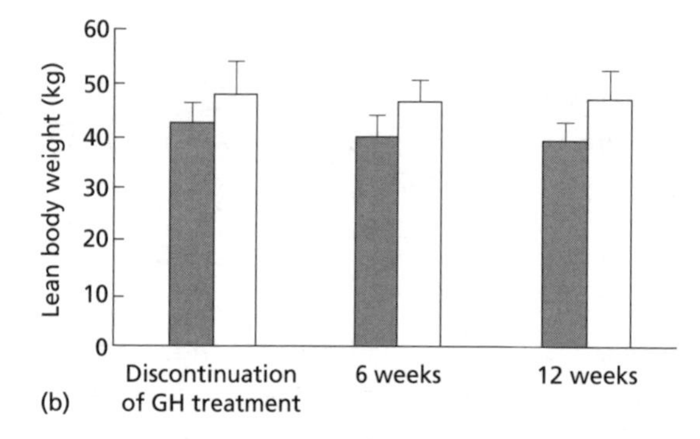

Fig. 15.3 (a) Fat weight and (b) lean body weight in six patients with growth hormone deficiency (GHD) (shaded bars) and 10 non-GHD patients (open bars). Reproduced with permission from Colle and Auzerie [15].

Table 15.1 Effects of two-year discontinuation of growth hormone (GH) replacement in 40 adolescent patients (age 16–21 years); 22 with isolated growth hormone deficiency, 18 with multiple pituitary deficiencies. From Johannsson *et al.* [16].

Effect
Serum IGF-I reduced: mean 649–299 μg/L
Body weight increased: mean 2.7 kg
Total body fat mass increased: mean 4.8 kg
Fat free mass decreased: mean 2.3 kg
Increased total cholesterol, LDL-C, and apolipoprotein B
No change in HDL-C
Decreased Lp(a)
Decreased fasting serum insulin

IGF-I: insulin-like growth factor-I; HDL-C: high-density lipoprotein cholesterol; LDL-C: low-density lipoprotein cholesterol.

Is cessation of GH replacement at completion of linear growth associated with a failure to achieve peak bone mass?

This issue has been addressed by two important studies. Kaufman *et al.* [20] examined bone mineral content by photon absorptiometry in 30 adult males with childhood-onset GHD who had discontinued GH therapy at least six months prior to the study, and documented relative osteopenia compared with matched controls. Similar deficits were observed in

Table 15.2 Echocardiographic findings in childhood-onset adult growth hormone deficiency after discontinuation of GH therapy for more than five years. From Amato *et al.* [17].

Variable	Controls ($n = 20$)	Patients ($n = 7$)	p
Heart rate (per min)	73 ± 10	70 ± 9	N.S.
Systolic blood pressure (mmHg)	120 ± 13	108 ± 5	N.S.
Diastolic blood pressure (mmHg)	70 ± 9	67 ± 8	N.S.
End-diastolic dimension (mm)	48 ± 4	45 ± 3	N.S.
End-systolic dimension (mm)	30 ± 4	31 ± 3	N.S.
Interventricular septal thickness (mm)	9.0 ± 1.0	7.4 ± 1.6	< 0.01
Posterior wall thickness (mm)	9.0 ± 1.0	7.4 ± 1.5	< 0.01
Left ventricular mass index	85 ± 15	62 ± 15	< 0.005
Fractional shortening (%)	38 ± 5	31 ± 2	< 0.01
Velocity of circumferential fibre shortening	1.1 ± 0.1	0.97 ± 0.07	< 0.02

N.S.: not significant.

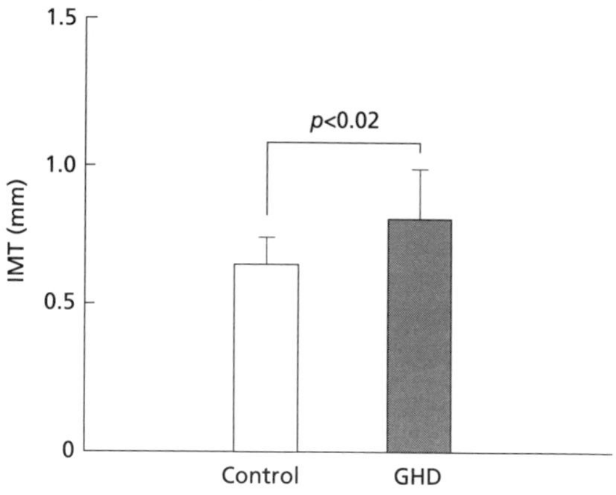

Fig. 15.4 Carotid intima–media thickness in 14 control individuals and 14 patients with childhood-onset growth hormone deficiency (GHD). Values represent the mean (SEM) thickness of both carotid arteries. Reproduced with permission from Capaldo *et al.* [18].

individuals with isolated GHD and multiple pituitary hormone deficiencies, indicating the major role of GH in determining bone mass. Furthermore, longitudinal follow-up was not associated with progressive bone loss, indicating that the reduction in bone mineral content was likely to have been due to a failure to achieve peak bone mass. De Boer *et al.* [21] reached similar conclusions based on the measurement of bone mineral density (BMD) by DEXA in 70 adult males with childhood-onset GHD. Importantly, they undertook careful corrections for the possible error in computing areal bone density in patients, who have significantly shorter stature than controls.

It should be borne in mind, however, that these assessments may simply reflect the inevitable suboptimal design of GH therapy in childhood in preceding decades, and therefore do not necessarily imply a similar deficit in BMD in the current generation of GHD patients. This notion is to some extent supported by the data provided by Saggese *et al.* [22], who demonstrated a significant advantage in BMD terms for a younger cohort of GHD patients compared with a group of patients treated initially with cadaveric GH; the study documented normal age-related BMD in the younger patients, and the authors concluded that GH therapy should be continued until the attainment of peak bone mass, irrespective of the height achieved. Clearly, these observations argue strongly that the assessment of the young person with GHD should include body composition and bone mineral density in addition to the achievement of final height.

How should GH therapy be recommenced after initial cessation?

Withdrawal of GH therapy in order to reassess GH reserve prompts the important question concerning the optimum method of dosing if treatment is to be recommenced. Although GH therapy has been traditionally based on body weight or surface area in children, the situation in adults has evolved into one of dose titration, most commonly based on serum IGF-I measurements [23]. This approach acknowledges the fact that there are substantial interindividual variations in susceptibility to GH on the basis of sex and other as yet unrecognized factors. The rationale underlying GH dose titration is discussed in Chapter 10, and is entirely applicable to therapy in the young adult. Reliable, age-related normative data for serum IGF-I within individual clinics are a prerequisite for this approach.

Timing of hand-over of adolescent patients from paediatric to adult endocrinologists

The strategy for hand-over will depend on a number of practical factors, which include:

- The geographical relationship between the paediatric and adult endocrine departments, i.e. same or different hospital buildings.
- The presence or absence of established clinical and/or academic liaison between the paediatric and adult department, i.e. are similar therapeutic protocols likely to be followed in both environments?
- The existence of other links, including training rotations for medical staff.
- The presence or absence of a transitional outpatient clinical environment.

Specific clinical factors should also be considered in deciding the timing of transfer, and these suggest that this will vary from one patient to another:

• Focus on metabolic consequences of GHD after cessation of linear growth is probably most appropriately undertaken in the adult clinic.
• Observation of the long-term effects on cardiovascular risk factors, bone mineral density and psychological well-being if GH replacement is withdrawn will be best conducted by adult endocrinologists.
• Relatively early transfer may increase the patient's and family's awareness of the continuing nature of the clinical problem.
• In young people with hypopituitarism, management of other aspects of hormonal deficiency, e.g. hypogonadism, will be most appropriately undertaken in the adult clinic, once secondary sexual development is complete.
• Early transfer, without adequate transitional arrangements, may compromise continuity of care and result in an increased rate of default from follow-up.
• If patients are transferred before completion of linear growth, an adequate auxological service and ensured data collection must be available in the adult clinic.
• Much depends on the approach and advice given to the family at the time of the initial presentation and treatment, i.e. a shift in emphasis from simply achieving maximum height to consideration of peak bone mass, as well as addressing the consequences of GHD throughout life.

Perhaps we should place the greatest emphasis on the manner in which transition is achieved rather than focusing simply on the timing of handover. Each unit must devise strategies for seamless transfer of clinical care. This, at the very least, must ensure that either the paediatrician or the adult physician should take responsibility for full reassessment of isolated GHD. It is clearly important that the small group of adolescents with evolving endocrinopathy should be identified.

Conclusions

The principles outlined above were endorsed recently by an international conference on the diagnosis and treatment of GH deficiency in children and adolescents held in Pisa, Italy, in March 1998. The following broad consensus on the principles of transition of care was reached:
• The overriding aim should be seamless transition. Re-testing of GH status should be arranged and the patient should be transferred on treatment, or have planned cessation with a clear plan for continuing surveillance.
• Patients should remain under paediatric care until the cessation of linear growth.
• Improved arrangements for liaison and mutual education between paediatric and adult endocrinologists are of paramount importance.

- It is inevitable that the burden of initiating and managing the transition process will fall substantially on the paediatric endocrine unit.
- In most instances, re-testing should be carried out by the paediatric endocrinologist.
- Those patients who have an apparently 'normal' GH reserve on re-testing require careful counselling and preferably long-term follow-up by either the paediatric or the adult endocrinologist.
- Children established on GH therapy require education about all aspects of GHD from early adolescence.
- Bone mineral density should be measured before discontinuation of GH on the achievement of final height, and consideration should be given to the continuation of therapy in those individuals with suboptimal bone mass.
- Young adults with documented GHD on re-testing should be offered continuation of GH therapy with appropriate monitoring, but recognizing the need for further randomized prospective trials in this age group.

References

1 Salomon F, Cuneo RC, Hesp R, Sönksen PH. The effects of treatment with recombinant human growth hormone on body composition and metabolism in adults with growth hormone deficiency. *N Engl J Med* 1989; **321**: 1797–803.

2 Jørgensen JOL, Pedersen SA, Thuesen L *et al.* Beneficial effects of growth hormone treatment in GH-deficient adults. *Lancet* 1989; **i**: 1221–5.

3 Bengtsson BÅ, Edén S, Lönn L *et al.* Treatment of adults with growth hormone (GH) deficiency with recombinant human GH. *J Clin Endocrinol Metab* 1993; **76**: 309–17.

4 De Boer H, Blok GJ, Van der Veen EA. Clinical aspects of growth hormone deficiency in adults. *Endocr Rev* 1995; **16**: 63–86.

5 Weaver JU, Monson JP, Noonan K *et al.* The effect of low dose recombinant human growth hormone replacement on regional fat distribution, insulin sensitivity and cardiovascular risk factors in hypopituitary adults. *J Clin Endocrinol Metab* 1995; **80**: 153–9.

6 Verhelst J, Abs R, Vandeweghe *et al.* Two years of replacement therapy in adults with growth hormone deficiency. *Clin Endocrinol* 1997; **47**: 485–94.

7 Savage MO, Besser GM. When and how to transfer patients from paediatric to adult endocrinologists: experience from St Bartholomew's Hospital, London. *Acta Paediatr Suppl* 1997; **423**: 127–8.

8 Monson JP. Conditions spanning paediatric and adult endocrine practice: the adult perspective. *Acta Paediatr Suppl* 1997; **423**: 124–6.

* 9 Wacharasindhu AM, Cotterill AM, Camacho-Hübner C, Besser GM, Savage MO. Normal growth hormone secretion in growth hormone insufficient children retested after completion of linear growth. *Clin Endocrinol* 1996; **45**: 553–6.

* 10 Nicholson A, Toogood AA, Rahim A, Shalet SM. The prevalence of severe growth hormone deficiency in adults who received growth hormone replacement in childhood. *Clin Endocrinol* 1996; **44**: 311–16.

* 11 Tauber M, Moulin P, Pienkowski C, Jouret B, Rochiccioli P. Growth hormone retesting and auxological data in 131 GH-deficient patients after completion of treatment. *J Clin Endocrinol Metab* 1997; **82**: 352–6.

12 Thorner MO, Bengtsson BÅ, Ho KY *et al.* The diagnosis of growth hormone deficiency in adults. *J Clin Endocrinol Metab* 1995; **80**: 3097–8.

* 13 Annatasio AF, Lamberts SWJ, Matranga AMC *et al.* Adult growth hormone

(GH)-deficient patients demonstrate heterogeneity between childhood onset and adult onset before and during human GH treatment. *J Clin Endocrinol Metab* 1997; **82**: 82–8.
*14 Rutherford OM, Jones DA, Round JM, Buchanan CR, Preece MA. Changes in muscle and body composition after discontinuation of growth hormone treatment in growth hormone-deficient adults. *Clin Endocrinol* 1991; **34**: 469–75.
15 Colle M, Auzerie J. Discontinuation of growth hormone treatment in growth hormone deficient patients: assessment of body fat mass using bioelectrical impedance. *Horm Res* 1993; **39**: 192–6.
16 Johannsson G, Albertsson-Wikland K, Bengtsson BÅ. Metabolic effects of discontinuation of GH treatment in adolescent patients. Paper presented at the 80th Annual Meeting of the Endocrine Society, 1998: OR31–1.
*17 Amato G, Carella C, Fazio S *et al.* Body composition, bone metabolism, heart structure and function in growth hormone deficient adults before and after growth hormone replacement therapy at low doses. *J Clin Endocrinol Metab* 1993; **77**: 1671–6.
18 Capaldo B, Patti L, Oliverio U *et al.* Increased arterial intima–media thickness in childhood-onset growth hormone deficiency. *J Clin Endocrinol Metab* 1997; **82**: 1378–81.
19 Markussis V, Beshyah SA, Fisher C *et al.* Detection of premature atherosclerosis by high-resolution ultrasonography in symptom-free hypopituitary adults. *Lancet* 1992; **340**: 1188–92.
*20 Kaufman JM, Taelman P, Vermeulen A, Vandeweghe M. Bone mineral status in growth hormone-deficient males with isolated and multiple pituitary deficiencies of childhood onset. *J Clin Endocrinol Metab* 1992; **74**: 118–23.
*21 De Boer H, Blok GJ, Van Lingen A *et al.* Consequences of childhood-onset growth hormone deficiency for adult bone mass. *J Bone Miner Res* 1994; **9**: 1319–26.
*22 Saggese G, Baroncelli GI, Bertelloni S, Barsanti S. The effect of long-term growth hormone (GH) treatment on bone mineral density in children with GH deficiency: role of GH in the attainment of peak bone mass. *J Clin Endocrinol Metab* 1996; **81**: 3077–83.
23 Drake WM, Kaltsas G, Korbonits M *et al.* Optimising growth hormone replacement by dose titration in hypopituitary adults. *J Clin Endocrinol Metab* 1998; **83**: 3913–19.

Part 4: Non-classical Uses of Growth Hormone

16: Is there a role for GH therapy in osteoporosis and ageing?

Christian Wüster

Introduction

Osteoporosis is the most frequently encountered metabolic bone disease. It is characterized by a reduction in bone mass, and is accompanied by biomechanical incompetence of the skeleton. This predisposes to fractures—typically of the vertebrae, hip, or radius. It is a complex, multifactorial, chronic disease that can progress without any symptoms for decades, until the loss of bone in an individual finally reaches a point at which fractures occur. Due to the enormous number of fractures and the consequent clinical symptoms and limitations in patients, osteoporosis is an enormous public health problem and impairs patients' quality of life. The costs of treating fractures are very high; hip fractures cost Germany DM 1 billion in 1992 [1]. In the United Kingdom alone, it is estimated that fractures of the hip, wrist and vertebrae are associated with annual costs to the Exchequer of close to £100 million (over $150 million dollars) [2]. Recent estimates of the annual costs of osteoporosis in Europe amount to ECU3500 million annually [3]. Services for the prevention and treatment of osteoporosis vary, and this has led to the development of European guidelines for prevention and treatment [4].

Since fracture rates increase with advancing age, this problem will be further aggravated in the future. This is due not only to the fact that the population is ageing, but also to the rising incidence of fractures caused by increasing risk factors. To date, many studies have been carried out showing that the risk of fractures increases with lower bone density. Often, other bone loss-inducing risk factors are added, including: advanced age, genetic risks (mother with hip fracture), no increase in weight since age 25, previous thyrotoxicosis, fracture after age 50, height at age 25 above 168 cm, anticonvulsant or benzodiazepine therapy, caffeine intake, less than four hours' physical activity per week, inability to stand up from a chair without using the arms, resting pulse of > 80/min and poor general health [5]. However, independent of these risk factors, bone density remains a predictive variable for fractures.

Although 60–90% of bone strength can be explained by bone mass and can thus be measured by densitometry, some other properties of bone strength seem to depend on other characteristics of the bone, such as elasticity, geometry, or structure [1]. These may be measurable using ultrasound techniques, which can therefore be used in parallel with conventional densitometry for predicting the fracture risk [1].

Osteoporosis is only one health problem seen more frequently with ageing. Others are the increased incidence of cardiovascular diseases, hyperlipidaemia, hypertension, increased obesity, decreased muscle mass and strength, and increased incidence of depression. Ageing is accompanied by several hormonal changes, especially decreases in the endogenous secretion of growth hormone (GH), sex hormones, dehydroepiandrosterone (DHEA), thyroid-stimulating hormone (TSH), and others [6–8]. The signs and symptoms associated with ageing have some similarity with those seen in patients with insufficiency in the above hormones (e.g. GH, Table 16.1). It is therefore challenging to see whether it may be possible to modify the signs and symptoms of ageing by substituting these hormones. This chapter will review the literature on the effects of GH in osteoporosis and in ageing, with particular emphasis on bone metabolism in the latter, not least because of the high morbidity of fractures in the ageing population.

Role of GH/IGF-I in the regulation of bone metabolism

The physiology of bone turnover is a very complex cycle occurring at all

Table 16.1 Signs and symptoms in 122 patients with growth hormone deficiency due to pituitary tumor, irradiation, or surgery. Adapted from Wüster *et al.* [112], with permission.

Signs and symptoms	Prevalence (%)
Tiredness, exhaustion and muscle weakness	53
Obesity	40
Hyperlipidaemia	77
Hypertriglyceridaemia	68
Hypercholesterinaemia	68
Hypertension	18
Signs of atherosclerosis (coronary heart disease, stroke)	14
Back pain	36
Decreased spinal bone density (<−1 SD)	57
Decreased radial bone density (<−1 SD)	73
Incidence of vertebral fractures	17

SD: standard deviation.

skeletal sites. Old bone is regularly replaced by new bone in well-defined quantities, the basic multicellular units (BMUs). Osteoclasts originating from haematopoietic stem cells are activated by an as yet unknown resorption stimulus. One might speculate that GH or parathyroid hormone (PTH), for example, may be required as systemic hormonal stimuli. Osteoclasts need about five to seven days to produce a resorption lacuna. Osteoblasts, which are derived from mesenchymal cells, move onto this freshly resorbed bone surface and produce new bone matrix, which subsequently mineralizes within 100 days. Some osteoblasts are incorporated into bone and become differentiated into osteocytes. They probably possess mechanoreceptors, responsible for the perception of physical and mechanical stress on bone. Skeletal growth factors (GFs) are stored within bone and released during the resorption process. They are needed to stimulate the differentiation of pre-osteoblasts into mature osteoblasts. During osteoblastic matrix deposition, GFs are again incorporated into bone. The amount of GFs deposited is dependent on the level of bone turnover, which is under hormonal control by GH and sex steroids, for example. GF activity is regulated by the presence of specific binding proteins and binding protein proteases, which are important for the control of the local concentrations of GFs within a BMU. The amount of transforming growth factor-β (TGF-β) released from bone originating from adult patients with GH deficiency of childhood onset after 1,25-vitamin D administration *in vitro* is diminished. However, this phenomenon is not reversible after GH treatment of the patients before taking the bone biopsy [9]. The release of cytokines such as interleukin-1 and interleukin-6, known to stimulate bone resorption, is increased after acute oestrogen deprivation in human bone marrow cells *in vitro* [10]. GFs may reflect a memory of previous hormonal disturbances within the skeleton. Together with possible variations at the local hormone receptor level, these changes might explain the changes in bone turnover and density seen with ageing or in hormone-deficiency states (menopause, somatopause).

GH is responsible for longitudinal bone growth via its direct stimulatory effect on chondrocytes and osteoblasts [11]. GH acts directly on osteoblasts via its own receptor [12] and via insulin-like growth factor-I (IGF-I) [13], which acts in an autocrine and paracrine fashion. This mechanism of action has been termed the 'dual effector theory' [14]. The GH receptor belongs to a receptor superfamily that includes the prolactin receptor and several cytokine receptors. The extracellular domain of the receptor circulates as GH-binding protein (GHBP), which may also regulate GH action [15]. Osteoclasts are stimulated by GH and/or IGF-I [16]. Nishiyama *et al.* [17] showed that bovine GH (bGH), 1–100 ng/mL, significantly stimulated bone resorption by pre-existing osteoclasts in stromal cell-containing mouse bone cell cultures, whereas it did not affect the bone-resorbing activity of isolated rabbit osteoclasts. GH also enhanced 1,25-dihydroxyvitamin D_3-induced osteoclast-like cell formation. Moreover, osteoclast-like cells newly formed from unfractionated bone cells in the presence of bGH possessed the ability to form pits on dentine slices. GH stimulated osteoclast-like cell formation from these haematopoietic blast cells in the absence of stromal cells in a concentration-dependent fashion, and these osteoclast-like cells formed pits on dentine slices in the presence of MC3T3-G2/PA-6 stromal cells. This study indicated for the first time that GH stimulates osteoclastic bone resorption through both direct and indirect actions on osteoclast differentiation and through indirect activation of mature osteoclasts, possibly via stromal cells, including osteoblasts. The osteoclastic stimulation by GH/IGF-1 is supported by *in vivo* data in mice [18], dogs [19] and humans (see below). Osteoblasts have GH [20,21] and IGF-I and IGF-II [22] receptors. They are stimulated to produce IGFs when GH is added *in vitro* [7,23,24], although these cells make greater amounts of IGF-II than IGF-I [25]. In contrast to the liver, where GH is the principal regulator of IGF-I production, osteoblastic IGF-I gene transcription is influenced by both paracrine and endocrine factors. The remainder of skeletal IGF production occurs in bone marrow stromal cells, which are rich sources of cytokines for haematopoietic progenitors and actively synthesize IGF-I and various insulin-like growth factor binding proteins (IGFBPs) [26]. Other IGF binding proteins may regulate IGF action; for example, IGFBP-4 inhibits the stimulatory effect of IGF-I on mouse osteoblasts, as described above [27]. GH, IGF-I and IGF-II have a potent stimulatory effect on bone cell growth in a dose-dependent manner [28]. GH-induced cell growth can be blocked by simultaneous addition of a specific monoclonal antibody to IGF-I. IGF-II is also a potent mitogen [29], stimulating mitogenesis even when high doses of IGF-I are also administered. This suggests that IGF-II regulates osteoblastic proliferation via the IGF type 2 receptor [30]. The role of IGF-I and IGF-II in the coupling of bone resorption and formation is extensively reviewed by Mohan and Baylink [31]. IGF-II stimulates type I collagen synthesis in human osteoblast-like cells [31]. IGF-I and IGF-II stimulate alkaline phosphatase activity [32]. Both IGFs act synergistically with 1,25-vitamin D to increase osteocalcin production [33].

The intraosseous concentrations of IGF-I, IGF-II, IGFBP-3 and IGFBP-5 in cortical bone decrease with age [34]. Another role of IGF-I becomes apparent in the regulation of bone remodelling by physical stress on bone. IGF-I mRNA expression is increased in osteocytes after the application of mechanical stress before initiation of bone formation [35].

GH has several other target organs apart from bone. These effects might indirectly influence bone metabolism. Its predominant action is an anabolic effect stimulating muscle mass, strength and exercise capacity. The same applies to the effects of GH on improving quality of life and cardiac function, which increases mobility and thus stimulates bone turnover. Interactions of GH with sex steroids are multiple. GH influences testicular and ovarian steroidogenesis and potentiates sex steroid effects on bone [36]. Furthermore, GH has enteral and renal effects that influence calcium metabolism. Whether GH has a daily regulatory effect on bone turnover is a matter of recent speculation. There is evidence that bone resorption is increased during the night and that this is preceded by the midnight GH peak [37]. Whether there is a direct link between bone resorption and GH secretion or merely an association requires further investigation. GH also affects the mechanical properties and biochemical composition of rat bones [38]. It increases the external diameter of long bones, whereas the internal diameters are either unaffected or increased [39]. It therefore improved the mechanical properties of the cortical femur and tibia, although bone density, bone collagen content and bone ash weight remained unchanged.

Pathophysiology of osteoporosis: role of GH/IGFs

Pubertal, menopausal and age effects on GH/IGF-1 serum levels

Patterns of GH secretion are highly dependent on age and sex in animals and humans. Age and sex have a strong influence on the frequency and amplitude of GH pulses and basal secretion; serum levels of IGF-I and IGFBP-3 are at least partly dependent on systemic GH secretion [40]. Characteristic changes are seen during puberty in rats [36]. GH secretion in male and female rats is similar after birth, but increases to a greater extent in males after puberty. At this time, a clear sexually differentiated pattern of secretion appears, with male rats displaying high-amplitude low-frequency pulses and female rats displaying pulses of high frequency but low amplitude [36]. The sexual dimorphism in this pattern can be altered by manipulating the gonadal steroid environment, suggesting that gonadal steroids may be important modulators of GH secretion. In humans, sexual differences during puberty seem to be less clear, and conflicting data have been reported. This might be due to differences in techniques (profiles vs. stimulatory tests) and the use of assays with different sensitivities. Generally, spontaneous and stimulated GH peaks are increased during puberty. In addition, GH secretion is higher in adolescents of Afro-Caribbean origin, a factor that might partly explain increased bone mass [41]. Studies investigating young males and females matched for age and body mass index have found higher integrated GH (IGHC) levels in women than in men [42,43]. Other secretory characteristics, including pulse amplitude, frequency and the fraction of GH secreted as pulses (FGHP) were similar in both sexes of young age.

IGHC, mean pulse amplitude and FGHP were found to be significantly lower in older subjects. The age-related fall in GH secretion was significantly greater in women than in men. Parameters of GH secretion correlated with serum oestradiol levels, but not with androgen levels in females.

GH secretion, as measured by changes in serum IGF-I, number of peaks, area under the curve and maximal amplitudes in GH pulsatility declines with age [44,45], a phenomenon referred to as the 'somatopause'. The characteristics of these age-dependent changes are shown in Table 16.2. This decline in endogenous GH secretion has been attributed to increased secretion of somatostatin [46], as has also been shown in the ageing rat [47]. This might be due to the general decrease in cholinergic activity with ageing. In addition, hypothalamic GHRH mRNA expression is decreased with age, as is the expression of pituitary GHRH receptors. The stimulated GH response to any stimulatory agent is decreased with age [48–53]. The same applies to the integrated 24-hour GH values [42,54–56] and to IGF-I concentrations in serum [57–60]. Serum IGFBP-3 levels have been shown to be closely related to 24-hour GH secretion in children [61]. However, Wüster *et al.* were unable to demonstrate an age-related decline in serum IGBFP-3 levels in healthy females (Fig. 16.1, lower panel) [62–64]. These measurements were only made up to an age of 65 years, so that it is possible that an age-related decline was missed in these subjects.

Reduced physical activity has been proposed as a mechanism for the low GH values in the elderly. However, it has been shown by Pyka *et al.* [65] that elderly people are unable to increase their endogenous GH secretion by exercise in comparison with healthy young subjects. In addition, elderly women participated in a vigorous exercise program for one year in this study, and did not demonstrate an increase in serum IGF-I levels compared to a non-exercising age-matched and sex-matched control group [66].

Table 16.2 Characteristics of the somatopause.

Characteristics
↓ central cholinergic tone → ↑ somatostatin
↓ hypothalamic GHRH mRNA expression
↓ expression of pituitary GHRH receptors
↓ GH response after GHRH
↓ IGF-I response after GH or GHRH
↓ number of GH pulses
↓ GH and GHBP serum concentrations
↓ GH half-life
↓ IGF-I and IGFBP-3 concentrations
↓ GH secretion after sports activity

GH: growth hormone; GHRH: growth hormone releasing hormone; IGF-I: insulin-like growth factor; IGFBP-3: insulin-like growth factor binding protein-3.

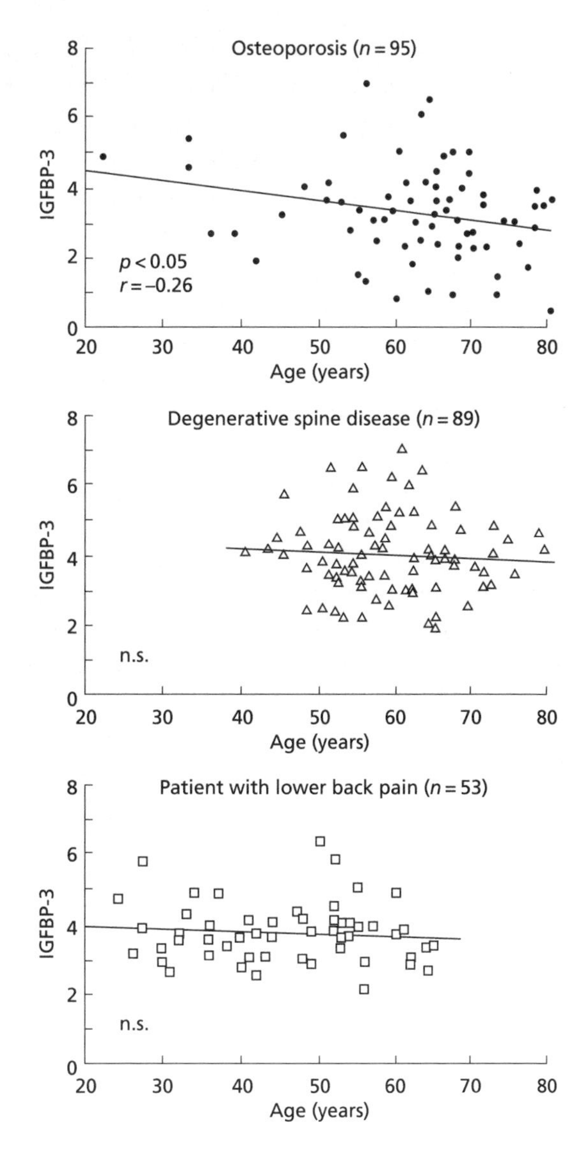

Fig. 16.1 Serum concentrations of insulin-like growth factor binding protein-3 (IGFBP-3) [62–64] in women with primary postmenopausal osteoporosis (upper panel); with degenerative spine disease (middle panel); and in control individuals (lower panel), relative to age. A significant age-related decline in serum IGFBP-3 was only seen in patients with established osteoporosis, as shown by the presence of vertebral fractures, but not in patients with osteoarthritis or patients with back pain but without spinal radiological changes. Reproduced with permission from Wüster *et al.* [7,62–64].

Thus, reduced exercise levels in the elderly do not explain the effects seen with the somatopause.

The menopause itself has also been implicated in the changes seen during the somatopause, as oestrogen has a stimulatory effect on GH secretion. Both serum IGHC and serum IGF-I decrease after the menopause [42,67,68]. Oral oestrogen replacement inhibits hepatic IGF-I synthesis and increases GH secretion through reduced feedback inhibition. However, reduced GH secretion in the menopause cannot be explained by oestrogen deficiency alone, since GH secretion is not restored by the attainment of physiological oestradiol levels using the transdermal route. As some signs of the somatopause are also seen in ageing men without any relation to prevailing testosterone levels, it is questionable whether sex steroid deficiency alone can explain the age-related decrease in GH secretion. Similar relationships between sex steroids and GH to those seen in puberty may be taking place in the somatopause. In puberty, oestrogens have an additive effect to GH on bone growth, whereas testosterone seems to have a potentiating effect [36]. Whether the reverse is seen in age-related changes remains to be investigated.

The GH–IGF-I axis undergoes a decline during life, so that in elderly individuals both spontaneous GH secretion and IGF-I levels are reduced. This is the result of changes in hypothalamopituitary regulation, as well as due to changes in lifestyle and nutrition with ageing. The GH response to all commonly used stimuli (with the exception of clonidine), such as GHRH, L-dopa, physostigmine, pyridostigmine, hypoglycaemia, and met-enkephalin is reduced with ageing. The somatotroph responsiveness to GHRH, when combined with arginine, which stimulates GH by suppressing endogenous somatostatin secretion, does not vary with age, suggesting that maximal GH secretory capacity is preserved in elderly subjects. It is hypothesized that due to impaired cholinergic function with ageing, the activity of somatostatin neurones is exaggerated, leading to a reduction in GH secretion. However, the GHRH–GH axis does not seem to be impaired in patients with Alzheimer's disease, despite their markedly impaired cholinergic activity.

Obesity is a very difficult problem, interfering with all cross-sectional studies, since adiposity is known to increase with ageing and increased fat tissue is known to decrease endogenous GH secretion [68, 69].

Changes in GH/IGF-I in patients with osteoporosis

The GH response to intravenous L-arginine was found to be reduced in patients with osteoporosis in comparison with patients with osteoarthritis. Wüster *et al.* have shown low serum IGF-I, IGF-II and IGFBP-3 as measured by radioimmunoassays in 98 women with postmenopausal osteoporosis compared to 59 normal individuals and 91 patients with osteoarthrosis or degenerative bone disease [64] (Fig. 16.1). Similar results were obtained in men with osteoporosis, and both serum IGF-I and IGFBP-3 concentrations were positively correlated with lumbar bone mineral density (BMD) in osteoporotic patients [70]. Comparable results have been found by others for IGF-I [71]. Patients with osteoarthrosis seemed to have higher values [64]. This is consistent with results from studies on IGF-II concentrations in bones from patients with osteoarthrosis [72]. Human marrow stromal osteoblast-like cells from patients with osteoporosis do not appear to have reduced responsiveness to *in vitro* stimulation with GH [73]. However, researchers in Loma Linda have shown an age-related decline of the content of growth factors and their binding proteins such as IGFBP-5, IGF-I and TGF-β [34]. It remains to be determined whether there is a difference between patients of the same age with and without osteoporosis. Low IGF-I levels are inversely correlated to osteoblastic surface in patients with idiopathic osteoporosis [74], and my own data do not support the hypothesis that a low serum IGF-I is due to diminished secretion of endogenous GH [75]. Therefore, one might explain the low serum IGF-I levels in osteoporosis patients with low turnover due to decreased osseous production of IGF-I and diminished osteoblastic activity, or simply by virtue of low BMD, bearing in mind the strong inverse correlation between serum IGF-I and BMD.

In conclusion, the GHRH–GH–IGF system is the major regulatory system of cell growth of any organ system. Important differences are seen between changes in local and systemic concentrations of growth factors, which are not always parallel. The local environment for IGFs is kept constant for as long as possible during the development of an age-related disease such as osteoporosis. Local activity of IGFs is regulated by specific binding proteins and IGFBP-specific proteases, and is thus potentially independent of short-term changes of the systemic hormonal regulation by GH or parathyroid hormone (PTH). It is hypothesized that the system is compromised only if the content of IGFs stored within the bone is reduced. This IGF content may be under systemic hormonal control and thus susceptible to long-term pathological changes such as those pertaining in patients with GH deficiency. Based on these assumptions, it has been hypothesized that diminished secretion of endogenous GH may underlie the pathogenesis of osteoporosis in some patients [6]. More recent studies have demonstrated that possibly reduced PTH secretion, rather than a decline in endogenous GH, might be responsible for the reduced content of intraosseous growth factors [17]. Clearly, osteoporosis is a multifactorial disease and a premature somatopause is probably just one factor contributing to the development of the disease in individual patients.

Acromegaly as a surrogate model for GH treatment in osteoporosis

The effects of GH treatment on bone metabolism in adult patients with GH deficiency (GHD) are described in Chapter 9.

Skeletal changes in patients with GH excess (e.g. acromegaly) have been described in a number of studies [75–86] with conflicting results, which may be explained by differences in vitamin D metabolism and gonadotrophin secretion. However, compared to osteoporosis, a disease that is often associated with hypogonadism, patients with acromegaly and secondary hypogonadism demonstrate the effects on bone mass that might be expected with GH treatment. Furthermore, eugonadal patients with active acromegaly do not suffer vertebral fractures. In acromegaly, increased bone turnover has been demonstrated by means of biochemical markers [76–79] and histologically [80]. Cortical BMD seems to be increased [81–83], whereas trabecular BMD is normal or low [79,83,85]. In one study using quantitative computed tomography (QCT) of the lumbar spine, trabecular BMD was elevated in one of 14 patients with active acromegaly [79]. Low trabecular BMD in patients with acromegaly may be due to previous hypogonadal phases without hormonal substitution. Interestingly, other parameters of bone metabolism, such as serum calcium, phosphate, PTH and 1,25-dihydroxyvitamin D are normal, despite active GH hypersecretion [79]. The old dogma of acromegaly being a cause of secondary osteoporosis [85] should be abandoned. Changes in bone metabolism during treatment of acromegaly have been described. However, changes in BMD have rarely been published [86]. Wüster examined five patients with active acromegaly on treatment with octreotide over five years, with achievement of normal IGF-I levels. Spinal BMD decreased in all five patients and normalized in three. All patients were eugonadal throughout the follow-up period [7].

Effects of GH treatment in animals, healthy young people and in old age

Andreassen *et al.* conducted a study treating aged rats with 2.7 mg/kg/d rhGH (1 mg = 3 IU) for 80 days. Significant increases in cortical bone volume, mineralizing surface/total surface, mineral apposition rate, and mineralized bone formation rate were found. Furthermore the transverse and midsagittal diameters were increased by GH. The compressive mechanical strength of the vertebral body specimens was increased, and this could be explained by formation and deposition of cortical bone [87]. However, treatment with 5 mg/kg/day of rhGH in glucocorticoid-treated rats (5 mg/kg/d prednisolone) did not inhibit loss of body weight, decrease in bone length and diameter, or decreased bone strength induced by glucocorticoid administration [88]. Elevated levels of GH in metallothionein promotor GH-transgenic mice increased the amounts of vertebral bone and tibia bone in young mice. Intact ovaries are a prerequisite for the stimulatory effect of elevated levels of GH, and the fact that ovariectomy decreased the stimulatory effect of elevated GH levels suggested that this GH effect was dependent on the presence of endogenous sex steroid secretion [89].

Denis *et al.* treated growing pigs with 40 g porcine GH/kg body weight or its vehicle twice daily for two months [90]. GH accelerated growth, with greater tibial and metacarpal weights, greater tibial length and diameter and greater tibial ash weight in GH-treated pigs than in controls. The similar values of apparent bone density in the two groups suggest adequate coupling between bone growth and mineralization in GH-treated pigs. Histomorphometric data for the distal metacarpal metaphysis indicated greater trabecular bone volume, osteoblastic surface, and mineral apposition rate in GH-treated pigs. The osteoclast surface, lacuna depth, and osteoid-related parameters in GH-treated and control pigs were similar. The plasma PTH of the two groups of pigs were similar throughout the experiment. These data and the elevated plasma alkaline phosphatase activity in GH-treated pigs suggest that GH specifically affects bone formation. GH had no effect on plasma 25-hydroxyvitamin D_3, but 1,25-dihydroxyvitamin D_3 (calcitriol) was higher in treated pigs throughout the experiment. This suggests that calcitriol may help adapt bone mineralization to accelerated bone formation during growth hormone treatment [90]. Effects of rhGH on bone formation are blunted in unloaded hypophysectomized rats (Fig. 16.2), indicating that the physiological stimulus of bone turnover is the predominant effect on bone formation.

GH has been given to young males [91] and to healthy elderly people [92] for short periods. The effects seen were similar to those seen in GHD. Rudman *et al.* [93] treated elderly men who had low IGF-I levels with 0.03 mg/kg of rhGH three times per week for one year. They reported on

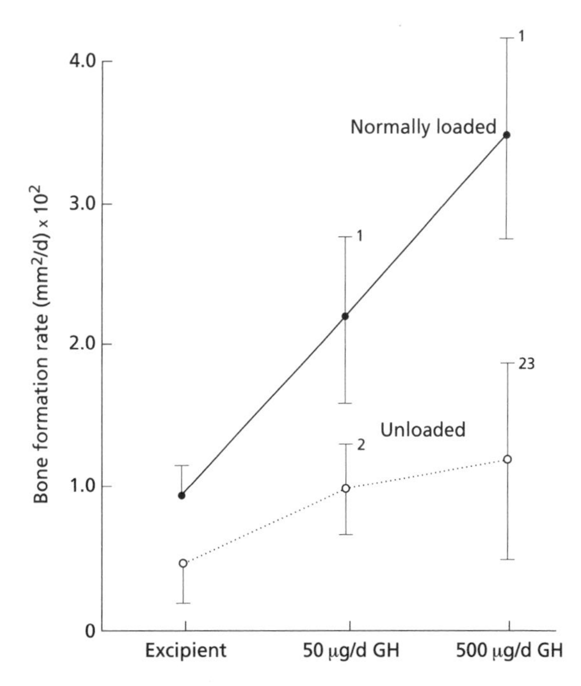

Fig. 16.2 Bone formation rate in hypophysectomized rats treated with recombinant human growth hormone (rhGH) if loaded and unloaded. Reproduced with permission from Halloran *et al.* [113].

the effects of GH on muscle and fat mass, and demonstrated a significant increase in lumbar bone mineral density that was not seen at other sites (hip and radius); this might have been due to a statistical problem, as multivariate analysis was not reported. It is doubtful whether this effect can be reproduced and whether it is maintained for a prolonged period.

Holloway *et al.* reported a longer study of GH treatment in 27 healthy postmenopausal elderly women, eight of whom took a stable dose of oestrogen throughout the study [94]. Thirteen women completed six months of treatment and 14 women completed six months in the placebo group. Side-effects prompted a 50% reduction in the original dose of rhGH (from 0.043 mg/kg body weight, or ≈0.3 mg rhGH/kg/week to 0.02 mg/kg/day), and led to several drop-outs in the treatment group. Although fat mass and percentage body fat declined in the treatment group, there was no significant effect on BMD at the spine or hip after six months or one year of treatment [94]. Although bone mass did not change, there were some changes in biochemical markers, particularly markers of bone resorption, the urinary pyridinolines. The effects on markers of bone formation were more variable; osteocalcin increased, but type I procollagen peptide levels did not change. For women taking oestrogen replacement therapy, indices of bone turnover were blunted.

Effects of GH treatment in osteoporosis

The potential rationale and limitations of GH treatment for osteoporosis are listed in Table 16.3; this therapy has not been particularly successful to date. As early as 1975, two patients with osteogenesis imperfecta and one patient with involutional osteoporosis were treated with GH [95], and increases in serum and histology markers of bone turnover were achieved. Subsequent studies employed GH with and without antiresorptive agents [96–98]. Aloia *et al.* administered between 2 and 6 U of GH per day for 12 months to eight patients with postmenopausal osteoporosis (the first six months of treatment featured low-dose GH; the last six months consisted of high-dose GH, 6 U/day). Radial bone mineral content dropped slightly and histomorphometric parameters did not change during treatment. However, the severity of back pain decreased considerably in several patients [96]. Daily GH injections (4 U/day) combined with alternating doses of calcitonin produced an increase in total body calcium but a decline in radial bone mass after 16 months [97]. In a separate trial, 14 postmenopausal women were given two months of GH followed by three months of calcitonin in a modified form of coherence therapy [98]. Total body calcium increased by 2.3% per year and there were few side-effects, but there were no changes in bone mineral density or histomorphometric indices.

Table 16.3 Potential benefits and limitations of growth hormone treatment for osteoporosis.

Theoretical benefits	Limitations
GH directly stimulates growth factors for bone such as IGFs	GH is not tissue-specific
GH can inhibit osteoblastic apoptosis via IGF-I	Long-term mitotic activity is unknown
GH stimulates bone formation and collagen production directly and via IGF-I	GH receptors are probably present on osteoclasts
GH receptors are found on osteoblasts	Type I IGF receptor is present on osteoclasts
GH is anticatabolic	GH suppresses endogenous GH in patients with osteoporosis
GH increases BMD in patients with GHD	There are no long-term studies on BMD in humans
GH increases bone size in patients with GHD	There are no studies on fracture rates in humans
GH leads to increased biomechanical competence of bones in rats	There are no studies showing that GH-induced increase in BMD is associated with increased biomechanical competence
GH increases muscle size and strength	Increased muscle strength might lead to increased mobility and increased falls, causing more fractures

BMD: bone mineral density; GH: growth hormone; GHD: growth hormone deficiency; IGF-I: insulin-like growth factor-I.

Dambacher *et al.* administered 16 U of rhGH every other day along with daily sodium fluoride to six women with postmenopausal osteoporosis [99]. On histomorphometric analysis, there was a significant increase in the number of osteoblasts and osteoclasts, but the bone mass was unchanged. Johannsen *et al.* conducted a placebo-controlled double-blinded cross-over trial of rhGH and IGF-I in 14 men with idiopathic osteoporosis [100]. In this seven-day trial with rhGH (2 IU/m^2), procollagen peptide and osteocalcin levels increased after treatment, as did urinary markers of bone resorption. The changes in osteocalcin were relatively small, however, and were not sustained after discontinuation of growth hormone treatment. Erdtsieck *et al.* treated 21 postmenopausal women who had osteoporosis with the amino bisphosphonate, pamidronate, for 12 months. During the initial six months, rhGH (0.0675 IU/kg, three times per week) was administered in a placebo-controlled fashion [101]. The bone mineral content (BMC) of the lumbar spine and femoral neck (on dual-energy X-ray absorptiometry) and BMC of the radius showed no change in the rhGH group. However, a consistent increase of about 5% at the lumbar spine and somewhat less in the distal forearm was reached from six months onwards in women treated with pamidronate. Compared to baseline values, the biochemical measurements of bone turnover showed a decrease of about 50% in the pamidronate group, but this was blunted in the group additionally treated with rhGH. The body composition measurements showed clear effects of rhGH administration: a decrease in fat mass of about 5% and an increase in lean body mass of about 3%. However, these effects disappeared after treatment with rhGH was stopped, and both fat mass and lean body mass returned to initial values. Thus, rhGH blunted both the pamidronate-induced accumulation of bone mineral mass and the reduction of biochemical markers of bone turnover.

Recent studies have combined PTH and GH treatment in males with osteoporosis [102]. Finally, Gonelli *et al.* (1997) [103] compared three sequential treatment regimens in a single-blind randomized study including 30 women with postmenopausal osteoporosis. The treatments given were (a) GH for seven days, calcitonin for 21 days and a drug-free period of 61 days; (b) GH for seven days, placebo for 21 days and a drug-free period of 61 days; and (c) placebo for seven days, calcitonin for 21 days and a drug-free period of 61 days. These cycles were repeated eight times over 24 months. GH was given at 12 IU/day (4 mg/day) and salmon calcitonin at 50 IU/day. A significant increase in the BMD of the lumbar spine (2.5% per year) was seen in the first group, but this was accompanied by a significant decrease in the BMD of the femoral shaft.

Conclusions

GH is a potent anabolic hormone for almost all systems including bone and calcium metabolism. This has previously been discussed in numerous excellent reviews by others [104–111]. Patients with adult GH deficiency have clinical features [112] (Table 16.1) that have some similarities with the signs and symptoms of ageing. Most of the signs and symptoms of GHD and hypopituitarism resolve with substitution of the necessary hormones including GH. The results of long-term studies on the effects of GH on bone mineral density and fracture rates are keenly awaited. Some patients with established osteoporosis and vertebral fractures seem to have low serum IGF-I levels, which are probably not due to diminished secretion of endogenous GH. A combination of the effects of the menopause and the somatopause as a reason for osteoporosis has been hypothesized, and is illustrated schematically in Figure 16.3. Treatment of osteoporosis with GH alone or in combination with calcitonin has not been successful in increasing BMD. However, studies of the effects of GH treatment on the vertebral fracture rate in large numbers of patients over several years are not available. Recent studies have shown the effect of exercise or loading in combination with GH treatment [113]. As with any anti-osteoporotic drug, rhGH does not stimulate bone formation in an unloaded state (Fig. 16.2). Thus, vigorous exercise in combination with muscle training and rhGH treatment is an option requiring further study.

As GH seems to reverse some consequences of the ageing process, such as muscle weakness and reduced exercise capacity, one might speculate that

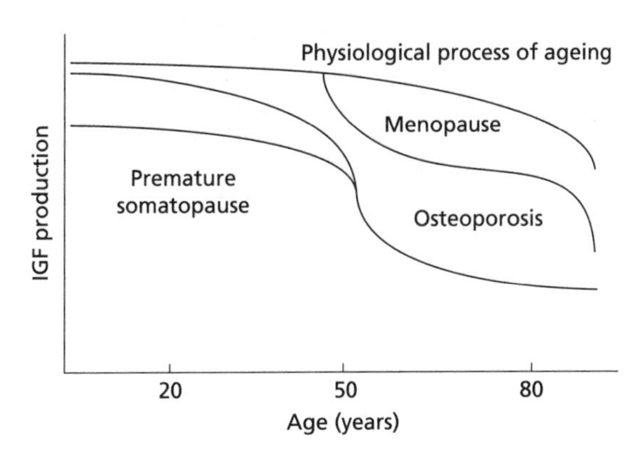

Fig. 16.3 Hypothetical model of the involvement of the somatopause in the pathogenesis of osteoporosis.

GH could prevent the occurrence of hip fractures in elderly people. It has been postulated that this might be achieved more cheaply by increasing physical activity. However, it has been shown that physical activity in old age does not lead to increased endogenous GH secretion. Thus, effective GH treatment against the negative symptoms of ageing is probably beneficial, but there are clearly ethical and financial considerations. Treatment with GH in this situation should only be undertaken by experienced physicians in the context of controlled clinical studies—not least because of the uncertain potential for increased mitogenesis in this age group.

Acknowledgements

I should like to thank my wife and children for their patience with me while I was in a different world, preoccupied with the writing of this chapter.

References

* 1 Wüster C, Heilmann P, Pereira-Lima J *et al.* Quantitative ultrasonometry (QUS) for the evaluation of osteoporosis risk: reference data for various measurement sites, limitations and application possibilities. *Exp Clin Endocrinol Diabetes* 1998; **106**: 277–88.

2 Dolan P, Torgerson DJ. The costs of osteoporotic fractures in women in the United Kingdom. *Osteoporos Int* 1998 [in press].

3 Report on osteoporosis in the European Community: building strong bones, preventing fractures. Summary report on osteoporosis in the European Community: action for prevention. Lyons: European Communities/ European Foundation For Osteoporosis, 1998.

** 4 Kanis JA. Assessment of fracture risk and its application to screening for postmenopausal osteoporosis: synopsis of a WHO report. *Osteoporosis Int* 1994; **4**: 368–81.

** 5 Cummings SR, Nevitt MC, Browner WS *et al.* Risk factors for hip fractures in white women. *N Engl J Med* 1995; **332**: 767–73.

6 Wüster C. Growth hormone and ageing. In: Scherbaum W, Rossmanith WG, eds. *Endocrinology of aging.*

Berlin: De Gruyter, 1995: 95–112.
7 Wüster C. Growth hormone, insulin-like growth factors and bone metabolism. *Endocrinol Metab* 1995; **2**: 3–12.
8 Rosen C, Wüster C. Growth hormone, insulin-like growth factors: potential applications and limitations in the management of osteoporosis. In: Marcus R, Feldman D, eds. *Osteoporosis*. San Diego: Academic Press, 1996: 1313–33.
9 Sterck JGH, Klein-Nulend J, Lips P, De Boer H, Burger EH. Osteoblasts derived from growth hormone-deficient patients show an impaired TGF-β release in response to 1,25-dihydroxyvitamin D_3. *Endocrinol Metab* 1994; **1** (Suppl): 113.
10 Bismar H, Diel I, Ziegler R, Pfeilschifter J. Increased cytokine secretion by human bone marrow cells after menopause or discontinuation of estrogen replacement [abstract]. *J Bone Miner Res* 1994; **9** (Suppl 1): A149.
11 Isaksson DGP, Lindahl A, Nilsson A, Isgaard J. Mechanism of the stimulatory effect of growth hormone on longitudinal bone growth. *Endocr Rev* 1987; **8**: 426–38.
12 Leung DW, Spencer SA, Cachines G *et al.* Growth hormone receptor and serum binding protein: purification, cloning and expression. *Nature* 1987; **330**: 537–43.
13 Ernst M, Froesch ER. Growth hormone-dependent stimulation of osteoblast-like cells in serum free cultures via local synthesis of insulin-like growth factor I. *Biochem Biophys Res Commun* 1988; **151**: 142–7.
14 Green H, Morikawa M, Nixon T. A dual effector theory of growth hormone action. *Differentiation* 1985; **29**: 195–8.
* 15 Carlsson B, Eden S, Nilsson A *et al.* Expression and physiological significance of growth hormone receptors and growth hormone binding protein in rat and man. *Acta Pediatr Scand Suppl* 1991; **379**: 70–6.
16 Mochizuki H, Hakeda Y, Wakasuki N *et al.* IGF-I supports formation and activation of osteoclasts. *Endocrinology* 1992; **131**: 1075–80.
17 Nishiyama K, Sugimoto T, Kaji H, *et al.* Stimulatory effect of growth hormone on bone resorption, osteoclast differentiation. *Endocrinology* 1996; **137**: 35–41.
18 Slootweg MC, Most WW, Van Beek E *et al.* Osteoclast formation together with interleukin-6 production in mouse long bones is increased by insulin-like growth factor-I. *J Endocrinology* 1992; **132**: 433–8.
19 Harris EH, Heaney RP, Jowsey J *et al.* Growth hormone: the effect of skeletal renewal in the adult dog, 1: morphometric studies. *Calcif Tissue Res* 1972; **10**: 1–13.
20 Barnard R, Ng KW, Martin TJ, Waters MJ. Growth hormone (GH) receptors in clonal osteoblast-like cells mediate a mitogenic response to GH. *Endocrinology* 1991; **128**: 1459–64.
21 Swolin D, Lövstedt K, Piros C *et al.* Expression of functional growth hormone receptors in cultured human osteoblast-like cells. *Endocrinol Metab* 1994; **1** (Suppl): 112.
22 Slootweg MC, Hoogerbrugge CM, De Porter TL, Duursma SA, Van Buul Offers SC. The presence of classical IGF type 1 and type 2 receptors on mouse osteoblasts: autocrine/paracrine growth effect of IGF? *J Endocrinol* 1990; **125**: 271–7.
23 Stracke H, Schulz A, Rossol R, Moeller D, Schatz H. Effect of growth hormone on osteoblasts: demonstration of somatomedin C/IGF-I in bone organ culture. *Acta Endocrinol (Copenh)* 1984; **107**: 16–24.
24 Wüster C. Osteoporose durch Mangel an Calcitonin und Wachstumshormon: Untersuchungen mittels Knochenzellkultur, Tiermodell und Osteodensitometrie [postdoctoral thesis]. Heidelberg: University of Heidelberg, 1993: 79–82.
25 Mohan S, Baylink DJ. Autocrine–paracrine aspects of bone metabolism. *Growth Genet Horm* 1990; **6**: 1–9.
26 Abboud SL, Bethel CR, Aron DC. Secretion of IGF-I and IGFBPS by murine bone marrow stromal cells. *J Clin Invest* 1991; **88**: 470–5.
27 Scharla SH, Strong DD, Mohan S, Baylink DJ, Linkhart TA. 1,25-Dihydroxyvitamin D_3 differentially regulates the production of insulin-like growth factor I (IGF-1) and IGF-binding protein-4 in mouse

osteoblasts. *Endocrinology* 1991; **129**: 3139–46.
28 Scheven BA, Hamilton NJ, Fakkeldji TM, Duursma SA. Effects of rhIGF-I and IGF-II and GH on the growth of normal human osteoblast-like cells and human osteogenic sarcoma cells. *Growth Regul* 1989; **1**: 160–7.
29 Mohan S, Baylink DJ. Bone growth factors. *Clin Orthop Rel Res* 1991; **263**: 30–48.
30 Mohan S, Linkart T, Rosenfeld R, Baylink DJ. Characterization of the receptor for IGF-II in bone cells. *J Cell Physiol* 1989; **140**: 169–76.
31 Mohan S, Baylink DJ. The role of IGF-II in the coupling of bone formation to resorption. In: Spencer EM, ed. *Modern Concepts of Insulin-Like Growth Factors*. New York: Elsevier, 1991: 169–84.
32 Schmid C, Steiner T, Froesch ER. IGF-I supports differentiation of cultured osteoblast-like cells. *FEBS Lett* 1984; **173**: 48–52.
33 Chenu C, Valentin-Opran A, Chavassieux P *et al.* IGF-I hormonal regulation by GH and 1,25-vitamin D and activity on human osteoblast-like cells in short-term cultures. *Bone* 1990; **11**: 81–6.
34 Nicolas V, Prewett A, Mohan S *et al.* An age-related decrease of insulin-like growth factor-binding protein-5 (IGFBP-5) in human cortical bone: implications for bone loss with aging [abstract]. *J Bone Miner Res* 1994; **9** (Suppl 1): A151.
35 Lean JM, Jagger CJ, Chambers TJ, Chow JWM. Increased insulin-like growth factor-I mRNA expression in osteocytes precedes the increase in bone formation in response to mechanical stimulation [abstract]. *J Bone Miner Res* 1994; 1994; **9** (Suppl 1): A86.
36 Jansson JO, Edén NS, Isaksson O. Sexual dimorphism in the control of growth hormone secretion. *Endocr Rev* 1985; **6**: 128–50.
37 Müller C, Wüster C, Seibel M, Knauf K. Ziegler R. Cosecretion of human growth hormone and bone markers. *Proceedings of the International Symposium Growth Hormone and Growth Factors in Endocrinology and Metabolism, Gothenburg, 21–22 October 1994*: D4.
38 Jørgensen PH, Bak B, Andreassen TT. Mechanical properties and biochemical composition of rat cortical femur and tibia after long-term treatment with biosynthetic human growth hormone. *Bone* 1991; **12**: 353–9.
39 Andreassen TT, Melsen F, Oxlund H. The influence of growth hormone on cancellous and cortical bone of the vertebral body in aged rats. *J Bone Mineral Res* 1996; **11**: 1094–102.
40 Ho KY, Evans WS, Blizzard RM *et al.* Effects of sex and age on the 24-hour profile of growth hormone secretion in man: importance of endogenous estradiol concentrations. *J Clin Endocrinol Metabol* 1987; **64**: 51–8.
41 Wright NM, Renault J, Willi S, Key LL, Bell NH. Greater growth hormone secretion in black compared with white adult males: possible factor in greater bone mineral density [abstract]. *J Bone Miner Res* 1994; **9** (Suppl 1): A113, Ps149.
42 Ho KY, Weissberger AJ. Secretory patterns of growth hormone according to sex and age. *Horm Res* 1990; **33** (Suppl 4): 7–11.
43 Weissberger AJ, Ho KY, Lazarus L. Contrasting effects of oral and transdermal routes of estrogen replacement therapy on 24-hour growth hormone (GH) secretion, insulin-like growth factor-I, and GH-binding protein in postmenopausal women. *J Clin Endocrinol Metab* 1991; **72**: 374–81.
44 Meites J. Neuroendocrine biomarkers of aging in the rat. *Exp Gerontol* 1988; **23**: 349–58.
* 45 Rudman D, Kutner MH, Rogers CM *et al.* Impaired growth hormone secretion in the adult population: relation to age and adiposity. *J Clin Invest* 1981; **67**: 1361–9.
46 Wehrenberg WB. Physiological role of somatocrinin and somatostatin in the regulation of growth hormone secretion. *Biochem Biophys Res Commun* 1975; **109**: 562–7.
47 Casad RC, Adelman RC. Aging enhances inhibitory action of somatostatin in rat pancreas. *Endocrinology* 1992; **130**: 2420–1.
48 Carlson HE, Gilin JC, Gorden P, Snyder F. Absence of sleep-related

growth hormone peaks in aged normal subjects and in acromegaly. *J Clin Endocrinol Metab* 1972; **34**: 1102–5.
49 Franchimont P. Effects of repetitive administration of growth hormone-releasing hormone on growth hormone secretion, insulin-like growth factor-I and bone metabolism in postmenopausal women. *Acta Endocrinol* 1989; **120**: 121–8.
50 Kalk WJ. Growth hormone response to insulin hypoglycemia in the elderly. *J Gerontol* 1973; **28**: 431–3.
51 Lang I, Schernthaner G, Pietschmann P *et al.* Effects of sex and age on growth hormone response to growth hormone-releasing hormone in healthy individuals. *J Clin Endorinol Metab* 1987; **65**: 535–40.
52 Muggeo M, Fedele D, Tiengo A, Molinari M, Crepaldi G. Human growth hormone and cortisol responses to insulin stimulation in aging. *J Gerontol* 1975; **30**: 546–51.
53 Shibaski T, Shizume K, Nakhara M *et al.* Age-related changes in plasma growth hormone response to growth hormone-releasing hormone in man. *J Clin Endocrinol Metabol* 1984; **58**: 212–14.
54 Finkelstein JW, Rolfwarg HP, Boyar RM, Kream J, Heltman L. Age-related changes in the twenty-four-hour spontaneous secretion of growth hormone. *J Clin Endocrinol Metabol* 1972; **35**: 665–70.
55 Florini JR, Prinz PN, Vittelo MV, Hintz RL. Somatomedin C levels in healthy young and old men: relationship to peak and 24-hour integrated levels of growth hormone. *J Gerontol* 1985; **40**: 2–10.
56 Zadik Z, Chalew SA, Gilula Z, Kowarski AA. Reproducibility of growth hormone testing procedures: a comparison between 24-hour integrated concentration and pharmacological stimulation. *J Clin Endocrinol Metab* 1990; **71**: 1127–30.
57 Pavlov EP, Harman SM, Merriam GR, Gelato MC, Blackman MR. Response of growth hormone (GH) and somatomedin-C to GH-releasing hormone in healthy aging men. *J Clin Endocrinol Metabol* 1986; **62**: 595–600.
58 Hattori N, Kurahachi H, Ikekubo K *et al.* Effects of sex and age on serum GH binding protein levels in normal adults. *Clin Endocrinol* 1991; **35**: 295–7.
59 Copeland KC, Colletti RB, Devlin JT, McAuliffe TL. The relationship between insulin-like growth factor-I, adiposity and aging. *Metabolism* 1990; **39**: 584–7.
60 Bennett AE, Wahner HW, Riggs BL, Hintz RL. Insulin-like growth factors I and II: aging and bone density in women. *J Clin Endocrin Metabol* 1984; **59**: 701–4.
61 Blum WF, Ranke MB. Use of insulin-like growth factor binding protein 3 for the evaluation of growth disorders. *Horm Res* 1990; **33** (Suppl 4): 31–7.
62 Wüster CHR, Blum W, Schlemilch S, Ranke MB, Ziegler R. Decreased serum levels of IGF-binding protein (IGFBP-3) in osteoporosis. *J Bone Miner Res* 1991; **6** (Suppl 1): 107.
63 Wüster CHR, Blum W, Schlemilch S, Ranke MB, Ziegler R. Decreased insulin-like growth factors 1 and 2 in sera of patients with osteoporosis. *Bone Miner* 1992; **17** (Suppl 1): 156.
64 Wüster CHR, Blum WF, Schlemilch S, Ranke MB, Ziegler R. Decreased serum levels of insulin-like growth factors 1 and 2 and IGF binding protein-3 in patients with osteoporosis. *J Intern Med* 1993; **234**: 249–55.
* 65 Pyka G, Wiswell RA, Marcus R. Age-dependent effect of resistance exercise on growth hormone secretion in people. *J Clin Endocrinol Metab* 1992; **75**: 404–7.
66 Mochizuki H, Hakeda Y, Wakatsuki N *et al.* Insulin-like growth factor-I supports formation and activation of osteoclasts. *Endocrinology* 1992; **131**: 1075–80.
67 Dawson-Hughes B, Stern D, Goldman J, Reichlin S. Regulation of growth hormone and somatomedin-C secretion in postmenopausal women: effect of physiological estrogen replacement. *J Clin Endocrinol Metab* 1986; **63**: 424–32.
68 Rudman D, Kutner MH, Rogers CM *et al.* Impaired growth hormone secretion in the adult population: relation to age and adiposity. *J Clin Invest* 1981; **67**: 1361–9.
69 Rudman D. Growth hormone, body

composition, and aging. *J Am Geriatr Soc* 1985; **33**: 800–7.
70 Ljunghall S, Johansson AG, Burman P *et al.* Low plasma levels of insulin-like growth factor 1 (IGF-1) in male patients with idiopathic osteoporosis. *J Intern Medicine* 1992; **232**: 59–64.
71 Nakamura T, Hosoi T, Mizuno Y *et al.* Clinical significance of serum levels of insulin-like growth factors as bone metabolic markers in postmenopausal women [abstract]. *Bone Miner* 1992; **17** (Suppl 1): 170.
72 Mohan S, Dequeker J, Van Den Eyned R *et al.* Increased IGF-I and IGF-II in bone from patients with osteoarthritis [abstract]. *J Bone Miner Res* 1991; **6** (Suppl 1): S131.
73 Kassem M, Brixen K, Mosekilde L, Eriksen EF. Human marrow stromal osteoblast-like cells do not show reduced responsiveness to *in vitro* stimulation with growth hormone in patients with postmenopausal osteoporosis. *Calcif Tissue Int* 1994; **54**: 1–6.
74 Reed BY, Zerwegh JE, Sakhaee K *et al.* Serum IGF-I is low and correlates with osteoblastic surface in idiopathic osteoporosis. *J Bone Miner Res* 1995; **10**: 1218–24.
75 Wüster CHR, Köppler D, Müller C *et al.* Normal GH, PICP and ICTP and decreased PTH: 24-hour secretion in osteoporosis. *Osteoporosis Int* 1996; **6** (Suppl 1): 102.
76 Halse J, Gordeladze JO. Urinary hydroxyproline excretion in acromegaly. *Acta Endocrinol (Copenh)* 1978; **89**: 483–91.
77 Halse J, Gordeladze JO. Total and non-dialyzable urinary hydroxyproline in acromegalics and control subjects. *Acta Endocrinol (Copenh)* 1981; **96**: 451–7.
78 De La Piedra C, Carboe Larranaga J, Castro N *et al.* Correlation among plasma osteocalcin, growth hormone and somatomedin C in acromegaly. *Calcif Tissue Int* 1998; **43**: 44–5.
79 Ezzat S, Melmed S, Endres D, Eyre D, Singer F. Biochemical assessment of bone formation and resorption in acromegaly. *J Clin Endocrinol Metab* 1993; **76**: 1452–7.
80 Halse J, Melsen F, Mosekilde L. Iliac crest bone mass and remodelling in acromegaly. *Acta Endocrinol (Copenh)* 1981; **97**: 18–22.
81 Riggs LB, Randall RV, Wahner HW *et al.* The nature of the metabolic bone disorder in acromegaly. *J Clin Endocrinol Metab* 1972; **34**: 543–51.
82 Aloia JF, Roginsky MS, Lowsey J *et al.* Skeletal metabolism and body composition in acromegaly. *J Clin Endocrinol Metab* 1972; **35**: 543–51.
83 Seeman E, Wahner WH, Offord KP *et al.* Differential effects of endocrine dysfunction on the axial and the appendicular skeleton. *J Clin Invest* 1982; **69**: 1302–9.
84 Diamond T, Nery L, Posen S. Spinal and peripheral bone mineral densities in acromegaly: the effects of excess growth hormone and hypogonadism. *Ann Intern Med* 1989; **111**: 567–73.
85 Schulz G, Manns M. Ätiologie, Diagnostik und Therapie der Osteoporose. In: Schild HH, Heller M, eds. *Osteoporose.* Stuttgart: Thieme, 1992: 39.
86 Bijlsma JWJ, Nortier JWR, Duursma SA *et al.* Changes in bone metabolism during treatment of acromegaly. *Acta Endocrinol (Copenh)* 1983; **104**: 153–9.
87 Andreassen TT, Jørgensen PH, Flyvbjerg A, Orskov H, Oxlund H. Growth hormone stimulates bone formation and strength of cortical bone in aged rats. *J Bone Miner Res* 1995; **10**: 1057–67.
88 Ørtoft G, Brüel A, Andreassen TT, Oxlund H. Growth hormone is not able to counteract osteopenia of rat cortical bone induced by glucocorticoid with protracted effect. *Bone* 1995; **17**: 543–8.
89 Sandstedt J, Tornell J, Norjavaara E, Isaksson OG, Ohlsson C. Elevated levels of growth hormone increase bone mineral content in normal young mice, but not in ovariectomized mice. *Endocrinology* 1996; **137**: 3368–74.
90 Denis I, Zerath E, Pointillart A. Effects of exogenous growth hormone on bone mineralization and remodeling and on plasma calcitriol in intact pigs. *Bone* 1994; **15**: 419–24.
91 Brixen K, Nielsen HK, Mosekilde L, Flyvbjerg A. A short course of recombinant human growth hormone treatment stimulates osteoblasts and

activates bone remodeling in normal human volunteers. *J Bone Miner Res* 1990; 5: 609–18.

92 Marcus R, Butterfield G, Holloway L *et al.* Effects of short-term administration of recombinant human growth hormone to elderly people. *J Clin Endocrinol Metab* 1990; **70**: 519–27.

* 93 Rudman D, Feller AG, Nagraj HS *et al.* Effects of growth hormone in men over 60 years old. *N Engl J Med* 1990; **323**: 1–6.

94 Holloway L, Butterfield G, Hintz RL, Gesundheit N, Marcus R. Effects of recombinant human growth hormone on metabolic indices, body composition, and bone turnover in healthy elderly women. *J Endocrinol Metab* 1994; **79**: 470–9.

95 Kruse HP, Kuhlencordt F. On an attempt to treat primary and secondary osteoporosis with human growth hormone. *Horm Metab Res* 1975; 7: 488–91.

96 Aloia JF, Zanzi I, Ellis K *et al.* Effects of growth hormone in osteoporosis. *J Clin Endocrinol Metab* 1976; **43**: 922–99.

97 Aloia JF, Vaswani A, Kapoor A, Yeh JK, Cohn SH. Treatment of osteoporosis with calcitonin with and without growth hormone. *Metabolism* 1985; **34**: 124–31.

98 Aloia JF, Vaswani A, Meunier PJ *et al.* Coherence treatment of postmenopausal osteoporosis with growth hormone and calcitonin. *Calcif Tissue Int* 1987; **40**: 253–9.

99 Dambacher MA, Lauffenberger T, Haas HG. Vergleich verschiedener medikamentöser Therapieformen bei Osteoporose (NaF, NaF + Vitamin D, 1,25(OH)$_2$D$_3$ und menschliches Wachstumshormon). Kurz- und Langzeituntersuchungen. *Akt Rheumatol* 1982; 7: 249–52.

100 Johannson AG, Lindh E, Blum WF *et al.* Effects of growth hormone and insulin-like growth factor-I in men with idiopathic osteoporosis. *J Clin Endocrinol Metab* 1996; **81**: 44–8.

** 101 Erdtsieck RJ, Pols HAP, Valk NK *et al.* Treatment of postmenopausal osteoporosis with a combination of growth hormone and pamidronate: a placebo-controlled trial. *Clin Endocrinol* 1995; **43**: 557–65.

102 Harms HM, König S, Wüstermann PR, von zur Mühlen A, Hesch RD. Knochenstoffwechselparameter bei Patienten mit Osteoporose unter Therapie mit humanem Parathormon-(1–38) (hPTH1-38) und rekombinantem Wachstumshormon (rhGH). In: Wüster C, Raue R, Ziegler R, eds. *Osteologie '92*. Heidelberg: Merges, 1992: 29.

103 Gonelli S, Cepollaro C, Montomoli M *et al.* Treatment of postmenopausal osteoporosis with recombinant human growth hormone and salmon calcitonin: a placebo-controlled study. *Clin Endocrinol (Oxf)* 1997; **46**: 55–61.

* 104 Rosen CJ, Donahue LR, Hunter SJ. Insulin-like growth factors and bone: the osteoporosis connection. *Proc Soc Exp Biol Med* 1994; **206**: 83–102.

* 105 Johansson A, Lindh E, Ljunghall S. Growth hormone, insulin-like growth factor I, and bone: a clinical review. *J Intern Med* 1993; **234**: 553–60.

** 106 Slootweg MC. Growth hormone and bone. *Horm Metab Res* 1993; **25**: 335–43.

* 107 Inzucchi SE, Robbins RJ. Clinical review 61: effects of growth hormone on human bone biology. *J Clin Endocrinol Metab* 1994; **79**: 691–4.

** 108 Parfitt AM. Growth hormone and adult bone remodelling. *Clin Endocrinol* 1991; **35**: 467–70.

** 109 Eriksen EF, Kassem M, Langdahl B. Growth hormone, insulin-like growth factors and bone remodelling. *Eur J Clin Invest* 1996; **26**: 525–34.

* 110 Ohlsson C, Bengtsson BA, Isaksson OG, Andreassen TT, Slootweg MC. Growth hormone and bone. *Endocr Rev* 1998; **19**: 55–79.

111 Bouillon R, ed. *GH and Bone.* London: OCC, 1998.

112 Wüster CHR, Slenczka E, Ziegler R. Erhöhte Prävalenz von Osteoporose und Arteriosklerose bei konventionell substituierter Hypophysenvorderlappeninsuffizienz: Bedarf einer zusätzlichen Wachstumshormonsubstitution? *Klin Wochenschr* 1991; **69**: 769–73.

* 113 Halloran BP, Bikle DD, Harris J *et al.* Skeletal unloading induces selective resistance to the anabolic actions of growth hormone on bone. *J Bone Miner Res* 1995; **10**: 1168–76.

17: What is the current status of GH therapy in catabolic adults?

Richard J.M. Ross and Richard C. Jenkins

Introduction

The rationale for the use of growth hormone (GH) therapy in catabolic illness is based on three assumptions: firstly, that improving the nutritional state of patients will improve outcome; secondly, that the anabolic actions of GH have a beneficial effect on nutritional state; and thirdly, that most catabolic illness is associated with GH resistance or deficiency.

Even relatively minor injuries, such as fracture of the tibia, are associated with a 14-day period of hypercatabolism and negative nitrogen balance [1], and this is prolonged in sicker patients [2]. This hypermetabolic response is most marked in patients with burns or in septic patients [3], who are resistant to nutritional support [4]. The magnitude of the hypermetabolic response is related to outcome [5], and nutritional state is an important predictor of survival in conditions such as liver and renal failure [6], and critical illness. Once illness is prolonged, protein loss results in muscle wasting, poor wound healing, an increased incidence of infection and delayed recovery [7].

GH seems a natural choice as an anabolic agent, as it promotes muscle protein synthesis and liberates metabolic fuels by lipolysis. In GH-deficient children, replacement therapy has a potent effect on skeletal growth, and in the GH-deficient adult, replacement therapy has a marked effect on body composition, in effect replacing fat with muscle (see Chapter 10).

Fasting, malnutrition, organ failure and acute illness are associated with an increase in GH secretion, a decrease in insulin-like growth factor-I (IGF-I) levels and resistance to the anabolic actions of GH [8]. This GH resistance may be an adaptive change, permissive to protein catabolism, in the sick and fasting patient [9]. In the chronic phase of protracted critical illness, it has been suggested that GH secretion is depressed and that there is a relative deficiency of GH [10].

GH has been used as a pharmacological agent in catabolic illness to overcome GH resistance, with the aim of increasing muscle mass and improving immune function in the hope that this will have a beneficial effect on outcome [11]. Numerous studies have demonstrated positive effects of GH on nitrogen economy [12]. However, there have been few studies that have shown this to be translated into a benefit in terms of patient outcome, and disturbingly a recent report suggests that GH may have deleterious effects in critically ill patients (personal communication). In this review, we consider the changes that occur in the GH–IGF-I axis in the catabolic patient, studies of GH therapy, and the potential side-effects of this treatment.

The GH–IGF-I axis in catabolic illness

The major problem in describing the metabolic changes in catabolic illness is in defining the term 'catabolic illness'. Virtually all 'illness' is associated with a period of hypercatabolism. Fasting alone, although not hypercatabolic in the initial phase, is on balance a catabolic state. Even if we were to pick out a single disease state such as liver failure, the degree of catabolism would depend on the individual patient's condition. A patient who has just suffered a variceal bleed will be metabolically very different from a patient with a similar degree of hepatic dysfunction, but no bleeding. This problem is even greater when we consider the 'critically ill', who represent a heterogeneous group of patients, often with multiple organ failure. We might try to subdivide these groups into those with acute or chronic phases of the illness; but this effort is confounded by the fact that an acute illness, such as an infection, may compound recovery during the chronic phase of the illness. The approach we have taken here is to look at a selected group of patients and try to draw some general conclusions.

In describing the changes in the GH–IGF-I axis during catabolic illness, it is important to define the components of this axis. GH is released in a pulsatile fashion from the anterior pituitary under the influence of hypothalamic factors, which include growth hormone-releasing hormone (GHRH), somatostatin and a possible endogenous GH secretagogue or u-factor. GH circulates attached to a binding protein, and acts through its own specific receptor, which occurs as multiple isoforms. GH then has various biological actions, which can be broadly divided into those that are anabolic, usually mediated by IGF-I, and those that are direct, such as lipolysis. The actions of IGF-I are modulated by at least six high-affinity binding proteins and possibly four low-affinity binding proteins. Insulin-like growth factor binding protein-3 (IGFBP-3) acts as the main intravascular store of IGF-I, and IGFBP-1 appears to modulate IGF-I bioavailability acutely. Levels of IGFBP-3 and IGFBP-1 are regulated by GH and insulin, respectively. To date, our understanding of the regulation of the IGFBPs and their influence on IGF bioactivity remains limited.

Fasting, malnutrition and anorexia are associated with a fall in IGF-I levels and an increase in pulsatile GH release [13,14]. This is associated with a fall in IGFBP-3 levels and increasing levels of IGFBP-1 and IGFHP-2. Elective abdominal surgery results in a more distinctive pattern in IGF-I response (Fig. 17.1). There is an acute fall in IGF-I, which is then prolonged,

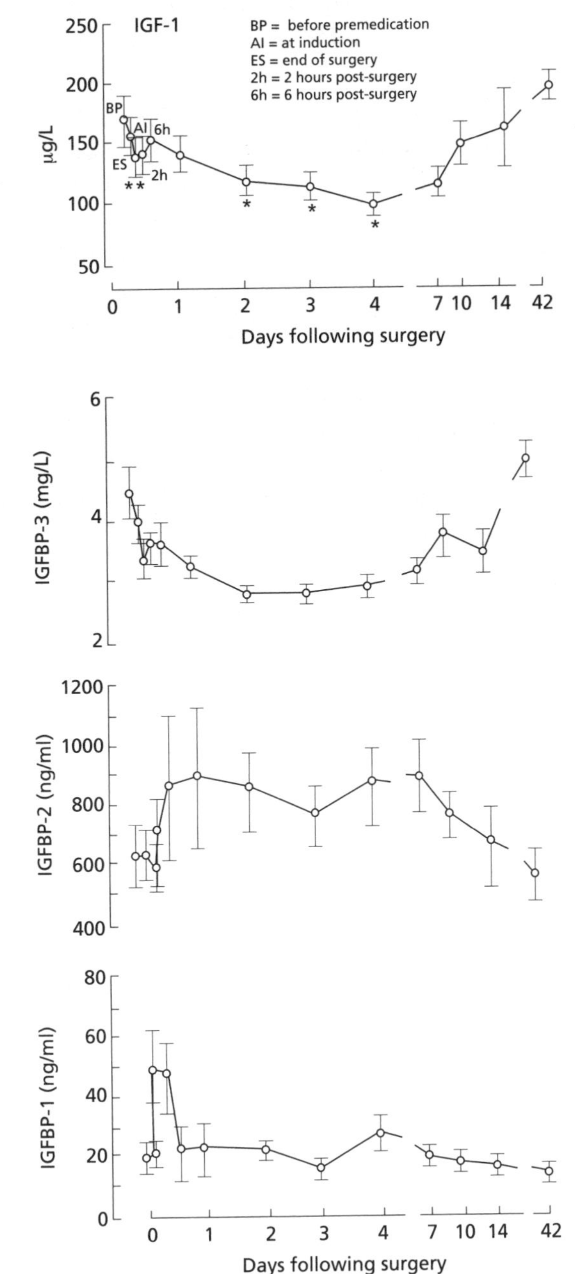

Fig. 17.1 Changes in insulin-like growth factor-I (IGF-I) and IGF binding proteins in 12 patients undergoing elective abdominal surgery. Reproduced with permission from Cotterill *et al.* [15].

with a nadir at four days and recovery only after two weeks [15]. Similar changes are seen in IGFBP-3, but IGFBP-1 rises acutely at the time of surgery. In addition to the fall in IGF-I, a protease appears that reduces the affinity of IGF-I for IGFBP-3. The changes seen in patients with organ failure are more difficult to interpret, as the individual organ may affect the production or clearance of GH, IGF-I, or binding proteins. In liver failure, GH levels are raised both due to increased secretion and decreased clearance [16], and IGF-I levels are low. It has been assumed that this is due to a lack of growth hormone receptor on the damaged liver. However, cirrhotic liver expresses GH receptor [17] and is transcriptionally active for most components of the IGF-I axis [18,19], and GH treatment increases IGF-I levels (Fig. 17.2) [20]. In renal failure, GH secretion is increased, and although IGF-I levels are often normal, secretion is reduced and there is retention of IGFBP-3. In cardiac failure, GH levels have been variously reported as high or low [21,22]. In critical illness, pulsatile GH secretion is reduced but basal GH levels are increased [10,23]. In patients with a prolonged intensive-care unit (ICU) stay, GH levels may be low [10]. A finding that can be generalized for most catabolic states is that IGF-I levels are low, fall in parallel to markers of nutrition, and reflect the severity of the illness [24]. Most illness associated with surgical stress or sepsis is associated with the production of an IGFBP-3 protease [25]. IGF-I clearance is increased with the severity of the illness, being short in postsurgical patients [26] and the critically ill [27]. Overall, the changes in most catabolic states can be summarized as acquired GH resistance [8].

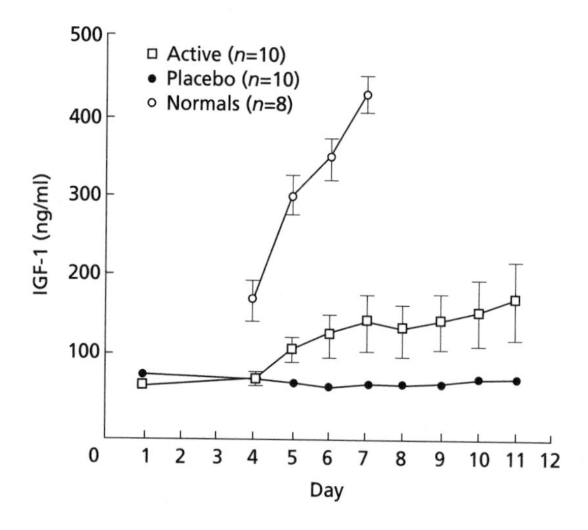

Fig. 17.2 Changes in insulin-like growth factor-I (IGF-I) levels in normal individuals or patients with chronic liver disease treated with growth hormone (GH) or placebo. Reproduced with permission from Donaghy *et al.* [20].

Clinical studies of GH therapy in catabolic illness

There have been many small studies investigating the use of GH as an anabolic agent in postoperative patients, following burns, glucocorticoid-induced catabolism, in acquired immune deficiency syndrome (AIDS), cancer, pulmonary disease, sepsis, trauma, renal failure, short bowel syndrome, and cardiac failure, as well as in the critically ill [12]. The majority of the studies have been in small groups of patients, using surrogate markers of outcome such as nitrogen balance. Below, we detail some of the studies that have attempted to analyse more clinically relevant outcome measures.

In postsurgical patients, GH treatment increased hand grip strength [28]. In the largest placebo-controlled study reported to date, GH treatment (8 IU/day) for eight days in 180 patients undergoing cholecystectomy reduced the wound infection rate (17%–3%) and reduced hospital stay from 12.5 to 9.6 days [29]. However, the incidence of wound infection and the duration of hospital stay were greater than would normally be expected for this patient group. Burns injury results in a severe catabolic state. Forty severely burned children were treated with 10–20 IU/m^2/day of GH in a double-blind, placebo-controlled study [30]. Donor-site healing times decreased with a higher GH dose, and length of admission was shortened—which for a child with a 60% burn could translate into a reduction from 46 to 32 days. However, the criticism of this study is that donor-site healing is a subjective assessment. In AIDS, a short-term study of seven days with GH (10 IU/m^2/day) in six human immunodeficiency virus (HIV)-positive patients resulted in weight gain [31], and longer studies have confirmed this effect [32]. In renal failure, GH in supraphysiological doses promotes linear growth in childhood [33].

Despite the many studies showing a benefit in nitrogen balance and the few studies reporting benefits in functional outcome, there are also reports demonstrating no benefit of GH in catabolic illness. In patients receiving ventilation for chronic lung disease, GH treatment had no benefit, despite a 250% increase in circulating IGF-I levels and improved nitrogen balance [34]. A similar result was found in patients requiring prolonged ventilation [35]. Combination of therapy with GH and IGF-I resulted in a transient increase in weight, but no improvement in muscle strength in AIDS patients [36,37].

Growth hormone in sepsis and the critically ill

Initial studies in sepsis suggested that patients were absolutely resistant to the anabolic actions of GH [38]. Subsequent studies suggested that GH therapy could improve nitrogen balance even in the early phase of severe sepsis [39]. For this reason, a large European multicentre trial of GH therapy in ICU patients was established. In 1997, the study was terminated because of increased mortality in the GH-treated vs. placebo-treated patients (42% vs. 18%) in a total of 532 patients (personal communication). This dramatic and very disturbing result has led to a great deal of reflection on the role of GH in catabolism. A full analysis of the study is not yet available, but from the interim analysis it seems clear that GH was the cause for the increased mortality. The mechanism seems less clear. Since the 1960s, there have been over 90 investigations of GH in catabolic states, including sepsis and critical illness, using similar doses of GH. From these publications, there was no evidence of an increase in mortality. The side-effects that had been reported include carpal tunnel syndrome, parotid tenderness, papilloedema [40], electrolyte disturbance [28,41], hypercalcaemia [42], sodium retention and insulin resistance.

One possibility for the effect in ICUs is that GH has a deleterious effect when there is associated endotoxaemia. Support for this hypothesis comes from two animal studies. GH potentiated the effect of endotoxaemia in the rat [43], and produced unfavourable effects on carbohydrate metabolism in septic piglets [44].

Summary

Hypercatabolism is a common response to many insults, and when it is prolonged it results in protein wasting. This is associated with changes in the GH–IGF-I axis, which in many conditions can be summarized as acquired GH resistance. Numerous small-scale and often uncontrolled studies have suggested that GH may improve the nutritional state in many different catabolic patient groups. There have been few controlled studies looking at important outcome measures, but among these there is some evidence to suggest that GH may benefit patients after surgery, those with burns patients and those with AIDS. Disturbingly, the largest study to date, in ICU patients, showed a doubling of mortality in GH-treated patients. The latter study forces us to question the original assumptions that the anabolic actions of GH will be of benefit in catabolism and that improving nutritional state will necessarily improve outcome. Future studies will need to examine the physiological basis for acquired GH resistance and test the hypothesis that its reversal will improve outcome.

Acknowledgements

We are grateful for support from the Northern General Hospital Research Committee, Special Trustees of the Former United Sheffield Hospitals, Trent

Regional Research Schemes, YCR, Serono Laboratories and Pharmacia and Upjohn.

References

1 Cuthbertson DP. Observations on the disturbance of metabolism produced by injury to the limbs. *Q J Med* 1931; **24**: 233–46.

2 Arnold J, Campbell IT, Samuels TA *et al.* Increased whole body protein breakdown predominates over whole body protein synthesis in multiple organ failure. *Clin Sci* 1993; **84**: 655–61.

3 Herndon DN, Wilmore DW, Mason AD. Development and analysis of a small animal model stimulating the human post-burn hypermetabolic response. *J Surg Res* 1978; **25**: 394–403.

4 Shaw JHF, Wildbore M, Rolfe RR. Whole body protein kinetics in severely septic patients: the response to glucose infusion and TPN. *Ann Surg* 1987; **205**: 288–94.

5 Herndon DN, Curreri PW, Abston S, Rutan TC, Barrow RE. Treatment of burns. *Curr Probl Surg* 1987; **24**: 341–97.

6 Mendenhall CL, Tosch T, Weesner RE *et al.* VA cooperative study on alcoholic hepatitis, 2: prognostic significance of protein–calorie malnutrition. *Am J Clin Nutr* 1986; **43**: 213–18.

7 Alexander JW. Nutrition and infection: a new perspective for an old problem. *Arch Surg* 1986; **121**: 966–72.

8 Ross RJM, Chew SL. Acquired growth hormone resistance. *Eur J Endocrinol* 1995; **132**: 655–60.

9 Ross RJM, Miell JP, Buchanan CR. Avoiding autocannibalism: consider growth hormone and insulin-like growth factor I. *Br Med J* 1991; **303**: 1147–8.

10 Van den Berghe G, De Zegher F, Bouillon R. Acute and prolonged critical illness as different neuroendocrine paradigms. *J Clin Endocrinol Metab* 1998; **83**: 1827–34.

11 Wilmore DW. Catabolic illness: strategies for enhancing recovery. *N Engl J Med* 1990; **325**: 323–56.

12 Jenkins RC, Ross RJM. Growth hormone therapy for protein catabolism. *Q J Med* 1996; **89**: 813–19.

13 Counts DR, Gwirtsman H, Carlsson LMS, Lesem M, Cutler GB. The effect of anorexia nervosa and refeeding on growth hormone-binding protein, the insulin-like growth factors (IGFs), and the IGF-binding proteins. *J Clin Endocrinol Metab* 1992; **75**: 762–7.

14 Ho KY, Veldhuis JD, Johnson ML *et al.* Fasting enhances growth hormone secretion and amplifies the complex rhythms of growth hormone secretion in man. *J Clin Endocrinol Metab* 1988; **81**: 986–75.

15 Cotterill AM, Mendel P, Holly JMP *et al.* The differential regulation of the circulating levels of the insulin-like growth factor binding proteins (IGFBP) 1, 2 and 3 after elective abdominal surgery. *Clin Endocrinol* 1996; **44**: 99–101.

16 Cuneo RC, Hickman PE, Wallace JD *et al.* Altered endogenous growth hormone secretory kinetics and diurnal GH-binding protein profiles in adults with chronic liver disease [abstract]. *Clin Endocrinol* 1995; **43**: 265–75.

17 Shen XY, Holt RIG, Miell JP *et al.* Cirrhotic liver expresses low levels of the full length and truncated growth hormone receptor. *J Clin Endocrinol Metab* 1998; **83**: 83.

18 Ross RJM, Chew SL, D'Souza LL *et al.* IGF-I and IGF binding protein genes in normal liver and cirrhosis. *J Endocrinol* 1996; **143**: 209–16.

19 Ross RJM, Rodriguez-Arnao J, Donaghy A *et al.* Expression of IGFBP-1 in normal and cirrhotic human livers. *J Endocrinol* 1994; **141**: 377–82.

20 Donaghy A, Ross RJM, Wicks C *et al.* Growth hormone therapy in cirrhosis: a double-blind, placebo-controlled, pilot study of efficacy and safety [abstract]. *Gastroenterology* 1997; **113**: 1617–22.

21 Anand IS, Ferrari R, Kalra GS *et al.* Studies of body water and sodium, renal function, hemodynamic indexes, and plasma hormones in untreated congestive cardiac failure. *Circulation* 1989; **80**: 299–305.

22 Giustina A, Lorusso R, Borghetti V *et al.*

Impaired spontaneous growth hormone secretion in severe dilated cardiomyopathy. *Am Heart J* 1996; **131**: 620–2.

23 Ross RJM, Miell J, Freeman E *et al.* Critically ill patients have high basal growth hormone levels with attenuated oscillatory activity associated with low levels of insulin like growth factor-I. *Clin Endocrinol* 1991; **35**: 47–54.

24 Hawker FH, Stewart PM, Baxter RC *et al.* Relationship of somatomedin-C/insulin-like growth factor I levels to conventional nutritional indices in critically ill patients. *Crit Care Med* 1987; **15**: 732–6.

25 Davies SC, Wass JAH, Ross RJM *et al.* The induction of a specific protease for insulin-like growth factor binding protein-3 in the circulation during severe illness. *J Endocrinol* 1991; **130**: 469–73.

26 Miell JP, Taylor AM, Jones J *et al.* Administration of human recombinant insulin-like growth factor-I to patients following major gastrointestinal surgery. *Clin Endocrinol* 1992; **37**: 542–51.

27 Yarwood GD, Ross RJM, Medbak S, Coakley J, Hinds CJ. Administration of human recombinant insulin like growth factor-I to critically ill patients on the intensive care unit. *Crit Care Med* 1997; **25**: 1352–61.

28 Jiang ZM, He GZ, Zhang SY *et al.* Low dose GH and hypocaloric nutrition attenuate the protein catabolic response after major operation. *Ann Surg* 1989; **210**: 513–25.

29 Vara-Thorbeck R, Guerrero JA, Rosell J, Ruiz-Requena E, Capitan JM. Exogenous growth hormone: effects on the catabolic response to surgically produced acute stress and on postoperative immune function. *World J Surg* 1993; **17**: 530–8.

30 Herndon DN, Barrow RE, Kunkel KR. Effect of recombinant human growth hormone on donor-site healing in severely burned children. *Ann Surg* 1990; **212**: 424–9.

31 Mulligan K, Grunfeld C, Hellerstein MK, Neese RA, Schambelan M. Anabolic effects of recombinant human growth hormone in patients with wasting associated with human immunodeficiency virus infection. *J Clin Endocrinol Metab* 1993; **77**: 956–62.

32 Schambelan M, Mulligan K, Grunfeld C *et al.* Recombinant human growth hormone in patients with HIV-associated wasting: a randomized, placebo-controlled trial. *Ann Intern Med* 1996; **125**: 873–82.

33 Lippe B, Yadin O, Fine RN, Moulton L, Nelson PA. Use of recombinant human growth hormone in children with chronic renal insufficiency: an update. *Horm Res* 1993; **40**: 102–8.

34 Suchner U, Rothkopf MM, Stanislaus G *et al.* GH and pulmonary disease: metabolic effects in patients receiving parenteral nutrition. *Arch Intern Med* 1990; **150**: 1125–30.

35 Pichard C, Kyle U, Chevrolet J. Lack of effects of recombinant GH on muscle function in patients requiring prolonged mechanical ventilation: a prospective, randomized, controlled study. *Crit Care Med* 1996; **24**: 403–13.

36 Lee PD, Pivavnik JM, Bukar JG *et al.* A randomized, placebo-controlled trial of combined IGF-I and low-dose GH therapy for wasting associated with human immunodeficiency virus infection. *J Clin Endocrinol Metab* 1996; **81**: 2968–75.

37 Waters D, Danska J, Hardy K *et al.* rhGH, IGF-I and combination therapy in AIDS-associated wasting: a randomized, double-blind, placebo-controlled trial. *Ann Intern Med* 1996; **125**: 865–72.

38 Dahn MS, Lange MP, Jacobs LA. Insulin-like growth factor I production is inhibited in human sepsis. *Arch Surg* 1988; **123**: 1409–14.

39 Voerman HJ, Strack van Schijndel RJM, Groeneveld ABJ *et al.* Pulsatile hormone secretion during severe sepsis: accuracy of different blood sampling regimens. *Metabolism* 1992; **9**: 934–40.

40 Koller EA, Stadel BV, Malozowski SN. Papilledema in 15 renally compromised patients treated with GH. *Pediatr Nephrol* 1997; **11**: 451–4.

41 Ziegler TR, Rombeau JL, Young LS *et al.* Recombinant human GH enhances the metabolic efficacy of parenteral nutrition: a double-blind, randomized, controlled study. *J Clin Endocrinol Metab* 1992; **74**: 865–73.

42 Knox JB, Demling RH, Wilmore DW, Sarraf P, Santos AA. Hypercalcaemia associated with the use of hGH in an adult surgical intensive care unit. *Arch Surg* 1995; **130**: 442–5.

43 Liao W, Rudling M, Angelin B. Growth hormone potentiates the in vivo biological activities of endotoxin in the rat. *Eur J Clin Invest* 1996; **26**: 254–8.
44 Balteskard L, Unneberg K, Mjaaland M *et al.* Treatment with growth hormone and insulin-like growth factor-I in septicemia: effects on carbohydrate metabolism. *Eur Surg Res* 1998; **30**: 79–94.

18: Is GH therapy indicated in the metabolic syndrome?

Gudmundur Johannsson, Per Björntorp and
Bengt-Åke Bengtsson

Introduction

Obesity is associated with decreased longevity and increased morbidity from cardiovascular disorders. Obesity is a powerful risk factor for type 2 diabetes mellitus, and like obesity, type 2 diabetes mellitus is also associated with increased cardiovascular disorders. Obesity has long been recognized as a cause of impotence and oligospermia in men, and of amenorrhea and reduced fertility in women. Endocrine aberrations associated with obesity are therefore well established. More recently, it has been established that abdominal fat predominance is an independent risk factor for type 2 diabetes mellitus in both sexes, as well as in several ethnic groups [1–3]. Moreover, abdominal fat distribution has been reported in prospective studies as being a strong independent risk factor for coronary heart disease, stroke and mortality. Therefore, the body fat distribution, together with obesity *per se,* is a strong predictor of the metabolic profile and health complications such as type 2 diabetes mellitus and cardiovascular diseases.

This chapter will focus on the endocrine aberrations associated with abdominal/visceral obesity, and its metabolic consequences. The possible pathogenic role of disturbances in the growth hormone (GH)–insulin-like growth factor-I (IGF-I) axis, and the place of hormonal interventions in the treatment of the metabolic syndrome are discussed.

The metabolic syndrome

The association of several risk factors (obesity, dyslipoproteinaemia, insulin resistance and hypertension) for the pathogenesis of type 2 diabetes mellitus and myocardial infarction has been known in the literature for many years, and was termed 'the metabolic syndrome' early on [4]. In 1988, Reaven introduced the term 'syndrome X' as the link between insulin resistance and hypertension [5].

Visceral adipose tissue

As was pointed out by Vague some 50 years ago, the type of adipose tissue distribution is associated with both endocrine perturbations and human disease [6]. It has now been suggested that a critical factor for the association between obesity, type 2 diabetes mellitus and cardiovascular morbidity is the mass of intra-abdominal fat [3,7]. This is probably an effect of the unique metabolic characteristics and anatomical localization of visceral adipose tissue.

The blood flow from visceral fat depot is drained via the portal vein to the liver, in contrast to other fat depots, which are drained to the systemic circulation. Visceral adipose tissue has a higher turnover rate of fat than other adipose tissue depots. The visceral adipose tissue therefore has a sensitive system for mobilization of free fatty acids (FFAs). The increased lipolytic activity of visceral fat, combined with its anatomical localization, means that the liver is exposed to higher concentrations of FFAs than any other organ. This might be the key to several important consequences [3,8]. Increased levels of FFAs stimulate hepatic gluconeogenesis and also attenuate the hepatic clearance of insulin from the pancreas. This will exaggerate the peripheral hyperinsulinaemia, which is also caused by increased insulin secretion, known to follow obesity in general. The secretion of very low-density lipoproteins (VLDLs) from the liver is increased in response to increased availability of FFAs in the portal circulation. As peripheral hyperinsulinaemia and hypertension are statistically interconnected [5], the consequences of enlarged visceral adipose tissue depots might be expected to cause peripheral hyperinsulinaemia, hyperglycaemia, hypertension and elevated levels of VLDLs, all of which are known to be important risk factors for type 2 diabetes mellitus and arteriosclerosis.

The lipolytic process is mainly regulated by catecholamines in human adipose tissue [2,3]. Furthermore, visceral adipocytes have a higher density of lipolytic β-adrenergic receptors, mediating lipolysis by the action of noradrenaline, and lower α_2-adrenoreceptor inhibition, than other fat cells. This is seen particularly in men and abdominally obese women, but not in normal women or obese women with gluteal femoral adipose tissue distribution. Insulin is an important regulator of lipoprotein lipase (LPL) activity and triglyceride synthesis, and has important interactions with β-adrenergic receptors. These effects of insulin may differ between subcutaneous and visceral adipocytes. In addition to these intrinsic characteristics of the visceral adipocytes, the surroundings of these cells are different from other adipocytes. Blood flow is higher than in other adipose tissues, which is of fundamental importance for both lipid uptake and mobilization, and in addition visceral adipose tissue contains greater catecholamine innervation than inguinal fat. The density of glucocorticoid as well as androgen receptors is also higher. The effect of cortisol is mainly to increase visceral fat mass by increasing the expression of LPL, while testosterone has the capability to decrease fat accumulation by inhibiting LPL and enhancing lipolysis by increasing the expression of β-adrenergic receptors [1]. Hypogonadal men therefore have abdominal obesity, which is reduced by testosterone replacement therapy [8]. It is, however, unclear why women with abdominal obesity and relative hyperandrogenism accumulate triglycerides in visceral fat depots. The β-adrenergic, α_2-adrenoreceptor and LPL activity in visceral and gluteal fat are therefore influenced by sex hormones and sexual dimorphism [3]. In summary, this means that visceral adipose tissue has unique metabolic characteristics, hormone receptor density, blood flow and innervation. This is of considerable importance in view of the effects of FFAs on the hepatic regulation of metabolism.

The drawback with this synthesis of data is that it does not explain why visceral adipose tissue is enlarged, and this phenomenon is, together with insulin resistance, the cornerstone of the expression of the metabolic syndrome. It has been suggested that the syndrome develops as a consequence of a neuroendocrine arousal, with multiple endocrine perturbations that direct excess fat to visceral depots. The visceral adiposity and the FFAs from these depots are responsible for the insulin resistance. The 'portal FFA mechanism' may then add to the generation of other risk factors, such as hyperinsulinaemia, lipoprotein abnormalities, and possibly hypertension [7]. Results from recent studies using new methodology to estimate the regulation of the hypothalamic–pituitary–adrenal (HPA) axis support this hypothesis [9]. These studies show clear perturbations of HPA axis regulation, based on environmental pressure such as mental and physical stress, toxins (alcohol and smoking), overeating and psychiatric traits. In susceptible individuals, the feedback control by central glucocorticoid receptors (GRs) is diminished, causing regulatory perturbations of the secretion of cortisol. The genetic basis seems to be a polymorphism of the GR gene, diminishing GR effects by a transcription defect, resulting either in an inefficient transcription and/or an abnormal transcript of the GR. No less than 13.7% of Swedish men are homozygous for this polymorphism [10]. This is the central abnormality, most likely followed by inhibition of other central endocrine axes, including both the gonadotrophic axis and GH secretion through known mechanisms [11]. In addition, it has recently been demonstrated, *in vitro*, that activity of 11β-hydroxysteroid dehydrogenase type 1 in omental adipocytes may enhance the local conversion of inactive cortisone to cortisol and that this phenomenon is positively regulated by glucocorticoids [12]. In contrast data from *in vivo* studies in hypopituitary patients have suggested an inverse relationship between total body fat mass and its distribution, and cortisone to cortisol conversion [13].

In summary, there is now considerable evidence that the metabolic syndrome may be a consequence of perturbations in the HPA axis due to environmental pressure, which are expressed in susceptible individuals with molecular genetic susceptibility in the feedback inhibitory mechanism exerted by central GRs. This, together with attenuation of the activity of the gonadotrophin and GH axes, may be responsible for the development of the metabolic syndrome, with visceral obesity, insulin resistance, dyslipoproteinaemia and hypertension. Attenuated GH secretion is an important component of this cascade of events, and, interestingly, the metabolic syndrome may apparently develop as a consequence of low GH secretion alone (reduced serum IGF-I concentration), i.e. without involvement of the HPA axis inhibition. This seems to have a prevalence of 5% in the middle-aged population of Swedish men [14]. This might be partly explained by the inhibitory effect of GH on 11β-hydroxysteroid dehydrogenase [15] so that GH deficiency may be associated with increased local concentration of cortisol within adipocytes.

Patients with acromegaly have reduced adipose tissue mass. After successful treatment that normalizes their GH secretion, they demonstrate an increase of predominantly visceral adipose tissue mass [16]. The reverse scenario is seen in adults with hypopituitarism and untreated GH deficiency, who have an increased amount of body fat mass with abdominal preponderance; GH administration results in a profound reduction of visceral adipose tissue and less marked effects on other adipose tissue depots [17]. These observations indicate that GH has profound effects on adipose tissue mass and distribution.

Metabolic aberrations

Striking similarities exist between the metabolic syndrome and untreated GH deficiency in adults [18]. The most central findings in both these syndromes are abdominal/visceral obesity and insulin resistance. Other features common to both conditions are high triglyceride and low high-density lipoprotein cholesterol (HDL-C) concentrations, an increased prevalence of hypertension, elevated levels of plasma fibrinogen and plasminogen activator inhibitor-1 (PAI-1) activity, premature atherosclerosis and increased mortality from cardiovascular diseases [19]. Due to the similarities between these two syndromes, undetectable and low levels of GH may be of importance for the metabolic aberrations observed in both these conditions.

Insulin resistance

Insulin resistance is a common condition and can be seen, for example, in type 2 diabetes mellitus, obesity and hypertension. The interrelationship between insulin resistance and these conditions, as well as the exact mechanisms underlying insulin resistance, have not yet been fully clarified. It recently became clear that even adults with GH deficiency have insulin resistance in peripheral tissues, as measured using the hyperinsulinaemic euglycaemic clamp technique [20]. Glucose disposal rate (GDR) in GH-deficient adults is found to be less than half than that of controls, both when calculated according to body weight and when corrected for amount of body fat. The decreased lean body mass and the increased abdominal obesity in GH deficiency [20] may be of importance for this finding, as the association between increased body fat mass and insulin resistance is stronger in the presence of abdominal obesity [21]. The attenuated GDR seen in abdominal obesity has in some studies been quantitatively similar to that seen in overtly hyperglycaemic type 2 diabetes mellitus [22]. Low levels of serum IGF-I may also contribute to insulin resistance, as IGF-I stimulates glucose transport in skeletal muscle. Other factors—such as different composition in skeletal muscle fibres, with a decrease in the slow-twitch, insulin-sensitive type I fibres and an increase in the fast-twitch, type Ib fibres, the degree of capillary rarefaction and decreased physical activity in adults with GH deficiency—may be of importance as well in healthy adults [2,3].

Dyslipoproteinemia

GH has important effects on the lipoprotein metabolism. For example, hypophysectomy in the rat changes the lipoprotein pattern from being predominantly HDL to becoming a pattern with predominately a low-density lipoprotein (LDL) peak, suggesting that the presence of GH is essential for maintaining a normal lipoprotein profile. Moreover, in response to GH, the serum low-density lipoprotein cholesterol (LDL-C) and apolipoprotein B concentrations decrease [20], probably as a result of the increased clearance of these lipoproteins through increased hepatic LDL receptor activity [23].

A common finding in both GH deficiency and the metabolic syndrome is high levels of serum triglycerides and low HDL-C concentrations. This may be associated with increased abdominal adiposity and insulin resistance in both conditions. However, although a dramatic reduction in visceral adipose tissue occurs in response to GH treatment in adults with GH deficiency, serum triglyceride concentration is not reduced, while the concentration of HDL-C is increased. This may be an effect of the lipolytic action of GH treatment, which may increase the flux of FFAs to the liver, which in turn may increase the synthesis and secretion of VLDLs from the liver. The LPL activity in adipose tissue is attenuated, and the post-heparin plasma LPL is not affected by GH treatment [24]. As serum triglyceride concentrations do not increase under conditions of increased VLDL secretion, the peripheral catabolism must be enhanced. Increased LPL activity in other tissues, such as muscle, is therefore likely [24]. Furthermore, the strong association between glucose/insulin homeostasis and VLDL metabolism [25] might be reflected in the response to GH. The unaffected triglyceride levels might thus be explained by essentially unchanged insulin sensitivity during more prolonged GH treatment in GH-deficient adults [20].

Fibrinolysis

Plasminogen activator inhibitor-1 (PAI-1), the fast-acting tissue plasminogen activator (t-PA) inhibitor, is the major regulator of fibrinolytic activity in plasma. Increased PAI-1 activity acts in a thrombogenic direction. Elevated PAI-1 activity has been associated with coronary artery disease [26]. High PAI-1 activity has previously been found in patients with hypertension, insulin resistance and abdominal obesity. In addition, elevated PAI-1 activity has been shown in GH-deficient adults as compared with healthy controls matched for age, sex and body mass index (BMI) [20].

Previous population-based studies have shown that fibrinogen is an independent risk factor for stroke as well as myocardial infarction—at least as important as blood lipids and blood pressure [27]. Obesity has been associated with both increased fibrinogen levels and increased PAI-1 activity. Also, the fibrinogen levels are higher in GH-deficient adults than in controls. Although matched for BMI, both fibrinogen levels and PAI-1 activity are higher in the patients, suggesting that other factors in addition to obesity *per se* are of importance. Both the elevated fibrinogen levels and PAI-1 activity may be linked to the abdominal and visceral obesity in these patients [20].

Blood pressure

Both GH deficiency in adults and the metabolic syndrome are associated with an increased prevalence of hypertension. The insulin resistance in the metabolic syndrome (syndrome X) has been linked with hypertension through increased activity of the sympathetic nervous system [5]. Direct evidence for this assumption is provided by an apparent parallel activation of hypothalamic centres regulating the sympathetic nervous system and the HPA axis, a 'hypothalamic arousal syndrome' [28]. Central arousal of the sympathetic nervous system is considered to be a major pathogenic pathway for essential hypertension [29].

In adults with hypopituitarism and untreated GH deficiency, augmented activity of the sympathetic nervous system has been demonstrated by direct intraneural recordings [30], linking this condition to increased prevalence of hypertension. In addition, GH deficiency has been found to be associated with low levels of nitric oxide (NO), a paracrine vasodilator produced in endothelial cells, which normalizes in response to GH treatment [31].

Endocrine aberrations

The endocrine abnormalities in the metabolic syndrome are described in the introduction, above. These disturbances include increased activity of the HPA axis and a blunted secretion of GH and sex steroids in both men and women. Cortisol is of particular interest, as it causes accumulation of abdominal/visceral adipose tissue and an increased release of FFAs. The role of cortisol in obesity in general has been controversial through the years. Several investigators have found decreased plasma cortisol levels in obese subjects, while others have reported an increased cortisol secretion. These studies have not distinguished central from peripheral obesity [1]. The HPA axis is regulated by feedback inhibition from central GRs, which, when occupied, attenuate the activity of the axis. Under normal circumstances, serum levels of cortisol are high in the morning and low in the evening, and morning values are effectively suppressed by dexamethasone. Findings supporting increased activity of the HPA axis in abdominal obesity are a flattened diurnal cortisol curve [9], associated with blunted dexamethasone suppression of cortisol secretion [32]. Previous studies in abdominally obese subjects have also shown increased cortisol secretion after corticotrophin-releasing hormone (CRH) or adrenocorticotrophic hormone (ACTH) challenge, as well as after mental and physical laboratory stress tests [1].

Two subgroups of men with abdominal obesity in terms of HPA axis activity have recently been described [9]: one group had adequate diurnal variability in levels of cortisol, and the other had a flattened diurnal cortisol curve, indicating poor regulation of the HPA axis. It is evident that in this latter subgroup of men with abdominal obesity, the relationship between stress-related cortisol secretion and hyperinsulinaemia, abdominal obesity, blood pressure and dyslipoproteinaemia is consistently stronger than in the former group. This latter group also demonstrated a strong negative relationship with serum testosterone and IGF-I. These findings confirm that poor regulation of the HPA axis is followed by decreased serum testosterone and serum IGF-I values [11].

Recent results from testosterone intervention studies in men with abdominal/visceral obesity [8] may strengthen the presumed pathogenic importance of low testosterone in the syndrome of visceral obesity. A physiological amount of testosterone in middle-aged men with abdominal/visceral adiposity induced improved insulin sensitivity, plasma lipid levels, and diastolic blood pressure, as well as a specific decrease in visceral adipose tissue mass. This might theoretically be explained by a direct effect of testosterone on adipose tissue. However, as testosterone treatment in men with hypogonadotrophic hypogonadism increases GH secretion [33], the observed effects could be explained by increased GH levels or by additional or synergistic effects by GH and testosterone on adipose tissue metabolism [34].

Growth hormone in abdominal obesity

With increased adiposity, GH secretion is blunted, with a decrease in the mass of GH secreted per burst, but without any major impact on GH secretory burst frequency [35]. Moreover, the metabolic clearance rate of GH is accelerated [36]. The serum IGF-I concentration is primarily GH-dependent, and influences GH secretion through a negative feedback system [37]. The serum levels of IGF-I are inversely related to the percentage of body fat [35]. In addition, the low serum IGF-I concentration in obesity is predominantly related to the amount of visceral adipose tissue and not to the amount of subcutaneous fat mass [38]. Serum free IGF-I concentration may, on the other hand, be increased in abdominal obesity [39], possibly as an effect of the concomitant insulin-induced suppression of serum IGF binding protein-1 levels. The relationship between regional fat distribution and GH secretion has only recently been considered. No significant correlation was found between the waist-to-hip ratio and 24-hour GH secretion rates in a study of 21 healthy men [40]. However, in healthy non-obese men and women, intra-abdominal fat mass had a strong negative exponential relationship with mean 24-hour serum GH concentrations that was independent of age, sex and physical fitness [41]. This indicates that, for each increment in intra-abdominal fat mass, there is a more than linear reduction in mean 24-hour GH concentration.

Low levels of GH may be of importance for the metabolic consequences and the maintenance of the obese condition. One trial demonstrated near-normalization of the 24-hour GH secretion and serum IGF-I in nine obese subjects after massive weight loss [42], whereas others have not found a normalization of the GH response to provocative testing in response to weight loss [43,44]. However, the amount of intra-abdominal fat was not considered in these studies. Thus, whether the multiple endocrine aberrations, including low GH secretion, in abdominal obesity are primarily responsible for, or are the consequence of, the obese condition remains to be elucidated. The link between visceral fat mass and these endocrine aberrations has been discussed [1,2].

Growth hormone treatment in abdominal obesity

As GH promotes lipolysis, it has been suggested that low levels of GH may be of importance for the maintenance of the obese condition. The calorigenic effects of GH in obese subjects have also long been recognized. Some trials have therefore addressed the question of whether GH administration, through its calorigenic and lipolytic action, might enhance weight loss during dietary restriction in obese individuals. Both short-term GH treatment [45] and several weeks of GH treatment [46] in combination with dietary restriction were unable to enhance the loss of body fat or body weight, compared with saline treatment. GH administration may, however, decrease the loss of lean body mass during dietary restriction [45]. Twelve weeks of GH or placebo treatment combined with dietary restriction and exercise in moderately obese middle-aged women resulted in a similar weight reduction and a significant reduction in fat mass in both groups. These results therefore suggest that GH is not useful in the induction or enhancement of weight loss in obese patients [47]. None of the above studies, however, measured central adiposity or fat distribution.

It has been found that GH treatment can improve several of the aberrations that GH-deficient patients share with those who have the metabolic syndrome. Thus, in adults with GH deficiency, the lipolytic effects of GH result in a preferential reduction in visceral adipose tissue [17]. Furthermore, GH reduces diastolic blood pressure, total cholesterol and LDL-C, and increases HDL-C concentrations. In addition, long-term GH treatment does not impair insulin sensitivity [20]. Against this background, in a nine-month randomized, double-blind, placebo-controlled trial, we have studied the effects of GH on the metabolic, circulatory and anthropometric aberrations associated with abdominal/visceral obesity and the metabolic syndrome [48].

The men who were studied were moderately obese, with a preponderantly abdominal localization of body fat. As a group, they had slight to moderate metabolic changes known to be associated with abdominal/visceral obesity, with serum IGF-I concentrations in the low normal range and moderate insulin resistance as judged from GDR values obtained during a euglycaemic hyperinsulinaemic glucose clamp, although none of them had overt diabetes. Nine months of GH treatment in these middle-aged men with abdominal/visceral obesity reduced their total body fat and resulted in a specific and marked decrease in both abdominal subcutaneous and visceral adipose tissue (Fig. 18.1). Moreover, insulin sensitivity improved (Fig. 18.2) and serum concentrations of total cholesterol and triglyceride decreased (Table 18.1). Diastolic blood pressure decreased (Fig. 18.2), while plasma fibrinogen increased slightly (Table 18.1).

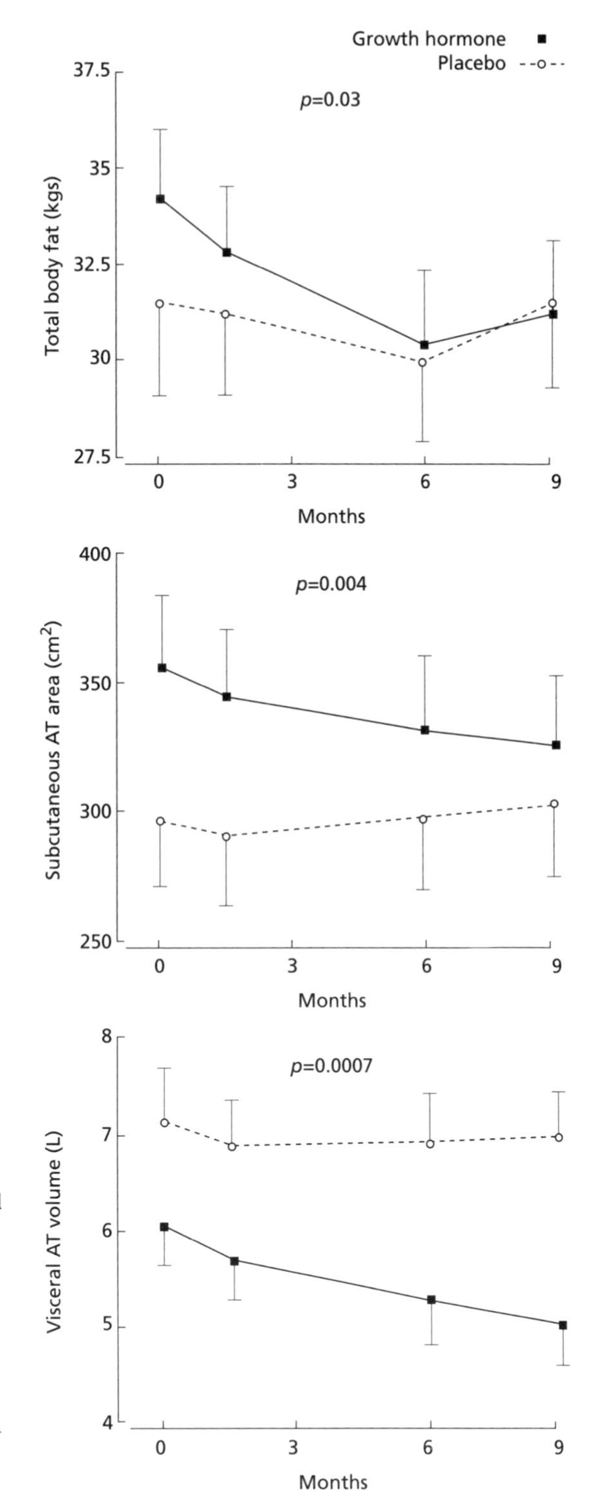

Fig. 18.1 Mean total body fat calculated from total body potassium, abdominal subcutaneous adipose tissue (AT) area at the level of L4–L5 and total volume of visceral AT, assessed with computed tomography during nine months of treatment with recombinant human growth hormone (rhGH) or placebo in 30 men with abdominal/visceral obesity. The horizontal bars indicate the standard error for the mean values shown, and *P* values denote the differences between the two groups using two-way ANOVA for repeated measurements. Reproduced with permission from Johannsson *et al.* [48].

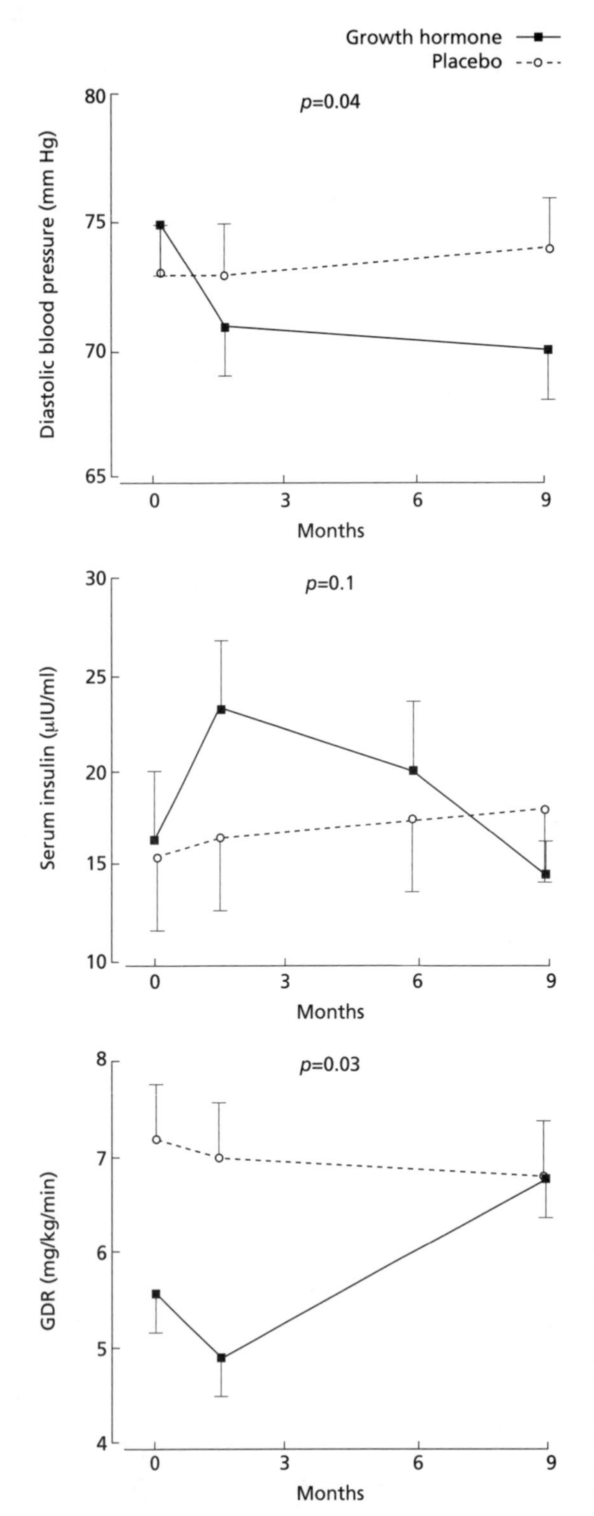

Fig. 18.2 Diastolic blood pressure, serum insulin and glucose disposal rate (GDR) assessed with a euglycaemic hyperinsulinaemic glucose clamp during nine months of treatment with recombinant human growth hormone (rhGH) or placebo in 30 men with abdominal/visceral obesity. The horizontal bars indicate the standard error for the mean values shown, and *p* values denote the differences between the two groups using two-way ANOVA for repeated measurements. Reproduced with permission from Johannsson *et al.* [48].

Table 18.1 Measurements of body composition, fasting blood glucose, glycosylated haemoglobin (HbA1c), total cholesterol (TC), triglycerides, free fatty acids (FFAs) and abdominal subcutaneous lipoprotein lipase (LPL) activity expressed as mU per gram of adipose tissue (AT) and plasma fibrinogen in 30 men, during nine months of growth hormone (GH) or placebo treatment.

Variable	Baseline	6 weeks	6 months	9 months	p^*
BMI (kg/m^2)					
GH	31.4 ± 0.7	31.6 ± 0.7	31.3 ± 0.7	31.1 ± 0.8	0.2
Placebo	30.5 ± 0.8	30.6 ± 0.7	30.5 ± 0.8	30.7 ± 0.8	
Fat free mass (kg)					
GH	67.5 ± 1.8	69.7 ± 1.9	71.1 ± 1.8	69.5 ± 2.2	0.4
Placebo	64.6 ± 1.4	65.2 ± 1.4	66.4 ± 1.3	65.3 ± 1.4	
Waist (cm)					
GH	111.8 ± 1.8	110.8 ± 1.8	107.6 ± 1.7	109.8 ± 1.9	0.002
Placebo	109.5 ± 2.5	109.4 ± 2.4	109.3 ± 2.3	111.0 ± 2.3	
Visceral AT (cm^2)					
GH	126 ± 15	121 ± 19	98 ± 14	99 ± 15	0.004
Placebo	163 ± 16	147 ± 13	142 ± 12	150 ± 13	
Blood glucose (mmol/L)					
GH	5.6 ± 0.1	5.9 ± 0.1	5.4 ± 0.3	5.6 ± 0.2	0.2
Placebo	5.5 ± 0.2	5.5 ± 0.2	4.9 ± 0.1	5.5 ± 0.2	
HbA1c (%)					
GH	5.2 ± 0.1	5.2 ± 0.1	5.4 ± 0.1	5.3 ± 0.1	0.8
Placebo	4.9 ± 0.1	4.9 ± 0.1	5.0 ± 0.1	5.1 ± 0.2	
TC (mmol/L)					
GH	6.1 ± 0.2	5.7 ± 0.2	6.1 ± 0.2	5.4 ± 0.3	0.006
Placebo	5.4 ± 0.3	5.6 ± 0.2	6.1 ± 0.3	5.5 ± 0.2	
Triglycerides (mmol/L)					
GH	2.09 ± 0.29	2.60 ± 0.42	2.31 ± 0.23	1.78 ± 0.23	0.02
Placebo	1.65 ± 0.13	1.80 ± 0.20	1.85 ± 0.22	2.05 ± 0.26	
FFAs (μmol/L)					
GH	0.77 ± 0.07	0.99 ± 0.11	0.78 ± 0.11	0.75 ± 0.07	0.5
Placebo	0.73 ± 0.05	0.73 ± 0.05	0.59 ± 0.06	0.67 ± 0.05	
LPL activity (mU/g AT)					
GH	287 ± 26	205 ± 15	N.D.	371 ± 65	0.2
Placebo	224 ± 31	228 ± 29	N.D.	263 ± 25	
Fibrinogen					
GH	3.13 ± 0.13	3.44 ± 0.10	3.49 ± 0.13	3.24 ± 0.19	0.04
Placebo	3.03 ± 0.11	2.73 ± 0.09	3.39 ± 0.28	2.84 ± 0.10	

N.D.: no data.

Although the daily GH dose used was lower than previously reported in trials involving healthy adults, the initial GH dose administered was apparently too high, as judged by the frequency of side-effects (eight patients in the GH-treated group and three in the placebo-treated group experienced clinical symptoms associated with fluid retention) and the initial high response in serum IGF-I concentrations [48]. Dose reduction resulted in serum IGF-I concentrations within the normal range, indicating that the GH dose during the latter part of the study was more physiological.

GH exerts direct insulin-antagonistic effects even when administered at physiological doses. GH has been considered to be the principal factor in the decrease in insulin sensitivity observed in the early morning, the so-called 'dawn phenomenon', and in the insulin resistance following hypoglycaemia. Thus, our observation of increased insulin sensitivity during prolonged GH treatment is unexpected, although not inexplicable. This improvement could be explained by the decrease in visceral adipose tissue mass induced by GH, followed by a decrease in FFA exposure to the liver counteracting the insulin-antagonistic effects of GH. Alternatively, as the major site of glucose disposal is in the skeletal muscle, the improvement in GDR in response to the more prolonged GH treatment might also be consequent on increased glucose transport in the skeletal muscle. This might be mediated through the IGF-I receptor and/or be an effect of an increased proportion of insulin-sensitive type I muscle fibres in response to the treatment.

The reduction in total cholesterol is conceivably an effect of enhanced hepatic LDL receptor activity in response to GH [23]. In healthy adults, short-term GH administration has been reported to increase serum triglyceride concentrations [49]. In this study, the serum triglyceride concentration also displayed an initial increase in response to GH treatment. This could be an effect of both an increased flux of FFA to the liver and a direct stimulatory effect of GH on the esterification of oleic acid into triglyceride and phospholipids in hepatocytes, which in turn enhances VLDL production from the liver [20]. However, after nine months of GH treatment, the serum triglyceride concentration had decreased again, probably as an effect of increased insulin-stimulated glucose uptake. GH treatment reduced diastolic blood pressure without affecting systolic blood pressure. This is in line with results from GH-deficient adults in whom GH administration decreased diastolic blood pressure, as an effect of reduced peripheral vascular resistance [20]. The mechanisms behind the reduction in peripheral vascular resistance might be indirect, through the reduced abdominal obesity and increased insulin sensitivity [5], or more direct, through the action on the vascular wall, with increased generation of NO [31].

This is the first trial clearly to demonstrate favourable effects of GH on the multiple disturbances associated with abdominal/visceral obesity. We therefore suggest that blunted GH secretion could be an important factor in the development of the metabolic and circulatory consequences of abdominal/visceral obesity.

GH treatment as compared with placebo in middle-aged men with abdominal obesity resulted in a reduction in serum leptin levels and an increase in basal metabolic rate (BMR) after six weeks of treatment, while after nine months of treatment, values were not different from baseline, although the GH treatment accomplished a sustained reduction in total body fat mass. No change in leptin gene expression from abdominal subcutaneous fat was noted in response to the GH treatment [50]. In studies in which a decrease in body fat was achieved with low-calorie diets, serum leptin concentrations and energy expenditure decreased [51]. It could therefore be speculated that long-term treatment with GH in men with abdominal obesity counteracts a decrease in serum leptin concentrations and BMR during body fat reduction, and therefore produces a new energy steady state through interactions between energy expenditure and serum leptin concentrations.

Is growth hormone treatment indicated in the metabolic syndrome?

The abnormal activity of the HPA axis, low levels of sex steroids and attenuated GH secretion in abdominal obesity suggest central neuroendocrine dysregulation. Whether this is of primary importance for the evolution of abdominal obesity or is a pathogenic co-factor remains to be elucidated. The finding that replacement therapy with testosterone and GH in men with abdominal obesity is able to diminish the negative metabolic consequences of the visceral obesity suggests that the low levels of these hormones are of importance for the metabolic aberrations associated with visceral/abdominal obesity.

Daily subcutaneous administration of recombinant human GH is cumbersome. The long-term compliance with such a treatment is expected to be low, and the cost is too high to make it feasible as a routine treatment in individuals with the metabolic syndrome. However, patients with severe forms of the syndrome are often treated with a combination of drugs in order to reduce blood pressure and blood lipids, and dietary interventions and exercise programs are prescribed. The total cost of these treatments together is not insignificant. Another reason for not using endocrine interventions as primary treatment for abdominal obesity is the uncertainty of the causal relationship between the neuroendocrine aberrations, including low GH levels, and the metabolic consequences of abdominal obesity. Until such an association has been established, administration of exogenous GH for the treatment of the metabolic syndrome will be questioned. Methods of increasing endogenous GH secretion would probably be more feasible.

Exercise and weight loss are two well-known methods of increasing endogenous secretion. Exercise does not, however, increase GH secretion to the same extent in older individuals as in young adults [52]. In addition, the GH response to chronic intermittent exercise is not as well demonstrated as the GH response to acute exercise. The administration of a GH secretagogue might increase endogenous GH secretion and possibly achieve results similar to those found with exogenous GH administration. The newly discovered GH-releasing peptides are substances with such potential. In a study of MK-677, an orally active non-peptidyl GH secretagogue, in obese but otherwise healthy men, an increase in GH secretion over a period of eight weeks was observed [53]. This was not, however, associated with a reduction in total body fat mass or visceral adiposity. This discrepancy with the study using exogenous GH administration [48] might be explained by an inadequate increment of serum IGF-I concentrations in response to MK-677, or by interactions with the neuropeptide Y-containing cells in the hypothalamus, or by altered appetite induced by the GH secretagogue.

Aerobic exercise training and weight loss affect the metabolic consequences of obesity in a different manner. Thus, aerobic exercise can improve insulin sensitivity without affecting abdominal adiposity and glucose tolerance, while weight loss reduces body fat mass and abdominal adiposity and improves glucose tolerance without affecting the insulin resistance in obese men [54]. The combination of both weight loss and exercise is therefore the preferable method of reducing visceral adiposity and improving the metabolic profile. These interventions may also increase the endogenous GH secretion, which may in turn be of importance for the correction of a hazardous metabolic profile in those with abdominal obesity.

Appropriate treatment for patients with abdominal obesity at present includes recommendations for increased physical activity, dietary adjustments, weight loss, and appropriate treatment of hypertension, hyperlipidaemia and glucose intolerance. These interventions have been shown to prevent cardiovascular morbidity and mortality in prospective trials. The endocrine disturbances are certainly of importance for the metabolic consequences of abdominal obesity. Further information about the association between metabolic and neuroendocrine aberrations in abdominal obesity is needed before we can determine the place of hormonal treatment in the metabolic syndrome.

References

* 1 Björntorp P. Visceral obesity: a 'civilization syndrome'. *Obes Res* 1993; **1**: 206–22.

** 2 Bouchard C, Després JP, Mauriège P. Genetic and nongenetic determinants of regional fat distribution. *Endocr Rev* 1993; **14**: 72–93.

* 3 Kissebah AH, Krakower GR. Regional adiposity and morbidity. *Physiol Rev* 1994; **74**: 761–811.

4 Herberg L, Bergmann M, Hennigs U, Major E, Gries FA. Influence of diet on the metabolic syndrome of obesity. *Isr J Med Sci* 1972; **8**: 822–3.

* 5 Reaven GH. Role of insulin resistance in human disease. *Diabetes* 1988; **37**: 1595–607.

6 Vague J. La differenciation sexuelle, facteur déterminant des formes de l'obésité. *Presse Méd* 1947; **55**: 339–41.

* 7 Björntorp B. 'Portal' adipose tissue as a generator of risk factors for cardiovascular disease and diabetes. *Arteriosclerosis* 1990; **10**: 493–6.

8 Mårin P, Holmäng S, Jönsson L *et al.* The effects of testosterone treatment on body composition and metabolism in middle-aged obese men. *Int J Obesity*

1992; **16**: 991–7.

* 9 Rosmond R, Dallman MF, Björntorp P. Stress-related cortisol secretion in men: relationships with abdominal obesity and endocrine, metabolic and hemodynamic abnormalities. *J Clin Endocrinol Metab* 1998; **83**: 1853–9.

10 Rosmond R, Chagnon YC, Holm G *et al.* A Bd I restriction fragment length polymorphism of the glucocorticoid receptor gene locus is associated with dysregulation of the hypothalamic–pituitary–adrenal axis. 1998.

* 11 Chrousos G, Gold P. The concept of stress and stress system disorders. *JAMA* 1992; **267**: 1244–52.

12 Bujalska IJ, Kumar S, Stuvart PM. Does central obesity reflect 'Cushing's disease of the omentum'? *Lancet* 1997; **349**: 1210–13.

13 Weaver JV, Taylor NF, Monson JP *et al.* Sexual dimorphism in 11β-hydroxysteroid dehyrogenase activity and its relation to fat distribution and insulin sensitivity; a study in hypopituitary subjects. *Clin Endocrinol* 1998; **49**: 13–20.

14 Rosmond R, Björntorp P. The interactions between hypothalamic–pituitary–adrenal axis activity, testosterone, insulin-like growth factor I and abdominal obesity with metabolism and blood pressure in men. *Int J Obes* 1998; **22**: 1184–96.

15 Gelding SV, Taylor NF, Wood PJ *et al.* The effect of growth hormone replacement therapy on cortisol-cortisone interconversion in hypopituitary adults: evidence for growth hormone modulation of extrarenal 11β-hydroxysteroid dehydrogenase activity. *Clin Endocrinol* 1998; **48**: 153–62.

16 Bengtsson BÅ, Brummer R, Edén S, Bosaeus I, Lindstedt G. Body composition in acromegaly: the effect of treatment. *Clin Endocrinol* 1989; **31**: 481–90.

** 17 Bengtsson BÅ, Edén S, Lönn L *et al.* Treatment of adults with growth hormone (GH) deficiency with recombinant human GH. *J Clin Endocrinol Metab* 1993; **76**: 309–17.

18 Bengtsson BÅ. The consequences of growth hormone deficiency in adults. *Acta Endocrinol* 1993; **128** (Suppl 2): 2–5.

19 Rosén T, Bengtsson BÅ. Premature mortality due to cardiovascular diseases in hypopituitarism. *Lancet* 1990; **336**: 285–8.

20 Bengtsson BÅ, Rosén T, Johansson JO *et al.* Cardiovascular risk factors in adults with growth hormone deficiency. *Endocrinol Metab* 1995; **2** (Suppl B): 29–35.

21 Kissebah AH, Peiris AN, Evans DJ. Mechanisms associating body fat distribution to glucose tolerance and diabetes mellitus: window with a view. *Acta Med Scand* 1988; **723**: 79–89.

** 22 Peiris AN, Struve MF, Mueller RA, Lee MB, Kissebah AH. Glucose metabolism in obesity: influence of body fat distribution. *J Clin Endocrinol Metab* 1988; **67**: 760–7.

** 23 Rudling M, Norstedt G, Olivecrona H *et al.* Importance of growth hormone for the induction of hepatic low density lipoprotein receptors. *Proc Natl Acad Sci USA* 1992; **89**: 6983–7.

24 Oscarsson J, Ottosson M, Johansson JO *et al.* Two weeks of daily injections and continuous infusion of recombinant human growth hormone (GH) in GH-deficient adults, 2: effects on serum lipoproteins and lipoprotein and hepatic lipase activity. *Metabolism* 1996; **45**: 370–7.

* 25 Reaven GM. Pathophysiology of insulin resistance in human disease. *Physiol Rev* 1995; **75**: 473–86.

26 Hamsten A, Wiman B, de Faire U, Blombäck M. Increased plasma levels of a rapid inhibitor of tissue plasminogen activator in young survivors of myocardial infarction. *N Engl J Med* 1985; **313**: 1557–63.

27 Wilhelmsen L, Svärdsudd K, Korsan-Bengtsen K *et al.* Fibrinogen as a risk factor for stroke and myocardial infarction. *N Engl J Med* 1984; **311**: 501–5.

28 Rosmond R, Björntorp P. Blood pressure in relation to obesity, insulin and the hypothalamic–pituitary–adrenal axis in Swedish men. *J Hypertens* 1998; **16**: 1721–6.

29 Julius S, Esler MD, Randall OS. Role of autonomic nervous system in mild human hypertension. *Clin Sci Mol Med* 1975; **48**: 243–52.

* 30 Sverrisdóttir YB, Elam M, Bengtsson BÅ, Johannsson G. Intense sympathetic nerve activity in adults with hypopituitarism and untreated growth hormone

deficiency. *J Clin Endocrinol Metab* 1998; **83**: 1881–5.

* 31 Böger RH, Skamira C, Bode-Böger SM *et al.* Nitric oxide may mediate the hemodynamic effects of recombinant growth hormone in patients with acquired growth hormone deficiency. *J Clin Invest* 1996; **98**: 2706–13.

32 Ljung T, Andersson B, Björntorp P, Mårin P. Inhibition of cortisol secretion by dexamethasone in relation to body fat distribution: a dose–respone study. *Obes Res* 1996; **4**: 277–82.

33 Liu L, Merriam GR, Sherins RJ. Chronic sex steroid exposure increases mean plasma growth hormone concentration and pulse amplitude in men with isolated hypogonadotropic hypogonadism. *J Clin Endocrinol Metab* 1987; **64**: 651–6.

* 34 Yang S, Xu X, Björntorp P, Edén S. Additive effects of growth hormone and testosterone on lipolysis in adipocytes of hypophysectomized rats. *J Endocrinol* 1995; **147**: 147–52.

* 35 Veldhuis JD, Liem AY, South S *et al.* Differential impact of age, sex steroid hormones, and obesity on basal versus pulsatile growth hormone secretion in men as assessed in an ultrasensitive chemiluminescence assay. *J Clin Endocrinol Metab* 1995; **80**: 3209–22.

* 36 Veldhuis JD, Iranmanesh A, Ho KKY *et al.* Dual effects in pulsatile growth hormone secretion and clearance subserve the hyposomatotropism of obesity in man. *J Clin Endocrinol Metab* 1991; **72**: 51–9.

37 Hartman ML, Clayton PE, Johnson ML *et al.* A low dose euglycemic infusion of recombinant human insulin-like growth factor I rapidly suppresses fasting-enhanced pulsatile growth hormone secretion in humans. *J Clin Invest* 1993; **91**: 2453–62.

38 Mårin P, Kvist H, Lindstedt G, Sjöström L, Björntorp P. Low concentrations of insulin-like growth factor-I in abdominal obesity. *Int J Obesity* 1993; **17**: 83–9.

39 Frystyk J, Vestbo E, Sklærbæk C *et al.* Free insulin-like growth factors in human obesity. *Metabolism* 1995; **44**: 37–44.

40 Iranmanesh A, Lizarralde G, Veldhuis JD. Age and relative adiposity are specific negative determinants of the frequency and amplitude of growth hormone (GH) secretory bursts and the half-life of endogenous GH in healthy men. *J Clin Endocrinol Metab* 1991; **73**: 1081–8.

** 41 Vahl N, Jørgensen JOL, Skjærbæk C *et al.* Abdominal adiposity rather than age and sex predicts mass and regularity of GH secretion in healthy adults. *Am J Physiol* 1997; **272**: E1108–16.

* 42 Rasmussen MH, Hvidberg A, Juul A *et al.* Massive weight loss restores 24-hour growth hormone release profiles and serum insulin-like growth factor-I levels in obese subjects. *J Clin Endocrinol Metab* 1995; **80**: 1407–15.

43 Jung RT, Campbell RG, James WPT, Callingham BA. Altered hypothalamic and sympathetic response to hypoglycaemia in familial obesity. *Lancet* 1982; i: 1043–6.

44 Kopelman PG, Pilkington TRE, White N, Jeffcoate SL. Evidence for existence of two types of massive obesity. *Br Med J* 1980; **281**: 82–3.

45 Clemmons DR, Snyder DK, Williams R, Underwood LE. Growth hormone administration conserves lean body mass during dietary restriction in obese subjects. *J Clin Endocrinol Metab* 1987; **64**: 878–83.

46 Snyder DK, Clemmons DR, Underwood LE. Treatment of obese, diet-restricted subjects with growth hormone for 11 weeks: effects on anabolism, lipolysis and body composition. *J Clin Endocrinol Metab* 1988; **67**: 54–61.

47 Thompson JL, Butterfield GE, Yesavage J *et al.* Effects of human growth hormone, insulin-like growth factor I, and diet and exercise on body composition of obese postmenopausal women. *J Clin Endocrinol Metab* 1998; **83**: 1477–84.

** 48 Johannsson G, Mårin P, Lönn L *et al.* Growth hormone treatment of abdominally obese men reduces abdominal fat mass, improves glucose and lipoprotein metabolism, and reduces diastolic blood pressure. *J Clin Endocrinol Metab* 1997; **82**: 727–34.

49 Marcus R, Butterfield G, Holloway L *et al.* Effects of short-term administration of recombinant human growth hormone to elderly people. *J Clin Endocrinol Metab* 1990; **70**: 519–27.

* 50 Karlsson C, Stenlöf K, Johannsson G *et al.* Effects of growth hormone treatment

on the leptin system: interactions with body composition and energy expenditure. *Eur J Endocrinol* 1998; **138**: 408–14.

51 Havel PJ, Kasim-Karakas S, Mueller W, Johnson PR, Gingerich RL. Relationship of plasma leptin to plasma insulin and adiposity in normal weight and overweight women: effects of dietary fat content and sustained weight loss. *J Clin Endocrinol Metab* 1996; **81**: 4406–13.

52 Pyka G, Wiswell RA, Marcus R. Age-dependent effect of resistance exercise on growth hormone secretion in people. *J Clin Endocrinol Metab* 1992; **75**: 404–7.

53 Svensson J, Lönn L, Jansson JO *et al.* Two-month treatment of obese subjects with the oral growth hormone (GH) secretagogue MK-677 increases GH secretion, fat-free mass, and energy expenditure. *J Clin Endocrinol Metab* 1998; **83**: 362–9.

54 Dengel DR, Pratley RE, Hagberg JM, Rogus EM, Goldberg AP. Distinct effects of aerobic exercise training and weight loss on glucose homeostasis in obese sedentary men. *J Appl Physiol* 1996; **81**: 318–25.

19: Does GH therapy have a role in the management of cardiac failure?

Jörgen Isgaard

Introduction

Recent experimental and clinical studies have suggested an important role for the growth hormone–insulin-like growth factor-I (GH–IGF-I) axis in the regulation of cardiac growth and function. Cardiac function in adult patients with GH deficiency has been shown to be compromised [1–3], and a recent study showing beneficial effects of GH treatment of patients with dilated cardiomyopathy has also led to the suggestion that GH may have a role in the future in the treatment of congestive heart failure (CHF) [4]. However, a recently published placebo-controlled study was not able to confirm beneficial effects of GH on haemodynamics in patients with idiopathic dilated cardiomyopathy [5]. Topics that will be reviewed and discussed in this chapter include observations that components of the GH–IGF-I axis are probably important for the development of cardiac hypertrophy in response to increased haemodynamic load. Effects of GH on intracellular calcium handling may facilitate the improvement of contractility observed in various experimental models and clinical studies. Moreover, a decrease in peripheral resistance and possible cardioprotective effects of GH and IGF-I may also be beneficial in situations with compromised heart function. However, potential adverse effects of GH treatment such as hyperglycemia, insulin resistance, hypertension and fluid retention must also be taken into account if GH is to be considered as a therapeutic agent in heart failure.

GH–IGF-I axis and cardiac growth

IGF-I

Several studies have demonstrated an increased left ventricle (LV) expression of IGF-I in different rat models of hypertension, including suprarenal aortic constriction, uninephrectomized spontaneously hypertensive rats, uninephrectomized, deoxycorticosterone-treated, saline-fed (DOCA salt) rats

[6] and in 2-kidney, 1-clip (2K1C) rats [7,8]. In the former two studies, IGF-I mRNA induction was shown to be accompanied by increased IGF-I protein immunoreactivity, which was most abundant in the inner layers of the left ventricle [7], where both tension and wall stress are high, with a gradual decline towards the epicardial surface [9]. No elevation of IGF-I was observed in the right ventricle (RV). The onset of IGF-I expression preceded or occurred in parallel with development of LV hypertrophy [7,8]. Taken together, these results indicate that hypertension and increased LV pressure load constitute a mechanical stimulus for a local growth response in the challenged ventricle, where locally synthesized IGF-I may play a role through autocrine–paracrine mechanisms. Results supporting the hypothesis that an increased wall stress may trigger a local induction of IGF-I expression have also been reported in experimental models in which the RV is predominantly challenged either through a surgically induced fistula between the abdominal aorta and the caval vein (ACF) [10] or through an induction of pulmonary hypertension [11].

IGF-I receptor (IGF-I-R) and binding proteins

Although data regarding up-regulation of IGF-I during cardiac hypertrophy in different rat models of pressure and volume overload are reasonably consistent, studies regarding the regulation of the IGF-R are more conflicting. During experimental 2K1C hypertension, a rapid and transient induction of IGF-I-R was found as early as two days after surgery [8]. In contrast, levels of IGF-I-R mRNA were reported to be stable over time in alternative rat models of experimental hypertension [6]. It should be noted, however, that in the latter study, the first measurements of IGF-I-R mRNA expression took place as late as two weeks after induction of hypertension. It was recently reported that both IGF-I-R and IGF-I expression, measured by reverse transcriptase polymerase chain reaction, were increased after two days in viable myocytes in rats subjected to experimentally induced myocardial infarction [12]. Hence, it appears that a rapid induction of IGF-I-R can occur during increased haemodynamic load and/or a regenerative process in the myocardium. It is reasonable to speculate that an initial elevation of IGF-I-R expression would enhance the response to locally available IGF-I and contribute to a hypertrophic response. However, whether the up-regulation of IGF-I-R gene expression in these conditions results in an increased number of functional receptors remains to be established.

So far, six binding proteins for IGF-I and IGF-II (IGFBP) have been isolated and characterized [13]. There is sparse information regarding the physiological role of IGFBPs in cardiac growth and development, or whether they are actually synthesized within cardiac tissue. It was recently demonstrated in an experimental model using urethral ligation and subsequent induction of IGF-I mRNA expression in the bladder, consistent with haemodynamic regulation of IGF-I in the heart, that gene expression of IGFBP-2 and IGFBP-4 was induced [14]. In pigs subjected to microembolization of coronary vessels, up-regulation of IGFBP-3 mRNA preceded elevated levels of IGFBP-5 mRNA in ischaemic tissue of the heart after injury [15]. Gene expression of IGFBP-6 was up-regulated at all time points, although levels of IGFBP-2 and IGFBP-4 mRNA were unchanged in ischaemic heart tissue in this experimental model [15].

GH receptor

The GH-receptor (GH-R) gene is expressed in the rat [16] and rabbit [17] heart at low levels around birth, and then rises to adult levels in the following months. Although basal levels of GH-R are lower in the heart than in the liver, GH-R expression in the heart appears to be relatively high compared to that in other tissues [16].

GH-R gene expression was shown to be elevated in the acutely overloaded RV of ACF-operated rats [10] and in the LV of 2K1C rats [8], suggesting a role for GH-R in compensatory growth of a haemodynamically challenged heart. However, since induction of GH-R expression preceded IGF-I mRNA in the volume overload model [10] but not in the 2K1C rats [8], it is unclear whether increased GH-R expression is definitely required for the upregulation of IGF-I message. Alternatively, it may be speculated that an increased expression of GH-R sustains the local synthesis of IGF-I in the heart. In summary, up-regulation of both IGF-I and GH-R expression is linked to increased haemodynamic load, and suggests a potential role for the GH–IGF-I axis during development of compensatory cardiac hypertrophy.

The hearts of patients with idiopathic dilated cardiomyopathy are characterized by ventricular dilation, thin ventricular walls and increased wall stress. A possible increase of ventricular mass with a subsequent reduction of wall stress by GH treatment in these patients would be beneficial, and this was observed in the study by Fazio and co-workers [4].

Cardiac effects of GH and IGF-I

An increasing amount of evidence indicates the importance of GH for normal cardiac function and structure. GH-deficient adults receiving GH substitution therapy showed improvement of systolic function and normalization of LV mass [2,3]. A number of experimental studies, both in normal rats [18] and in rats with surgically induced myocardial infarction [19–21], show that GH and IGF-I, alone or in combination, improve cardiac function. Moreover,

GH and IGF-I were still effective in improving cardiac function when added as an adjunct treatment to post-infarction rats receiving angiotensin-converting enzyme (ACE) inhibition [22]. However, an absence of positive effects of GH on impaired cardiac function after experimental myocardial infarction [23] and ventricular pacing in dogs [24] has also been reported. Interestingly, it was recently demonstrated that GH, apart from improving cardiac function, also attenuates left ventricular remodelling in rats after myocardial infarction, apparently mostly due to reduction of LV dilatation and additional hypertrophy of the non-infarcted myocardium [25]. Until recently, studies regarding GH treatment in heart failure were limited to case reports [26–28] in which GH administration dramatically improved cardiac function. In two of these reports [26,27], the patients were GH-deficient. In a small open study of seven patients with idiopathic dilated cardiomyopathy and CHF without GH deficiency who received GH treatment for three months, considerable improvement of cardiac function was reported [4]. More recent studies have demonstrated beneficial effects in patients with CHF due to both ischaemic and idiopathic dilated cardiomyopathy, with improvements in haemodynamics when GH was added both as a maintenance therapy and as a short-term infusion [29,30]. However, concern has also been raised regarding increased levels of circulating IGF-I [31], which may contribute to a subacromegalic condition with increased risk for hypertension, hyperinsulinemia, insulin resistance and hyperlipidemia. A possible risk for arrhythmias during prolonged GH treatment has also been pointed out [32]. Hence, careful dose titration of GH and safety monitoring of patients with congestive heart failure appears to be mandatory in future clinical studies. So far, two placebo-controlled studies with GH as adjunct therapy in patients with CHF have been reported, although neither study was able to confirm previously reported improvements in systolic function and lowering of wall stress. In a study by Osterziel and co-workers [5], 50 patients with idiopathic dilated cardiomyopathy were treated with recombinant human growth hormone (rhGH) for three months. A significant increase in LV mass was reported, which correlated with changes in serum IGF-I concentrations (Fig. 19.1), although systolic function and wall stress were not affected by GH treatment. In another placebo-controlled three-month study, rhGH treatment of patients with CHF of various aetiologies was reported to be safe and well tolerated, without serious adverse effects, although no significant improvement in cardiac function was seen [33]. Possible explanations for the absence of cardiovascular effects in these placebo-controlled trials may be that both studies were small and that the treatment time may have been too short. Moreover, the patients in these placebo-controlled studies were on a more optimal conventional therapy for CHF, including higher doses of ACE inhibitors, than in the study previously described

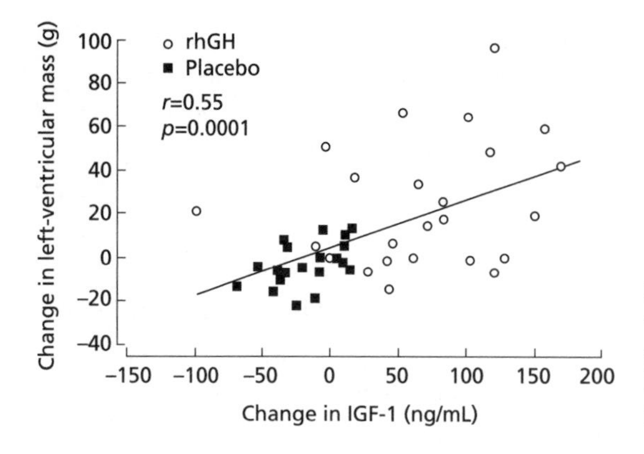

Fig. 19.1 Relation between changes in left ventricular mass and changes in serum insulin-like growth factor-I (IGF-I) concentrations. Reproduced with permission from Osterziel *et al.* [5].

by Fazio and co-workers [4]. A comparison of some of the aspects of study design and baseline characteristics is shown in Table 19.1.

The precise mechanisms of action of GH and IGF-I on the heart are less clear, and there are few *in vitro* studies addressing GH–IGF-I effects on cardiomyocytes. In papillary muscle from rats exposed to very high GH plasma levels due to a GH-secreting tumour, it was found that the maximum Ca^{2+}-activated force per cross-sectional area was increased, suggesting a GH-induced elevated Ca^{2+} responsiveness of the myofilaments [34]. In a model with

Table 19.1 A comparison of some aspects of study design and baseline characteristics of patients in recent trials with growth hormone (GH) added to treatment for congestive heart failure.

	Fazio *et al.* [4]	Osterziel *et al.* [5]	Isgaard *et al.* [33]
Placebo controlled	No	Yes	Yes
Duration of treatment	3 months	3 months	3 months
GH dose	4 IU every second day	2 IU daily	2 IU daily
Aetiology of CHF	IDCM	IDCM	IDCM ($n = 13$), IHD ($n = 8$), other ($n = 1$)
Number of patients	7	50	22
Mean age (years)	46 ± 9	54 ± 10	60 ± 2
Ejection fraction at baseline	34	26	29
NYHA function class at baseline	2.7	3.2	3.0
Number of patients on ACE inhibitors	7	48	20
Mean dose of ACE inhibitors as % of target dose	25	85	76

ACE: angiotensin-converting enzyme; CHF: congestive heart failure; IDCM: idiopathic dilated cardiomyopathy; IHD: ischaemic heart disease; NYHA: New York Heart Association.

isolated buffer-perfused hearts from rats treated with high doses of GH or IGF-I for four weeks, systolic function was improved, and the findings supported the notion of an increased maximal response to Ca^{2+} [35]. However, whether the effects seen are due to stimulation of cardiomyocytes rather than interstitial cells remains to be investigated. In addition to direct effects on the heart, a lowering of arterial blood pressure peripheral resistance by GH may also be beneficial for haemodynamics [3,22]. It may be speculated that this effect is more rapid, and could be an important factor in studies demonstrating acute cardiovascular effects of GH [29].

In vitro studies, with addition of IGF-I to cultured neonatal rat cardiomyocytes, have demonstrated changes associated with hypertrophy, including increments in cell size, protein synthesis and induced expression of myosin light chain-2 and troponin I [36]. In adult rat cardiomyocytes, it has been reported that IGF-I stimulates myofibril development [37] and increases isometric force and free cytosolic Ca^{2+} [38].

Apoptosis, or programmed cell death, has been reported to occur in the myocardium of patients both after ischaemic injury [39] and in conditions with dilated cardiomyopathy [40]. Some of the genes involved in the apoptotic process have been identified, including the Fas receptor, which has been shown to mediate signals for apoptosis [41,42]. Up-regulation of the Fas receptor has been shown to be associated with apoptosis after myocardial infarction in rats [43]. It has been reported that administration of IGF-I and GH prior to cardiac ischaemia attenuates the increased rate of myocyte apoptosis [44,45].

Summary

There is ample evidence to support a role for the GH–IGF-I axis in the regulation of cardiac growth, structure and function. GH may act directly on the heart, or through circulating IGF-I. Moreover, GH has been found to regulate local production of IGF-I in the heart. Both the GH receptor and IGF-I receptor are expressed in cardiac tissue. Hence, the IGF-I receptor can theoretically be activated through locally produced IGF-I acting via autocrine–paracrine mechanisms, or via circulating IGF-I exerting its effects as an endocrine agent. During conditions of pressure and volume overload, increased systolic wall stress triggers an induction of gene expression of IGF-I, GH-R and possibly IGF-I-R, implying a potential role for the GH–IGF-I axis in the development of adaptive hypertrophy of the heart and vessels. Cardiovascular effects of GH in clinical studies include beneficial effects on contractility, exercise performance and a decrease in peripheral resistance, and these effects, as well as possible adverse effects, are summarized in Table 19.2. Experimental studies suggest an increased Ca^{2+} responsiveness as one possible

Possible beneficial effects	Possible adverse effects
Increased left ventricular mass	Fluid retention
Decreased peripheral resistance	Hypertension
Increased contractility	Hyperglycaemia
Cardioprotective effects	Hyperinsulinaemia
	Cardiac hypertrophy
	Arrhythmias

Table 19.2 Summary of possible beneficial and adverse effects of growth hormone on cardiovascular parameters.

underlying cause for the improvement in contractility, although effects of GH and IGF-I on apoptosis may possibly also play a more long-term role for cardiomyocyte survival. Adverse effects of GH treatment, such as hyperglycaemia, insulin resistance, hypertension and fluid retention, must also be taken into consideration if GH is to be considered as a potential therapeutic agent in the treatment of heart failure. However, these effects are probably at least partially dose-dependent, and studies on heart failure patients treated with GH require careful monitoring and dose titration. In summary, it is clear that further basic and clinical studies are required to gain insight into the GH and IGF-I mechanisms of action and to monitor long-term effects when GH is administered as substitution therapy or as an agent in the treatment of congestive heart failure.

References

* 1 Merola B, Cittadini A, Colao A, *et al.* Cardiac structure and functional abnormalities in adult patients with growth hormone deficiency. *J Clin Endocrinol Metab* 1993; 77: 1658–61.

* 2 Amato G, Carella C, Fazio S *et al.* Body composition, bone metabolism and heart structure and function in growth hormone (GH)-deficient adults before and after GH replacement therapy at low doses. *J Clin Endocrinol Metab* 1993; 77: 1671–6.

* 3 Caidahl K, Edén S, Bengtsson BÅ. Cardiovascular and renal effects of growth hormone. *Clin Endocrinol (Oxf)* 1994; **40**: 393–400.

** 4 Fazio S, Sabatini D, Capaldo B *et al.* A preliminary study of growth hormone in the treatment of dilated cardiomyopathy. *N Engl J Med* 1996; **334**: 809–14.

* 5 Osterziel KJ, Strohm O, Schuler J *et al.* Randomised, double-blind, placebo-controlled trial of human recombinant growth hormone in patients with chronic heart failure due to dilated cardiomyopathy. *Lancet* 1998; **351**: 1233–7.

* 6 Donohue TJ, Dworkin LD, Lango MN *et al.* Induction of myocardial insulin-like growth factor I gene expression in left ventricular hypertrophy. *Circulation* 1994; **89**: 799–809.

* 7 Wåhlander H, Isgaard J, Jennische E, Friberg P. Left ventricular insulin-like growth factor I increases in early renal hypertension. *Hypertension* 1992; **19**: 25–32.

* 8 Guron G, Friberg P, Wickman A *et al.* Cardiac insulin-like growth factor I and growth hormone receptor expression in renal hypertension. *Hypertension* 1996; **27**: 636–42.

* 9 Mirsky J. Elastic properties of the myocardium: a quantitative approach with physiological and clinical applications. In: Berne RM, ed, *Handbook of*

Physiology, vol. 1. American Physiological Society, Bethesda, USA, 1979: 497–531.

* 10 Isgaard J, Wåhlander H, Adams MA, Friberg P. Increased expression of growth hormone receptor mRNA and insulin-like growth factor I mRNA in volume overloaded hearts. *Hypertension* 1994; **23**: 884–8.

* 11 Russell-Jones DL, Leach RM, Ward JPT, Thomas CR. Insulin-like growth factor-I gene expression is increased in the right ventricle hypertrophy induced by chronic hypoxia in the rat. *J Endocrinol* 1993; **10**: 99–102.

* 12 Reiss K, Meggs LG, Li P *et al.* Upregulation of IGF-I IGF-I receptor and late growth related genes in ventricular myocytes acutely after infarction in rats. *J Cell Physiol* 1994; **158**: 160–8.

* 13 Delafontaine P. Insulin-like growth factor I and its binding proteins in the cardiovascular system. *Card Vasc Res* 1995; **30**: 825–34.

* 14 Chen Y, Arner A, Bornfeldt KE, Uvelius B, Arnqvist HJ. Development of smooth muscle cell hypertrophy is closely associated with increased gene expression of insulin-like growth factor binding protein-2 and -4. *Growth Regul* 1995; **5**: 45–52.

* 15 Kluge A, Zimmermann R, Weihrauch Mohri M *et al.* Coordinate expression of the insulin-like growth factor system after microembolisation in porcine heart. *Cardiovasc Res* 1997; **33**: 324–31.

* 16 Mathews LS, Engberg B, Norstedt G. Regulation of rat GH receptor gene expression. *J Biol Chem* 1989; **264**: 9905–10.

* 17 Ymer SI, Herington AC. Developmental expression of the growth hormone receptor gene in rabbit tissues. *Mol Cell Endocrinol* 1992; **83**: 39–49.

** 18 Cittadini A, Strömer H, Katz SE *et al.* Differential cardiac effects of growth hormone and insulin-like growth factor-I in the rat: a combined in vivo and in vitro evaluation. *Circulation* 1996; **93**: 800–9.

* 19 Yang R, Bunting S, Gillett N, Clark R, Jin H. Growth hormone improves cardiac performance in experimental performance. *Circulation* 1995; **92**: 262–7.

* 20 Duerr RL, Huang S, Miraliakbar HR *et al.* Insulin-like growth factor-1 enhances ventricular hypertrophy and function during the onset of experimental cardiac failure. *J Clin Invest* 1995; **95**: 619–27.

* 21 Isgaard J, Kujacic V, Jennische E *et al.* Growth hormone improves cardiac function in rats with experimental myocardial infarction. *Eur J Clin Invest* 1997; **27**: 517–25.

** 22 Jin H, Yang R, Gillett N *et al.* Beneficial effects of growth hormone and insulin-like factor-1 in experimental heart failure in rats treated with chronic ACE inhibition. *J Cardiovasc Pharmacol* 1995; **26**: 420–5.

* 23 Shen YT, Wiedmann RT, Lynch JJ, Grossman W, Johnson RG. GH replacement fails to improve ventricular function in hypophysectomized rats with myocardial infarction. *Am J Physiol* 1996; **271**: H1721–7.

* 24 Shen YT, Woltmann RF, Appleby S *et al.* Lack of beneficial effects of growth hormone treatment in conscious dogs during development of heart failure. *Am J Physiol* 1998; **274**: H456–6.

* 25 Cittadini A, Grossman J, Napoli R *et al.* Growth hormone attenuates early ventricular remodeling and improves cardiac function in rats with large myocardial infarction. *J Am Coll Cardiol* 1997; **29**: 1109–16.

* 26 Cuneo RC, Wilmshurst P, Lowy C, McGauley G, Sönksen PH. Cardiac failure responding to growth hormone. *Lancet* 1989; **i**: 838–9.

* 27 Frustaci A, Perrone GA, Gentiloni N, Russo MA. Reversible dilated cardiomyopathy due to growth hormone deficiency. *Am J Clin Pathol* 1992; **97**: 503–11.

* 28 O'Driscoll JG, Green DJ, Ireland M, Kerr D, Larbalestier RI. Treatment of end-stage cardiac failure with growth hormone. *Lancet* 1997; **349**: 1068.

* 29 Volterrani M, Desenzani P, Lorusso R *et al.* Hemodynamic effects of intravenous growth hormone in congestive heart failure. *Lancet* 1997; **349**: 1067–8.

* 30 Beer N, Tortoledo F, Beer R, Pinedo M. Beneficial effects of growth hormone in patients with congestive heart failure [abstract]. *Circulation* 1997; **96** (Suppl 1): I521, A2920.

* 31 Turner H, Wass JAH. Growth hormone in the treatment of dilated cardiomyopathy [letter]. *N Engl J Med* 1996; **335**: 672.

* 32 Frustaci A, Gentiloni N, Russo MA. Growth hormone in the treatment of dilated cardiomyopathy [letter]. *N Engl J Med* 1996; **335**: 672–3.

* 33 Isgaard J, Bergh CH, Caidahl K *et al.* A placebo-controlled study of growth hormone in patients with congestive heart failure. *Eur Heart J* 1998; **19**: 1704–11.

** 34 Mayoux E, Ventura-Clapier R, Timsit J *et al.* Mechanical properties of rat cardiac skinned fibers are altered by chronic growth hormone hypersecretion. *Circ Res* 1993; **72**: 57–64.

** 35 Strömer H, Cittadini A, Douglas PS, Morgan JP. Exogenously administered growth hormone and insulin-like growth factor-I alter intracellular Ca^{2+} handling and enhance cardiac performance: *in vitro* evaluation in the isolated isovolumic buffer-perfused rat heart. *Circ Res* 1996; **79**: 227–36.

* 36 Ito H, Hiroe M, Hirata Y *et al.* Insulin-like growth factor-I induces hypertrophy with enhanced expression of muscle-specific genes in cultured rat cardiomyocytes. *Circulation* 1993; **87**: 1715–21.

* 37 Donath MY, Zapf J, Eppenberger-Eberhardt M, Froesch ER, Eppenberger HM. Insulin-like growth factor I stimulates myofibril development and decreases smooth muscle I-actin of adult cardiomyocytes. *Proc Natl Acad Sci USA* 1994; **91**: 1686–90.

* 38 Freestone NS, Ribaric S, Mason WT. The effect of insulin-like growth factor-1 on adult rat cardiac contractility. *Mol Cell Biochem* 1996; **163/164**: 223–9.

** 39 Saraste A, Pulkki K, Kallajoki M *et al.* Apoptosis in human acute myocardial infarction. *Circulation* 1997; **95**: 320–3.

** 40 Narula J, Haider N, Virmani R *et al.* Apoptosis in myocytes in end-stage heart failure. *N Engl J Med* 1996; **335**: 1182–9.

* 41 Itoh N, Yonehara S, Ishii A *et al.* The polypeptide encoded by the cDNA for human cell surface antigen Fas can mediate apoptosis. *Cell* 1991; **66**: 233–43.

* 42 Nagata S, Golstein P. The Fas death factor. *Science* 1995; **267**: 1449–56.

* 43 Kajstura J, Cheng W, Reiss K *et al.* Apoptotic and necrotic myocyte cell deaths are independent contributing variables of infarct size in rats. *Lab Invest* 1996; **74**: 86–107.

* 44 Buerke M, Mohara T, Skurk C *et al.* Cardioprotective effect of insulin-like growth factor-I in myocardial ischemia followed by reperfusion. *Proc Natl Acad Sci (USA)* 1995; **92**: 8031–5.

* 45 Buerke M, Prufer D, Ibe W *et al.* Human growth hormone exerts cardioprotective effects in murine reperfusion injury [abstract]. Paper presented at the 70th Scientific Session of the American Heart Association, Orlando, Florida, November 1997: A2116.

Index

Please note: page numbers in *italic* refer to figures; page numbers in **bold** refer to tables.